Civil, Infrastructure and Construction Engineering

Civil, Infrastructure and Construction Engineering

Edited by **Sarah Crowe**

Willford Press

New York

Published by Willford Press,
118-35 Queens Blvd., Suite 400,
Forest Hills, NY 11375, USA
www.willfordpress.com

Civil, Infrastructure and Construction Engineering
Edited by Sarah Crowe

International Standard Book Number: 978-1-68285-280-4 (Hardback)

Printed in China.

Contents

Preface

The purpose of the book is to provide a glimpse into the dynamics and to present opinions and studies of some of the scientists engaged in the development of new ideas in the field from very different standpoints. This book will prove useful to students and researchers owing to its high content quality.

Civil engineering combines elements of architecture and engineering to create sound structures. This book outlines the processes and applications of construction engineering in detail. The various sub-fields of civil, infrastructural and construction engineering along with technological progress in these disciplines that have future implications are glanced at in this book. Many concepts such as types of construction, surveying, property management, etc. have been elucidated thoroughly. This book is a resource guide for experts as well as students.

At the end, I would like to appreciate all the efforts made by the authors in completing their chapters professionally. I express my deepest gratitude to all of them for contributing to this book by sharing their valuable works. A special thanks to my family and friends for their constant support in this journey.

Editor

An Alignment Model for the Research Higher Degree Supervision Process Using Repertory Grids – Reflections on Application in Practice in Built Environment Research

1

Guillermo Aranda-Mena (RMIT University, Australia)

Rod Gameson (University of Salford, United Kingdom)

Abstract

From *PhD-doing* to *PhD-done* is not as from *A to B*. This research aims to improve the understanding of the complex relationship between doctoral research and the supervisory process. The research model developed utilises *repertory-grids* and discussion is presented in relation to the nature of factors contributing to successful completions. The paper is argumentative in nature and provides insights to a complex process which is largely untested. A supervisory model previously presented by the authors (Aranda-Mena & Gameson, 2004) is tested in two areas: (1) the research process and (2) personality factors. The research increases common understanding of what it takes to complete a PhD, and the supervisory challenges in such a long process. It is concluded that supervision is a key factor in completing a PhD, and in developing the intellectual, analytical and research skills expected of PhD graduates. The authors call for more research, both theoretical and, in particular, empirical, in this important area.

Keywords: Built environment, Doctorate, Research supervision

Introduction

An awareness and understanding by both student and supervisor of the research process that lies ahead is critical to the success of any research higher degree studies (Boud and Lee, 2009), given an increase in the types (e.g. traditional PhD, Professional Doctorate, DSc, DBA) and modes (e.g. full-time, part-time, online, split-site) of higher degree studies now available (Park, 2005; University of Salford, No Date). For example, Denicolo *et al* (2010) report a 71% increase, from 2000-01 to 2005-06, in students studying for professional doctorates in the United Kingdom (UK). A key driver for facilitating successful and timely completions is the input of such data to national research assessment exercises such as: the UK's 'Research Excellence Framework [REF]' (HEFCE, 2011), and the Australian 'Excellence in Research for Australia [ERA]' (Commonwealth of Australia, 2010).

Previous research has identified the issue of research supervision as being a critical factor in PhD completions (Baird, 1995; Barnes & Austin, 2009; HEFCE, 2011; Wichmann-Hansen *et al*, 2012). A major Australian report on PhD supervision, conducted at the Australian National University [ANU], (Cullen *et al*, 1994, p.108), concluded that, "*...the identification of effective supervisory practice was best accomplished not through simple aggregation of existing best practice, but rather through the deconstruction of supervisory practice and through the identification of those aspects of supervisory practice which would most benefit from strengthening, elaboration and change.*" The report also identifies three main issues requiring attention:

- *"Supervision should be conceptualised to encompass a broad view of PhD education which includes more than one-to-one interaction of a student and a supervisor;*
- *Programs for staff and students to improve practice can and should be designed to contextualise the generic process of supervision with attention to disciplinary and usual human variation, and,*

- *There is a need to go beyond individual supervisory interactions and restructure practice to ensure that responsibility for quality is shared and co-ordinated."* (Cullen *et al*, 1994, p.108-109).

The objective of this research is to critically review theoretical issues pertaining to the research supervisory process and the student – supervisor relationship, and identify key factors. A principal conclusion of Haksever & Manisali's (2000) research of construction management and engineering students, was that the development of a framework to, '*promote dialogue*', between students and supervisors would contribute to improved outcomes, which could include: improved contributions to knowledge and more highly developed student research skills. Such a framework, in terms of the proposal of a supervisor-student alignment model, is also described and discussed.

Doing a PhD

There has been an upward trend in the number of students commencing PhD studies in the UK. Data published by HEFCE (2011) shows a significant increase of 62% in students commencing PhDs from 14,045 in 1996/97 to 22,790 in 2009/10. HEFCE (2011) also reports an increase of 71% in PhD starters in the '*engineering / technology / building / architecture*' subject area over the same period.

A key factor to consider is how PhD 'starters' translate into PhD 'completions'. HEFCE (2005, p. 4) state that, *"by 2000-01, after five years, 57 per cent of PhD students who began their studies on a full-time course, and 19 per cent starting on a part-time course had completed. By 2002-03, after seven years, the completion rates were 71 per cent and 34 per cent for full-time and part-time starters respectively."* This data shows that, after seven years of study, 29% of full-time students, and 66% of part-time students had not completed their studies. A major Australian study by Sinclair (2004, p.v) states that, *"Sixty-four per cent of PhD candidates supervised over the 1990–97 period were conferred with the award of Doctor of Philosophy."* This result is similar to another Australian study conducted by Martin *et al* (2001) which estimated that only 65% of doctoral candidates completed their degrees. A HEFCE (2012) report presents projected rates of qualification of students on postgraduate research degrees. For students who commenced their studies in 2009/10, 67.6% are predicted to complete within 7 years, and 75.6% within 25 years. Therefore, it is predicted that 32.4% of these students will not complete their studies by 2016/17.

A survey by Harman (2003) of Australian research-intensive universities discovered low student satisfaction levels due to poor quality of supervision. Specific issues identified by students included: quality and effectiveness of supervision, interpersonal skills of the supervisor and help provided in designing their projects. Harman (2003) also identified the issue of an expansion in PhD enrolments in Australia since the early 1990's, suggesting that universities had not kept pace with the needs of a growing body of research students, leading to increasing levels of student dissatisfaction. A survey of UK PhD students, reported by Haksever & Manisali (2000), found that 30% of the cases cited, relating to non-completion of construction management and engineering students, were specifically due to problems with supervision; in particular 'direct research-related help', defined as, *"…critical analysis of work, help with methodological problems, precise direction and help with the management of the project"* (Haksever & Manisali, 2000, p.21).

In summary therefore, the data presented in this section identifies a problem with level of 'completions' of PhD research, with a key contributory factor being the quality of research supervision.

The Research Process

There are many books that purport to give guidance on research processes and methods, both in general terms (i.e. Howard and Sharp, 1983; Phillips and Pugh, 1987; Jankowicz, 1995) and, more recently, built environment related texts (i.e. Fellows and Liu, 1997; Runeson and Skitmore, 1999; Knight and Ruddock, 2008; Farrell, 2011). Such books are now increasingly supplemented by resources available to postgraduate researchers online (i.e. Vitae, 2011).

The stages of research could be regarded as sequential, starting with theory development and ending with conclusions. This could be considered as a 'traditional' model of research, aligning with Hughes's (1994) construction management contextualised view of a PhD process of, *"...consisting of three trimesters, the first to define and contextualize the problem [the theory], the second to do the field work [the test] and the third to write the thesis [concluding]."* These three components of research could be more appropriately represented as three points on a *'circle of research'* where, given different research approaches and/or methodological positions, the starting point of the research process may not always be the '*theory*' component (Marshall and Rossman, 1989). One example of an alternative would be the '*grounded theory'* approach where, as Strauss and Corbin (1990: p23) state, "*One does not begin with the theory, then prove it. Rather, one begins with an area of study and what is relevant to that area is allowed to emerge.*" Therefore new theory may emerge from a '*test and conclude'* exercise (Hunter and Kelly, 2008).

For the purpose of this paper the authors, whilst acknowledging that other methodological propositions exist, will present and discuss a 'traditional' research model of: building theory, testing theory and concluding.

Building Theory

There is a Chinese proverb, attributed to Lao-tzu, which says that, *"A journey of a thousand miles begins with a single step."* The first research 'step' involves an initial review of the underpinning theories which might have a relation to the questions under scrutiny. The theory of the research process, as a starting point, involves the selection of a topic and a research area. With some initial ideas in mind this first research '*step*' should be a comprehensive literature search and critical review. In the past this step could prove to be difficult in terms of access to information being readily available. The growth in information technology (i.e. online journals, web-based articles and reports) has led to a proliferation in information available to researchers, requiring well-developed skills in information searching; particularly during this first step of the research to ensure that recent and relevant literature is discovered (Neuman, 1997). Students no longer have to wait 3 to 6 months for inter-library loans, or print extracts from borrowed PhDs from microfilm. Journal publishers often make papers available online before they are published in hard copy. With the internet being the main domain of information for students there is a potential risk of being overwhelmed by information, and coming across erroneous or misleading information.

Jankowicz (1995, p.26-28) proposes that researchers should ask themselves a number of questions when selecting and evaluating a research topic: Does it interest you? How much do you know about it already? How difficult is it likely to be? These three questions would be hard to answer without the assistance of a mentor or supervisor because of factors such as: the inexperience / capability of the student, and not utilising the experience of the supervisor (Barnes & Austin, 2009). Hughes (1994) suggests that the first task for construction management students, in consultation with their supervisors, is to agree a detailed specification for the research, and goes on to cite Kane (1985, p.15) who contends that, *"The most difficult hurdle to overcome in doing research is not in learning the techniques or doing the actual work or even writing the report. The biggest obstacle, surprisingly, lies in figuring out what you want to know."*

Numerous authors have put forward suggestions as to how to focus research at this first '*step*' of the process. Jankowicz (1995, p.34) suggests, *"working up and narrowing down..."* a topic by moving from an '*area*' (i.e. construction management), to a '*field*' (i.e. legal issues) to an '*aspect*' (i.e. dispute resolution). Kane (1985) states that a '*research statement*' should be produced, Howard & Sharp (1983) propose a '*topic analysis*', Leedy (1985) tenders the need for a '*problem statement*' and Black (1993) contends the need for clear '*research questions and hypotheses*'. Often an outcome of this step will be a 'research/thesis proposal' which, *"...justifies and describes the proposed study."* (Griffith & Watson, 2008, p.185).

Testing Theory

Having clearly defined a research scope and developed a research direction the next '*step*' is to define a satisfying research question, aim, objectives and, if appropriate, hypotheses. Given the wide range of types of research projects conducted in the field of construction management numerous different research methodologies, ranging from the natural to social sciences, may be utilised (Dainty, 2008). If the research involves the collection of empirical data the selection of appropriate valid, reliable and objective methodologies needs to be addressed (Black, 1993; Cresswell & Miller, 1997). Perry (1998) draws a distinction between '*explanatory*' research, which is quantitative (i.e. considering what the precise relationship between variables is), and can be considered, according to Mostyn (1985, p.116), to be, *"...to determine what, where, when, and how many,* rather than *why.",* and '*exploratory*' research, which is of a qualitative nature (i.e. looking at what variables are involved).

A key consideration at this 'test' stage is to ensure that any research is conducted in an ethical manner. This is of particular importance when primary, non-publicly available data, such as questionnaires and interviews, are being conducted and/or where any data collection methods may impact upon humans. Therefore ethical approval procedures must be followed (University of Salford, 2010b; RMIT, 2011a). When considering this step in the research process at postgraduate level, it can be seen as another '*check-point*' allowing further discussion between student and supervisor, and providing an indicator of the student's progress so far. Most Universities now have, as a minimum, a requirement for periodic assessment / review of a student's progress (i.e. RMIT, 2007; University of Salford, 2010a), notwithstanding the importance of regular student – supervisor meetings throughout the whole PhD process.

Concluding

The final, and critical, step in the research process is to reach an outcome in the form of conclusions. This involves comparing "*theory*" with "*test*" to determine if the research aim and objectives have been achieved, and, if the research has hypotheses, if they are proven, or not. Black (1993, p.175) provides a list of, "*...common sources of misleading conclusions...*", which includes: ignoring data which is contrary to the research hypothesis; conclusions extended to a larger population than the one investigated and conclusions beyond the stated research boundaries (i.e. aim and objectives).

This is also the end of the '*writing-up*' of the research where it is critical that students clearly communicate what they have done in the research. Hughes (1994), commenting on PhDs in construction management, states that, for many students, the most daunting task is writing-up their work. A logical structure and presentation of the research is essential, '*telling the story*' to the reader, using typographically and grammatically correct conventions (see good and bad examples in: Truss, 2003). The Australian model of examination, based upon reading and assessment of the thesis by two or three external examiners, with no oral examination which is also common in other parts of the world such as Europe (Tinkler & Jackson, 2000), places significant importance upon how students write up their research. The defence of the thesis which takes place in an oral examination needs to, where no oral occurs, be made very explicitly in words, and other visual media, on paper.

PhD supervision as a Non-Linear Process

Traditionally, both the literature and University printed guidance on the research process provides prescriptive guidance (University of Salford, 2010a; RMIT, 2011b). However, there are a limited number of sources that acknowledge the complex process of thesis development in higher degrees. The more typical literature, including Universities' own guides, often prove to be rigid and highly prescriptive about supervisor, supervisee and the thesis or object of study (Lawson, 2000). For example, Phillips & Pugh (1987) suggest steps and sequential timetables. However, in reality, the situation often proves to be somewhat different. The process of doing a PhD is hardly one that goes from *a* to *b* to *c*. According to Grant (2003) the process is iterative, complex and unstable and, for the same reasons, it would seem sensible to develop more prescriptive approaches towards supervision. Figure 1 illustrates a model of the complexity and non-linearity in the supervision process. This model was first proposed by Grant (2003) and provides a potentially powerful tool to explicitly externalise the situation. Figure 1 also shows direct and indirect links in the '*supervisor-student-thesis'* relationship. The relationship between supervisor and student adjoins a third element of *'knowledge'* which in the figure is represented by the thesis. Grant (2003) states that the relationship between these three elements is in constant change. Such an arrangement is affected through power relationships that work through actions; these actions have an effect upon the elements and highlight more subtle difficulties dealing with personality, cultural and power differences (Lee, 2007; 2008). Power relations can be a consequence of aspects of institutional position and can also vary across disciplines, institutions and nations (Hofstede, 1986).

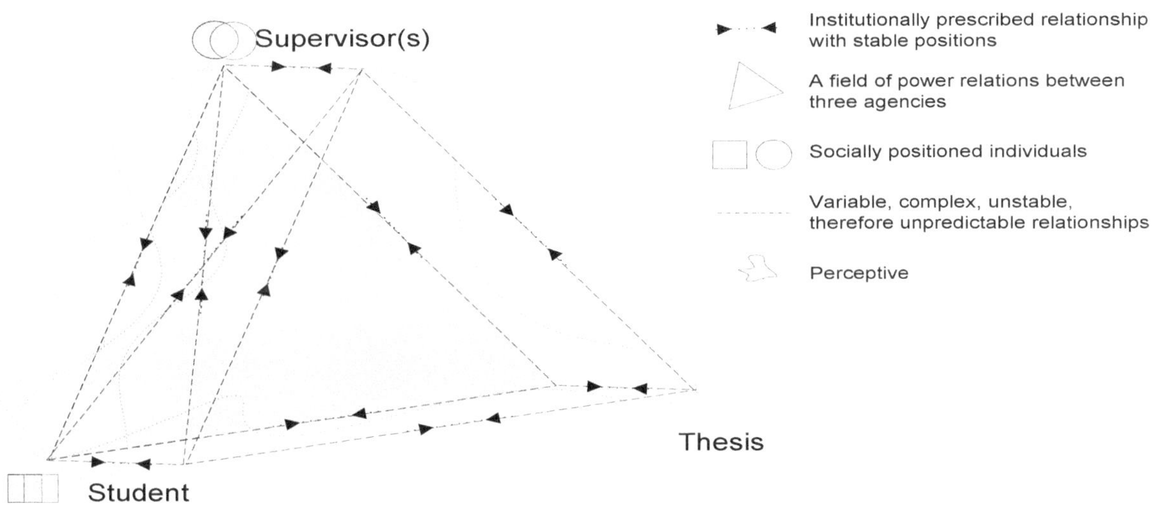

Figure 1 Student-supervisor multilayer relationship (Based on Grant, 2003)

Even if bright as undergraduates, neophyte higher degree researchers need to develop their skills further when conducting higher degree research. They can often be insecure, inexperienced, and demand close supervision. They may see the thesis as a culturally prescribed artefact and it may take some time before a student can comprehend the implications of doing postgraduate research (Johnson *et al*, 2000). Gurr (2001) suggests that the role of supervising would typically involve feedback, time, money, networks, and recognition to the student by an established authority in academia. Hughes (1997) disagrees with this and argues that the supervisor does not necessarily need to be an expert in all areas/fields and that the researcher should also seek support from within the department, the university and the wider research community. Whitelock *et al* (2008) suggest that promoting creativity in PhD supervision is important, but tensions and dilemmas within the student-supervisor relationship need to be resolved.

The dimensions of supervisory styles summarised in Table 1 were originally developed by Cullen *et al* (1994) to pinpoint the styles across individuals and academic disciplines. For instance, supervisors from computing, economics, physics and engineering disciplines tend to provide a closer style than those from arts, history or sociology. This relates to Hughes's (1994) discussion on styles in supervision in built environment research. He argues that building and built environment research should seek the support of core disciplines such as economics, law, cybernetics, psychology amongst others. If building-related research is to grow and be beneficial to industry and the wider scientific community, it must establish itself as a *'field of application'*. If this is the case, it could exacerbate the complexities to apply standard codes of practice because areas of research supervision could span across many fields.

Style:	**Close**	**Hands off**
Meetings:	Regular, frequent	Irregular
Project:	Collaborative	Individual
Relation to supervisor's research:	Closely related	Unrelated
Joint Publication:	The norm	Uncommon
Mentorship:	The norm	Rare

Table 1 Supervisory styles (Cullen et al, 1994)

In the traditional sense, the supervisor-student relationship has been described by FAUSA (1979, p.21) as, "*...the most important channel of intellectual inheritance between one generation and the next.*", with good supervision being central to thesis completion (Lawson, 2000). On the other hand it is generally expected that research students should develop their own ideas and test their own hypotheses – and not that of their supervisors (Harman, 2003). A graduate from a higher degree by research (especially in PhDs) is expected to be academically competent and independent. Achieving this is, arguably, the most challenging test in the supervision process (Haksever & Manisali, 2000).

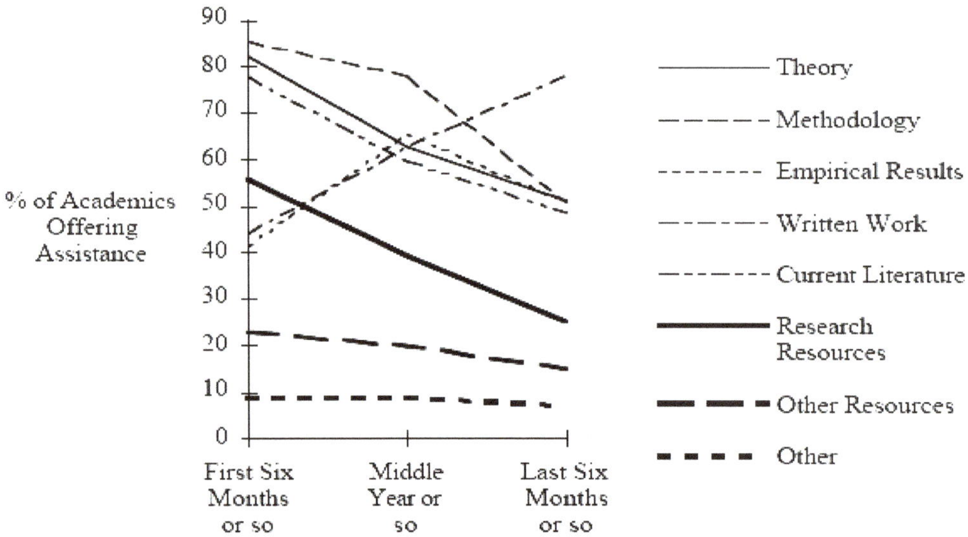

Figure 2 Supervisory assistance in research phases (Cullen et al, 1994, p.91)

Figure 2 presents the supervision process in terms of skills and expectations. Supervisees typically demand supervisory expertise in most of these areas (e.g. theory, methodology, writing skills, etc.). In reality it is very unlikely that a single supervisor possesses all of these skills. This often applies in social science related research where students are expected to develop critical thinking not just validating findings but defending their more philosophical standings such as ontological and epistemological positions (Heath, 2002).

Aligning the Players

This section introduces a model which, whilst being relatively simple, provides explicit value for academic supervision across the wide range of disciplines. The model seeks to raise awareness of the norms, expectations and standards within their discipline and be able to assess their own plans and actions to ensure compliance with them (Gurr, 2001). The aim is ultimately about teaching students to be their own supervisors (Phillips & Pugh, 1987).

Gurr's (2001) model operates to promote a *'plan-act-review'* learning cycle. In the increasingly common arrangement where associate supervisors or supervisory panels are involved, additional parties can be invited to join subsequent discussions. Figure 3 illustrates a dynamic alignment model which could be used in dialogue between a student and a supervisor to establish a shared understanding through reflection. The model is aimed at facilitating the production of a successful PhD thesis. A flexible approach is required on the part of the supervisor, an approach that should be informed by open discussion with the student (Vilkinas, 2002). This may then lead to establishing an appropriate supervisory style. Anderson (1988) developed a model comprising of the following four styles of supervision:

- "**Direct active**: Characterised by initiating, criticising, telling and directing;
- **Indirect active**: Characterised by asking for opinions and suggestions, accepting and expanding supervisee's ideas, or asking for explanations and justifications of supervisee's statements;
- **Indirect passive**: Characterised by listening and waiting for supervisees to process ideas and problem solving, and
- **Passive**: Characterised by having no input and not responding to supervisee's input." (Gurr, 2001, p.86).

Other emerging and increasingly popular modes of higher degrees by research include the Professional Doctorate (Chynoweth, 2006) and the embedded practice PhD. Examples of both of these modes, in the built environment, are currently delivered at RMIT University Schools of Architecture, Business Graduate Studies and School of Property, Construction and Project Management (RMIT, No Date). The University of Salford also delivers a Professional Doctorate where, *"...candidates are encouraged to generate explicit academic knowledge from their wealth of existing tacit professional knowledge*." (University of Salford, 2009). For these modes of doctoral studies the student-supervisor relationship takes on a different challenge (Walker, 2008).

In these cases the candidate is normally someone already established in industry and may have a supervisory committee of practitioners and academics. The candidates are request to report and present to the academic institution with progress reviews and presentations. Most of the time this would be done face-to-face but there are cases where such reviews may be conducted 'virtually' via video and/or teleconference. A particular challenge here is the lack of face-to-face contact time. There have been a number of successful completions, with outcomes including conference and journal paper publications; for example Bourne (2005) and Holzer (2009). Therefore, there is a need to find and use different and more innovative techniques to facilitate the student-supervisor relationship such as wiki-webs and social media, especially when geographic collocation is not possible.

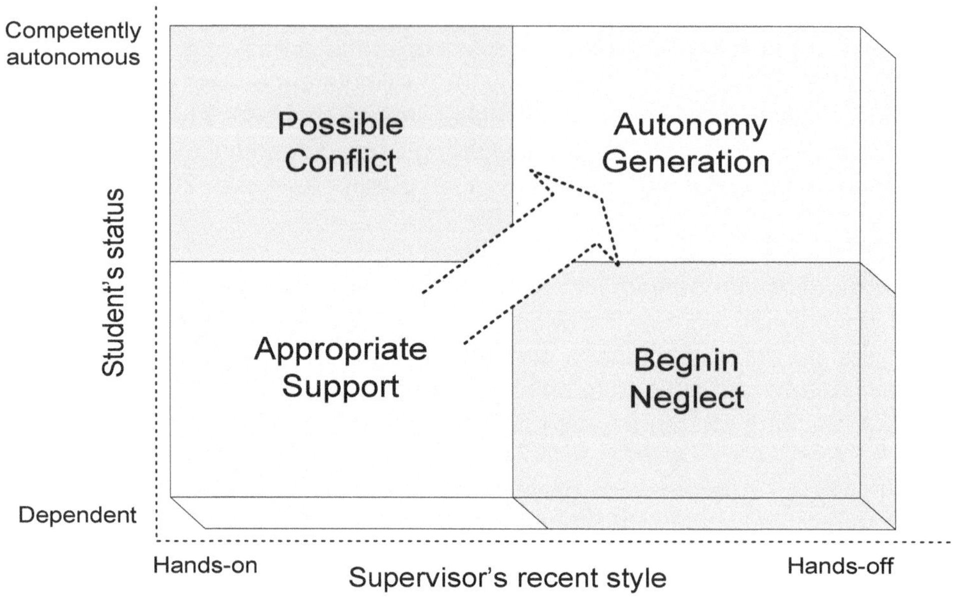

Figure 3 Dynamic alignment model (Gurr, 2001, p.87)

The research to date has not identified any existing, detailed models which could be used to capture perceptions of both students and supervisors during the PhD's life-cycle. Bradbury-Jones (2007) presented a systematic approach for exploring subjectivity during doctoral studies, but only from a student's perspective. Therefore this research now proposes using the *'Repertory Grid'* method to develop a model (Denicolo & Pope, 2001). The repertory grid technique is based upon the principles of *'personal construct psychology'* (Kelly, 1955). The use of an alignment model in conjunction with repertory grids provides a means of judging the degree of, and framework for working towards, effective supervision.

Mapping-out the Process

A comprehensive explanation of the personal construct theory and technique is beyond the scope of this paper. However, further information concerning the use and implementation of the technique can be found in Aranda & Finch (2003). A repertory grid technique (RGT) is proposed to facilitate the implementation of Gurr's (2001) model. Denicolo & Pope (2001) identify the value of repertory grids to explicitly indicate thinking in the supervision process, and to track and record perceptual and emotive changes, often a *tacit* phenomenon. RGT is therefore used to facilitate conversation, unearth insights and provide focus in the supervisory relationship.

Repertory Grid Technique (RGT) systems are traditionally developed through *triad elicitation* (e.g. Denicolo & Pope, 2001). This means that a respondent or interviewee is asked to find similarities and differences of three elements at a time. Those similarities and differences become what are known as a,' *bipolar construct system'*. A series of bipolar constructs outline *'attributes'* to a concept. In the context of this research such a concept would be *'supervision'*. The numerical or hierarchical grading of such construct systems follows a similar procedure to that of the Likert scales (Allport, 1935) or Osgood's (1954) semantic differential. Ratings are assigned to a bi-polar construct system; this would facilitate a relational structure between elements and construct polarity (ie. quality – quantity).

Mapping out: Repertory-Grids

The following repertory grids were originally discussed in 2004 (Aranda-Mena & Gameson, 2004) in relation to the perception and changes during the student-supervisor relationship. The repertory grid shown in figure 4 illustrates a *focused* construct system and its relationship and gap analysis as to what is perceived as *effective supervision* by both, the supervisor and the student. The grids provide a mechanism for further discussion, reflection and development of a new model.

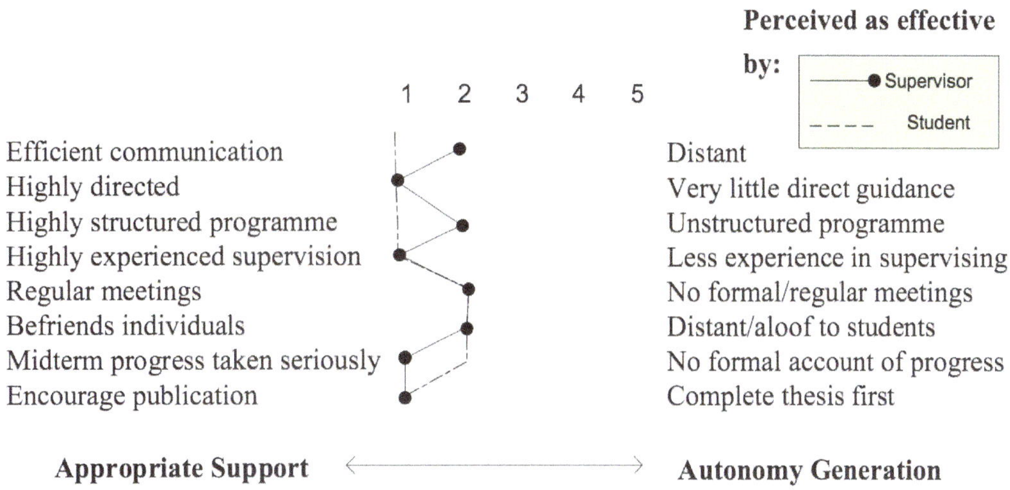

Figure 4 Initial rating (i.e. topic definition)

At this initial stage, the student perceives appropriate support and the supervisor also considers that the supervisee needs it. However, according to Gurr (2001) close support should only take place at an initial stage - otherwise there is a high chance for this causing conflict (i.e. see Figure 3). At this initial stage the student is more receptive and the supervisor is rather paternalistic. The grid technique is for continuous use by both the supervisor and the student and is to provide a means for discussion over differences and expectations so as to monitor changes over time. The grid provides insight into what the student and the supervisor perceive to be *'effective supervision'*.

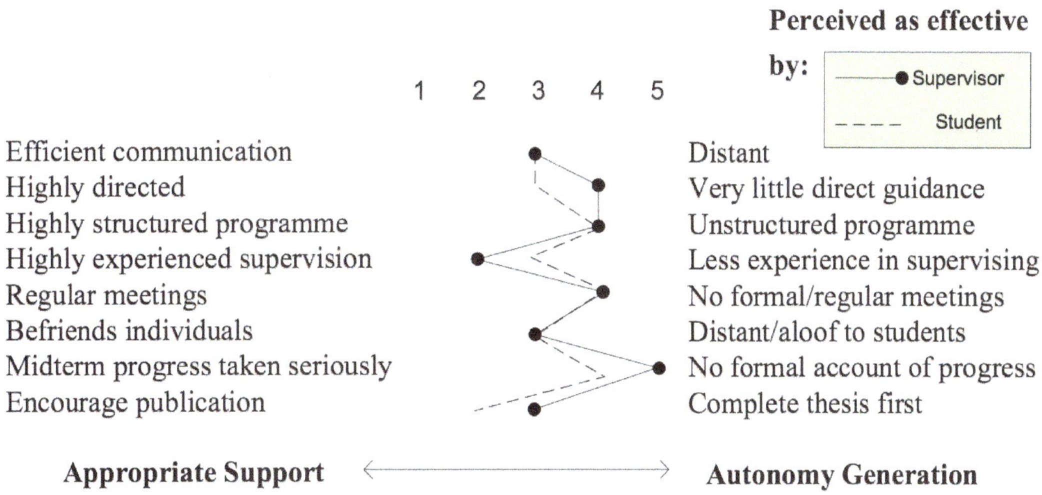

Figure 5 Progress rating (i.e. theory development)

The repertory grid shown in figure 5 reveals how the student now perceives an appropriate supervisory style as the research develops. Certainly they are experiencing a more demanding stage with less direct guidance; the distance between the two has increased. At this stage the student has started building confidence and is more in tune with the situation.

The grid in figure 6 illustrates how the supervisor has moved towards a '*hands-off*' style, which he or she perceives to be the effective one at this stage. The student has some anxieties but with the assistance of the grids, they can both discuss the current situation.

The supervisor may make it clear that it is time for the student to take *'authorship'* of their work from this point onwards, which is expected in higher degrees. The student feels somehow neglected; however, it has been important to be aware of such expectations of the supervisor. Note that new bipolar constructs can emerge at any point.

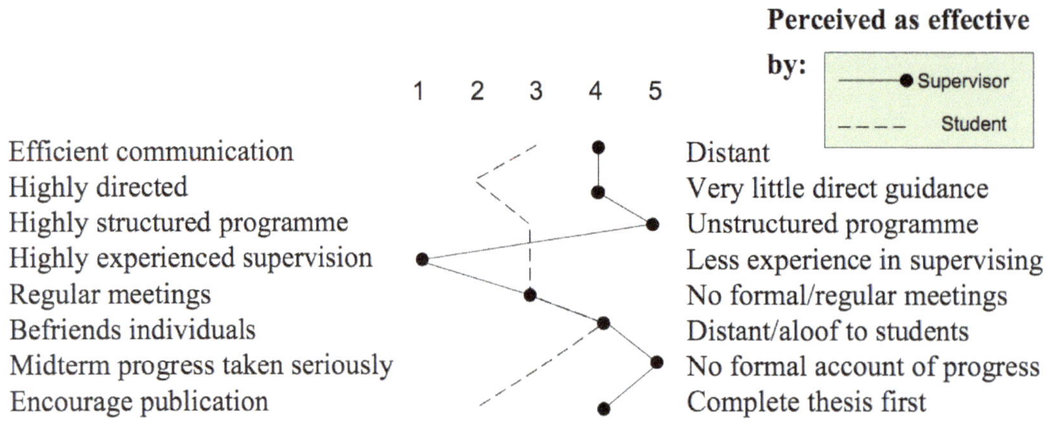

Figure 6 Further progress rating (i.e. data analysis / testing)

A breakthrough arises at this stage, as the student realises that he or she need to embrace or take control and direction of their project. The supervisor is still there but the relationship is somehow unstructured and there is no formal account of progress. The student frequently falls into procrastination and shifts attention to other activities such as tutoring and community work.

Figure 7 illustrates the situation towards thesis completion where the relationship seems to have improved. The student has overcome anxiety and insecurity and keeps better progress records by taking full control over the project. Although more confident, the student is always experiencing anxiety but this might remain until after the final examination.

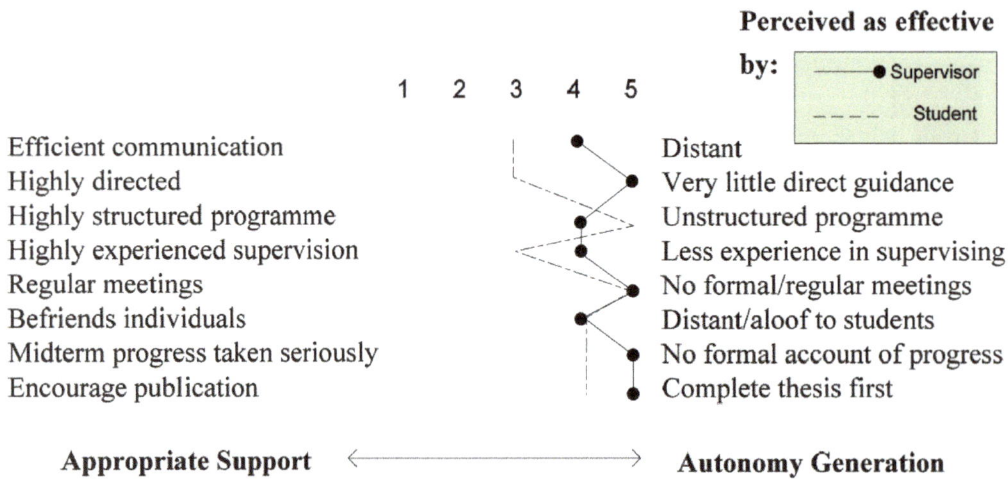

Figure 7 Rating towards completion (concluding)

The four grids above have been used to illustrate the personal perceptions of the supervisory process from both a student and a supervisor perspective. The vertical lines indicate 'perceived as effective' as seen/reflected by each party. The profile line of a co-supervisor could have been introduced here. It is contended that a co-supervisor could act as a bridge across the current existing supervisory gap. Arguably co-supervisors can only be effective if fully engaged in the process from the outset of the research.

Mapping out: Cluster Analysis

Having reflected upon the original model, developed in 2004, (Aranda-Mena & Gameson, 2004), and conducted further research to identify relevant literature published since then, such as Gatfield's (2005) study of PhD supervisory styles, a more complex repertory grid has been developed to provide further insights into research supervision encompassing a range of modes such as single and multiple supervisors.

The model, as shown in figure 8, is a repertory grid comprising of *elements*, relating to supervision, and *bi-polar constructs*.

The 'Elements' (i.e. supervision/supervisory support) are:

A) Single supervisor
B) Two supervisors
C) One industry and one academic supervisor
D) School academic board
E) School PhD students
F) Ghost supervisor
G) Two academic and one industry
H) On-line interest groups such as CNBR

Emerging *'constructs'* (i.e. attributes to supervision/supervisory support) after triads have been self-administered by the authors. Emerging constructs by comparing and contrasting elements form A to H are as follow:

1.- **Relationship**: [Left construct] Formal—*[Right construct] Informal*
2.- **Communication**: [L]Argumentative—*[R]Discussions*
3.- **Expertise**: [L]Specialised—*[R]Generic*
4.- **Engagement**: [L]No ownership over the project—*[R]Process Driven*
5.- **Vision**: [L]Here and now—*[R]Long term vision*
6.- **Opinion**: [L]Valued—*[R]Not Valued*
7.- **Proximity**: [L]Physical/Psychological Closeness—*[R]Remote*
8.- **Line of communication**: [L]Three way relationship—*[R]Unidirectional*
9.- **Line of communication**: [L]Unidirectional one-to-one – *[R]One-to-many*
10.- **Line of communication**: [L]Unidirectional strong – *[R]Unidirectional fragile/tenuous*
11.- **Line of communication**: [L]Sessional triangle – *[R]Sessional unidirectional*

The resulting above Figure 8 indicates the relationship between supervision/supervisory support in a cross tabulated grid with attributes on the left and right columns indicating a relationship across all attributes with all elements (i.e. 8 elements from A to H) and 11 bipolar constructs (22 individual tags either at the left (1) or right (5) sides of the grid). The shading indicates if a particular element is likely to relate to a Left to Right construct. For instance, *Element A*, Single Supervisor, is associated with the following constructs: 7R, 8R, 5R, 4R, 10L, 1L, 3L, 2L, 9L, 11R and 6R. The shadings also form clusters indicating similarities between elements and constructs. Such similarities are also indicated with the tree-shaped *dendritic* diagrams. The above grid indicates that supervision support as G, B and C (i.e. different modes of the supervisor and the co-supervisor) is seen as an effective supervisory mode. The contrasting situation is indicated by the proximity of A with F suggesting that the *single supervisor* is performing as a nowhere to be seen *ghost supervisor!*

The *nodal points* in the dendritic diagrams at the top and right-hand side of the grid indicate the proximity of elements or constructs on both the X-axis and the Y-axis. For instance, constructs 4 and 10 are the closest matching at 80, whereas construct *5 (Vision)* is the single most isolated construct with a match of 50 with any other construct. Constructs *7 (Proximity)* and 8 (*Line of communication)* are isolated with the rest of all other constructs matching at 30. Constructs *4 (Engagement)* and *10 (Line of communication)* are also closely aligned with a matching factor of 80.

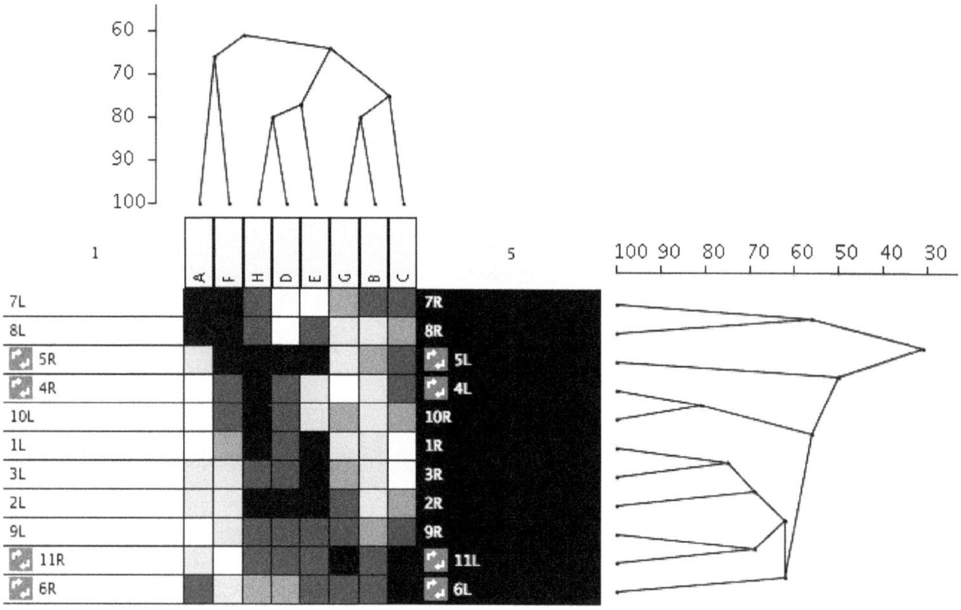

Figure 8 Repertory Grid: 8 elements by 11 constructs

The dendritic diagram at the top of the grid indicates *nodal points* or matching closeness factors of supervisors in three groups: (1) *A,F* (Single supervisor and Ghost supervisor) at 67 pointing out at deficiencies supervisory situation. A second supervisory group (2) *H,D,E* (On-line interest groups such as CNBR, School academic board, School PhD students) at 80 indicating what would be a supportive environment (both, rational and emotional) for the PhD candidate. A third group (3) *G,B,C* (Two supervisors, One industry and one academic supervisor and Two academic and one industry) at 78 create an ideal supervisory mode, which could be suggested to be a most appropriate model for future student, supervisor, co-supervisor relationships.

Conclusions

This research concludes that to improve both thesis completion rates and the quality of research a model for student-supervisor interaction and satisfaction is essential. A technique to facilitate such a process in therefore proposed. The Repertory Grid Technique (RGT) can provide insights into the student-supervisor relationship by explicitly revealing views and perceptions. It can also provide stronger objectivity in the progress of the research and its assessment processes. As the result of utilising the RGT the research contends that this would contribute to improved communication in dealing with student-supervisor demands and expectations. It is also hypothesised that the level of anxiety in the supervisory process would be reduced if clear and transparent communication emerges.

This, in turn, can contribute to stronger objectivity in the progress of the research and its assessment processes. For example, Heath (2002) found that the differences between a student and a supervisor can be diminished by the frequency of meetings. This research argues that the *'quality'* of such meetings is the most important factor.

This paper has put forward three main themes for consideration. The first describes the challenges in the current PhD research process and highlights the need to increase timely completions. Secondly, aspects relating to the research process are critically reviewed.

Finally, a model, initially conceptualised in 2004, has been extended and further developed using the RGT as a self administered reflective tool drawing upon a critical review of literature and the authors' experiences in supervising PhDs.

Although this research has been contextualised in the construction management domain the principles, and the RGT model, could be utilised for PhD's in other built environment related disciplines, such as economics and the social sciences, and in many other academic disciplines. The next logical stage of the research is to test this model by conducting an empirical, quantitative study.

References

Allport, G. W. (1935) *Attitudes: a handbook of social psychology*. Worcester, Mass: Clark University Press

Anderson, J. (1988) *The supervisory process in speech language pathology and audiology,* Boston: College Hill Press & Little Brown

Aranda, G. & Finch, E. (2003) 'Using repertory grids to measure changes in risk-taking behaviour', *Journal of Construction Research – Special Issue in Construction Safety*, **4** (1), 101-114

Aranda-Mena, G. & Gameson, R. (2004) 'An alignment model for the research higher degree supervision process using repertory grids', *Australian Universities Building Educators Association (AUBEA) Conference*, Newcastle, Australia, July, 98-113

Baird, L. L. (1995) 'Helping graduate students: a graduate adviser's view', In Pruitt-Logan, A.S. & Isaac, P.D. (eds.), *Student services for the changing graduate student population; New Directions for Student Services*, 72, San Francisco, CA: Jossey-Bass. 25-32

Barnes, B.J. and Austin, A.F. (2009) 'The role of doctoral advisors: A look at advising from the advisor's perspective', *Innovation in Higher Education*, **33** (5), 297-315

Black, T.R. (1993) *Evaluating social science research: an introduction*, London: Sage Publications

Boud, D. & Lee, A. (2009) *Changing practices is doctoral education*, London: Taylor & Francis

Bourne, L. (2005) *Project relationship management and the stakeholder circle,* Unpublished professional doctorate dissertation, RMIT University, Melbourne, Australia

Bradbury-Jones, C. (2007) 'Exploring research supervision through Peshkin's I's: the yellow brick road, *Journal of Advanced Nursing*, **60** (2), 220–228

Chynoweth, P. (2006) 'Professional doctorates in the built environment: how higher education is responding to the changing needs of industry and the professions', *Built Environment Education Annual Conference (BEECON 2006): Where Industry and Practice, the Professions and Education Meet*, London, September, 15

Commonwealth of Australia (2010, March 16) *The excellence in research for Australia [ERA] initiative* [online] http://www.arc.gov.au/era/

Cresswell, J.W. & Miller, G.A. (1997) 'Research methodologies and the doctoral process', *New Directions in Higher Education*, **99**, 33-46

Cullen, D. J., Pearson, M., Saha, L.J. & Spear, R.H. (1994) *Establishing effective PhD supervision,* Department of Employment Education and Training [DEET], Australian Governmental Publishing Service

Dainty, A. (2008) 'Methodological pluralism in construction management research', In Knight, A. & Ruddock, L. (eds.) *Advanced Research Methods in the Built Environment,* West Sussex: Wiley-Blackwell, 1-13

Denicolo, P. M. & Pope, M. (2001) *Transformative professional practice: personal construct approaches to education and research,* London: Whurr Publishers, London

Denicolo, P., Fuller, M. & Berry, D. with Raven, C. (2010) *A review of graduate schools in the UK,* Staffordshire: UK Council for Graduate Education

Farrell, P. (2011) *Writing a built environment dissertation: practical guidance and examples,* West Sussex: Wiley-Blackwell

Federation of Australian University Staff Associations [FAUSA] (1979) *Report on research in universities*, Federation of Australian University Staff Associations, Melbourne, Australia

Fellows, R. & Liu, A. (1997) *Research methods for construction*, London: Blackwell Science

Gatfield, T. (2005) 'An investigation into PhD supervisory management styles: development of a dynamic conceptual model and its managerial implications', *Journal of Higher Education Policy and Management*, **27** (3), 311-325

Grant, B. (2003) 'Mapping the pleasures and risks of supervision', *Discourse; Studies In The Cultural Politics Of Education,* **24** (2), 175-190

Griffith, A. & Watson, P. (2008) 'Managing the thesis', In Knight, A. & Ruddock, L. (eds.) *Advanced Research Methods in the Built Environment*, West Sussex: Wiley-Blackwell, 183-192

Gurr, G. M. (2001) 'Negotiating the rackety bridge: a dynamic model for aligning supervisory style with research student development', *Higher Education Research & Development,* **20** (1), 81-92

Haksever, A.M. & Manisali, E. (2000) 'Assessing supervision requirements of PhD students: the case of construction management and engineering in the UK', *European Journal of Engineering Education*, **25** (1), 19-32

Harman, G. (2003) 'PhD student satisfaction with course experience and supervision in two Australian research-intensive universities', *Prometheus,* **21** (3), 317-333

Heath, T. (2002) 'A quantitative analysis of PhD students' views of supervision', *Higher Education Research & Development,* **21** (1), 41-53

Higher Education Funding Council for England [HEFCE] (2005) *PhD research degrees - entry and completion*, Issues Paper 2005/02, HEFCE

HEFCE (2011) *PhD study - trends and profiles: 1996-97 to 2009-10*, Issues Paper 2011/33, HEFCE

HEFCE (2011, March 15) *Research Excellence Framework [REF]*, [online] http://www.hefce.ac.uk/research/ref/

HEFCE (2012) *Rates of qualification from postgraduate research degrees*, Issues Paper 2012/10, HEFCE

Hofstede, G. (1986) 'Cultural differences in teaching and learning', *International Journal of Intercultural Relations,* **10**, 301-320

Holzer, D. (2009) *Sense-making across collaborating disciplines in the early stages of architectural design*, PhD (Embedded Practice), RMIT University, Melbourne, Australia.

Howard, K. & Sharp, J.A. (1983) *The management of a student research project.* Aldershot: Gower Press

Hughes, W. (1994) 'The PhD in construction management', In Skitmore, M. & Betts, M. (eds.) *Proceedings of the 10th annual conference of the Association of Researchers in Construction Management [ARCOM]*, Loughborough University, UK, 1, 76-87

Hughes, W. (1997) 'Construction management research: a field of application', *Australasian Universities Building Education Association [AUBEA] - Third International Electronic Forum on Research and Education for Construction,* 1, 6

Hunter, K. & Kelly, J. (2008) 'Grounded theory', In Knight, A. & Ruddock, L. (eds.) *Advanced Research Methods in the Built Environment*, West Sussex: Wiley-Blackwell, 86-98

Jankowicz, A.D. (1995) *Business research projects* (2nd ed.), London: Chapman and Hall

Johnson, L., Lee, A. & Green, B. (2000) 'The PhD and the autonomous self: gender, rationality and postgraduate pedagogy', *Studies in Higher Education*, **25** (2), 135-147

Kane, E. (1985) *Doing your own research: how to do basic research in the social sciences and humanities*, London: Marion Boyars

Kelly, G. (1955) *The psychology of personal constructs*, Vols. 1 and 2, Norton: New York

Knight, A. & Ruddock, L. (eds.) (2008) *Advanced research methods in the built environment*, West Sussex: Wiley-Blackwell

Lawson, M. (2000) *Being a PhD supervisor*, Prepared by the Staff Development Training Unit, The University of Newcastle, Australia

Lee, A. (2007) 'Developing effective supervisors: Concepts of research supervision', *South African Journal of Higher Education*, **21** (4), 680-693

Lee, A. (2008) 'How are doctoral students supervised? Concepts of doctoral research supervision', *Studies in Higher Education*, **33** (3), 267-281

Leedy, P.D. (1985) *Practical Research* (3rd ed.) New York: Macmillan

Marshall, C. & Rossman, G.B. (1989) *Designing qualitative research*, California: Sage Publications

Martin, Y., Maclachlan, M. & Karmel, T. (2001) *Postgraduate completion rates*, Department of Education, Training and Youth Affairs [DETYA], Canberra

Mostyn, B. (1985) 'The content analysis of qualitative research data: a dynamic approach', In Brenner, M., Brown, J. & Canter, D. (eds.) *The research interview: uses and approaches*, London: Academic Press, 115-145

Neuman, W.L. (1997) *Social research methods: Qualitative and quantitative approaches* (3rd ed.) Boston: Allyn and Bacon

Osgood, C. E. (1954) 'A blind analysis of a case of multiple personality using the semantic differential', *Abnormal Psychology,* **49**, 579

Park, C. (2005) 'New variant PhD: The changing nature of the doctorate in the UK', *Journal of Higher Education Policy and Management*, **27** (2), 189–207

Perry, C. (1998) *Thesis resource paper: a structured approach to presenting theses* [online] http://www.uq.net.au/action_research/art/cperry.html accessed 2011, March 19

Phillips, E.M. & Pugh, D.S. (1987) *How to get a PhD*. Milton Keynes: Open University Press

Royal Melbourne Institute of Technology [RMIT] *The doctor of project management (DPM) @ RMIT University* [online] http://dhtw.tce.rmit.edu.au/pmgt/dpm.htm accessed 2011, March 22

RMIT (2007) *Higher degrees by research: policies and procedures information booklet* [online] http://mams.rmit.edu.au/6rp3m7h6fawnz.pdf accessed 2011, March 24

RMIT (2011a) *Ethical issues and responsible practice policy – higher degrees by research* [online] http://www.rmit.edu.au/browse;ID=qdvjsqchwsmf1 accessed 2011, March 23

RMIT (2011b) *Requirements of research degrees policy - higher degrees by research* [online] http://www.rmit.edu.au/browse;ID=04nydchcbbaf1 accessed 2011, March 24

Runeson, G. & Skitmore, M. (1999) *Writing research reports: a practical guide for students of the built environment.* Australia: Deakin University Press

Sinclair, M. (2004) *The pedagogy of 'good' PhD supervision: a national cross-disciplinary investigation of PhD supervision*, Australia: Commonwealth of Australia

Strauss, A. & Corbin, J. (1990) *Basics of qualitative research: grounded theory procedures and techniques.* California: Sage Publications

The Quality Assurance Agency for Higher Education [QAA] (2001) *The framework for higher education qualifications in England, Wales and Northern Ireland* [online] http://www.qaa.ac.uk/academicinfrastructure/fheq/ewni/default.asp accessed 2011, March 25

Tinkler, P. & Jackson, C. (2000) 'Examining the doctorate: institutional policy and the PhD examination process in Britain', *Studies in Higher Education*, **25** (2), 167-180

Truss, L. (2003) *Eats, shoots and leaves: the zero tolerance approach to punctuation*, London: Profile Books

University of Salford, *Research Degrees – School of the Built Environment* [online] http://www.sobe.salford.ac.uk/courses/research-degrees accessed 2011, March 15

University of Salford (2009) *Professional doctorate* [online] http://www.professionaldoctorate.co.uk/ assessed 2011, March 18

University of Salford (2010a) *Code of practice for the conduct of postgraduate research degree programmes* [online] http://www.governance.salford.ac.uk/cms/resources/uploads/File/AQA/PG_Research_Degrees_Code_of_Practice.pdf accessed 2011, March 17

University of Salford (2010b) *Governance and ethics* [online] http://www.rgc.salford.ac.uk/page/forms accessed 2011, March 21

Vilkinas, T. (2002) The PhD process: the supervisor as manager, *Education + Training*, **44** (3), 129-137

Vitae (2011) *Postgraduate researchers* [online] http://www.vitae.ac.uk/researchers/1218/Postgraduate-researchers.html accessed 2011, March 17

Walker, D.H.T. (2008) Collaborative academic/practitioner research in project management: theory and models, *International Journal of Managing Projects in Business*, **1** (1), 17-32

Whitelock, D., Faulkner, D. & Miell, D. (2008) Promoting creativity in PhD supervision: tensions and dilemmas, *Thinking Skills and Creativity*, **3**, 143-153

Wichmann-Hansen, G.M., Bach, L.W., Eika, B. & Mulvany, M.J. (2012) 'Successful PhD supervision: a two-way process', In Castano, M.A.R.B. & Guner-Akdogan, G. (eds.) *Mentoring in Academia and Industry: Researching, Teaching, and Learning Triangle,* Springer New York, **10** (2), 55-64

Motivation-Driven Learning and Teaching Model for Construction Education

2

Imriyas Kamardeen, (University of New South Wales, Australia)

Abstract

Quality learning outcomes are correlated with students' motivation to learn. Lecturers need to design courses that enthuse, inspire and motivate their students. But, this is a fundamental challenge facing many lecturers. A new motivation-driven learning and teaching model was developed to help lecturers in this regard. Its operationalisation, implementation and evaluation were conducted in a first year course in Construction Management degree through action research. Study findings suggest that the new model can help lecturers to improve overall teaching quality and student learning experience as it facilitates effective course delivery and stimulates student motivation to learn.

Keywords: Construction education, Pedagogy, Motivation, Student learning experience, Curriculum design, Teaching quality

Introduction

Motivation is an important factor in student learning; it is positively correlated with students' willingness to learn, high-level cognition, creativity and performance (Liu et al 2012). Ramsden (2003) maintained that quality learning outcomes are most likely to occur when students adopt a deep approach to learning, and active engagement and interactions are two important aspects in promoting deep approached to learning. However, self-motivation of students is a prime requirement for their active engagement and interactions. Freeman & Blayney (2005) argued that these are more difficult to achieve in large classes, which have become the norm in many university courses due to increasing student numbers and limited resources.

Learning motivation is an essential issue that needs considerations when designing a course (Liu et al 2012). A vital part in the role of a lecturer should be to enthuse, inspire and motivate their students to learn (Kember & McNaught 2007). Gavrin & Novak (1999) critiqued that motivating students is one of the fundamental challenges facing lecturers. Many students work hard, but all too often the motivation is a grade rather than understanding. Many active learning methods can partially obviate this problem, however, once they leave the classroom, time pressure (often combined with poor study skills) can sap a student's motivation. As such, developing and implementing a motivating course has become a challenging task for university lecturers. To this end, this paper addresses the research question of "how can an effective motivation-driven learning and teaching strategy be developed and implemented in construction education?"

In addressing the research question, first a thorough literature review was conducted to identify factors that influence students' motivation to learn. Then a new motivation-driven learning and teaching model was developed based on the literature review findings. Following that an action research approach was adopted to operationalise and implement the model, and to validate its effectiveness/suitability to enhance student learning. Discussions in the paper are laid out in a logical order to reflect the above stages. First, the literature review findings on the factors that influence students' motivation are explored. Second, the new model of motivation-driven learning and teaching is introduced. Third, the process involved in operationalizing and implementing the model in a course is expounded followed by the analysis of its effectiveness, using a formal online student survey. Finally, conclusions are drawn.

Critical Success Factors for Motivation-Driven Learning and Teaching

Scholarly articles by researchers in higher education were reviewed for the purpose of identifying critical success factors for motivation-driven learning and teaching, and findings are summarised in here.

Kember & McNaught (2007) identified five strategies to motivate students to learn:

- *Encouraging students* – students should be told that they have done well
- *Displaying enthusiasm as a teacher by conveying a passion/fascination for the discipline and by showing sufficient interest in students* - to get to know and establish a positive relationship with students
- *Interesting and enjoyable classes* – aim to make lessons interesting by entertaining teaching and engaging students
- *Relevant materials* – students should perceive that what is taught is relevant. Relevance can be established by using real life examples, drawing cases from current issues, giving local materials/examples and relating theories to practice
- *Using a variety of teaching approaches involving active learning* – learning occurs more through what students do than through what the lecturer says. Use a different teaching approach for lesson after lesson to avoid learning becoming monotonous

They further suggested that a variety in teaching approaches, involving active learning, is a feature of quality teaching, which naturally results in an innovative form of teaching. In addition to the most common types of university teaching: lectures, seminars and tutorials, several other forms can be used, which include: (1) projects, (2) student presentations, (3) case-based teaching, (4) problem-based learning, (5) role-play, (6) reflective journals, (7) experiential learning, (8) peer-tutoring, (9) games and simulations and (10) computer-enhanced learning. Additionally, recently emerged adaptive e-tutorials technique is claimed to create a more advanced interactive learning environment that provides instant feedback to students on their learning progress and allows repeated attempts, thereby improving students' motivational level in learning (Beldagli & Adiguzel 2010).

In a similar vein, Kember et al (2008) discussed findings of their interview survey with students from nine faculties in three universities on factors motivating and de-motivating student learning. Eight characteristics of a teaching and learning environment that motivate student learning in the order of importance are: (1) establishing relevance, (2) establishing interest, (3) allowing choice of courses/tasks so that interest can be followed, (4) learning activities, (5) teaching for understanding, (6) assessment of learning activities, (7) close teacher-student relationship and (8) sense of belonging between classmates. The most common reason given for a lack of motivation is that teaching is purely abstract or confined to theory and lacking applications. Without seeing applications which put the theory in context it becomes hard for students to grasp the meaning. They further suggested that lecturers map their courses to take account of these factors in building curriculums.

Edelson & Joseph (2004) recognised a student's interest in a subject as a powerful motivator of learning as it provides four benefits:

- *Natural appeal* – interest describes a natural draw to certain activities
- *Mastery goal orientation* – interest leads to mastery goal orientation. When a learner perceives that certain knowledge or skills are useful to the pursuit of an interest, he or she is drawn to master the knowledge or skills, not just to demonstrate them

- *Persistence and effort* – when a student is interested in a task, he or she is likely to expend more effort and persist longer at that task
- *Strongly connected knowledge* – individual interest carries an important cognitive benefit for learning because students have rich knowledge structures to build upon in their areas of interest

They further argued that learning materials, activities and social structures need to be designed carefully to capitalise on interest. Three pillars of the interest-driven learning framework proposed by them are:

- Creating relevance for learning materials
- Strengthening interest through the use of a variety of motivation
- Usefulness – when a student sees the usefulness and applications of acquired new knowledge to real world, that recognition of usefulness creates an inherent motivation to learn

In identifying teaching strategies for interest-driven learning, they included Goal-based Scenarios, Anchored Instruction, Concept Oriented Reading Instruction and Project-based Science. They further criticised that activities should be designed to establish relevance for students' areas of concern. For example, in adopting the Project-based Science strategy to teach air quality, learners can be asked to investigate the air quality in their own community. This will require them to learn the chemistry of air and its impact on the health of people, and to develop inquiry and data analysis skills.

In a structured fashion, Keller & Suzuki (2004) proposed the ARCS model (acronym for **a**ttention, **r**elevance, **c**onfidence and **s**atisfaction) to assist educators in designing motivational tactics that can be integrated with teaching/learning strategies. The model suggests four conditions for motivational design of instruction:

- *Attention* – a lesson must gain and sustain the learner's attention by incorporating a variety of tactics to gain learner attention by the use of interesting graphics, animation, any kind of event and stimulating inquiries, and varying approaches and pace are important for this purpose
- *Relevance* – it is necessary for learners to perceive the instructional requirements to be consistent with their goals, compatible with their learning styles and connected to their personal interests and experiences
- *Confidence* – this is accomplished by helping students establish positive expectancies for success and attribute their successes to their own abilities and efforts
- *Satisfaction* – this is necessary in order for learners to have positive feelings about their learning experiences. This means extrinsic reinforcements such as positive rewards and recognitions must be used

In exploring structured lectures, Jenkins (1992) suggested a number of tactics to encourage student learning, viz.: (1) lectures should be limited to short bursts, (2) after or before each short burst of lecturing, brief learning tasks should be introduced, involving the application of principles and discussions between students in twos or threes, (3) the learning tasks would work best if they are varied in format and process, engaging, open-ended, but possible to tackle and discuss in a few minutes, (4) it is necessary to provide handouts, course guides or guides to reading so that course material is accessible to students outside class, and (5) good social relationships between students, and between students and the lecturer certainly help. If students work in small groups in projects, it will help if they sit in the same grouping in the lecture.

Developing a Motivation-Driven Learning and Teaching Model

The review above identified six essential pillars for a motivation-driven learning and teaching practice, including: (1) establishing relevance to the real world in designing course materials, (2) adopting relevant assessments and providing balanced feedback promptly, (3) utilising a mix of active learning strategies, (4) creating an interactive classroom/learning atmosphere, (5) recognising students' efforts and achievements constantly and openly, and (6) building close student-lecturer relationships. A range of tactics were identified in literature as means to achieve each of these pillars and a preliminary model of motivation-driven learning and teaching was developed. Then, a critical review of the model was conducted with a senior academic in the author's discipline who has won the Vice-Chancellor's Teaching Award, the Dean's Teaching Award and several school level commendations for teaching quality. The peer approved of all the six pillars and their corresponding tactics. However, he suggested another important element that has to be a part of all these pillars; continual improvement. The effectiveness of tactics used for each pillar of the model in different courses should be measured and used for continual improvement of teaching. He further suggested a couple of techniques for measuring the effectiveness: maintaining a teaching logbook to record what worked well and what did not in every class and maintaining a logbook of qualitative student feedback that states students comments and the author's responses. With these, the model was finalised as illustrated in Figure 1. The model can work as an overall framework for teaching practice; however its operationalisation in individual courses is influenced by the context of the course, course aims, expected student learning outcomes and the student cohort. The next section discusses the research method adopted in validating the model.

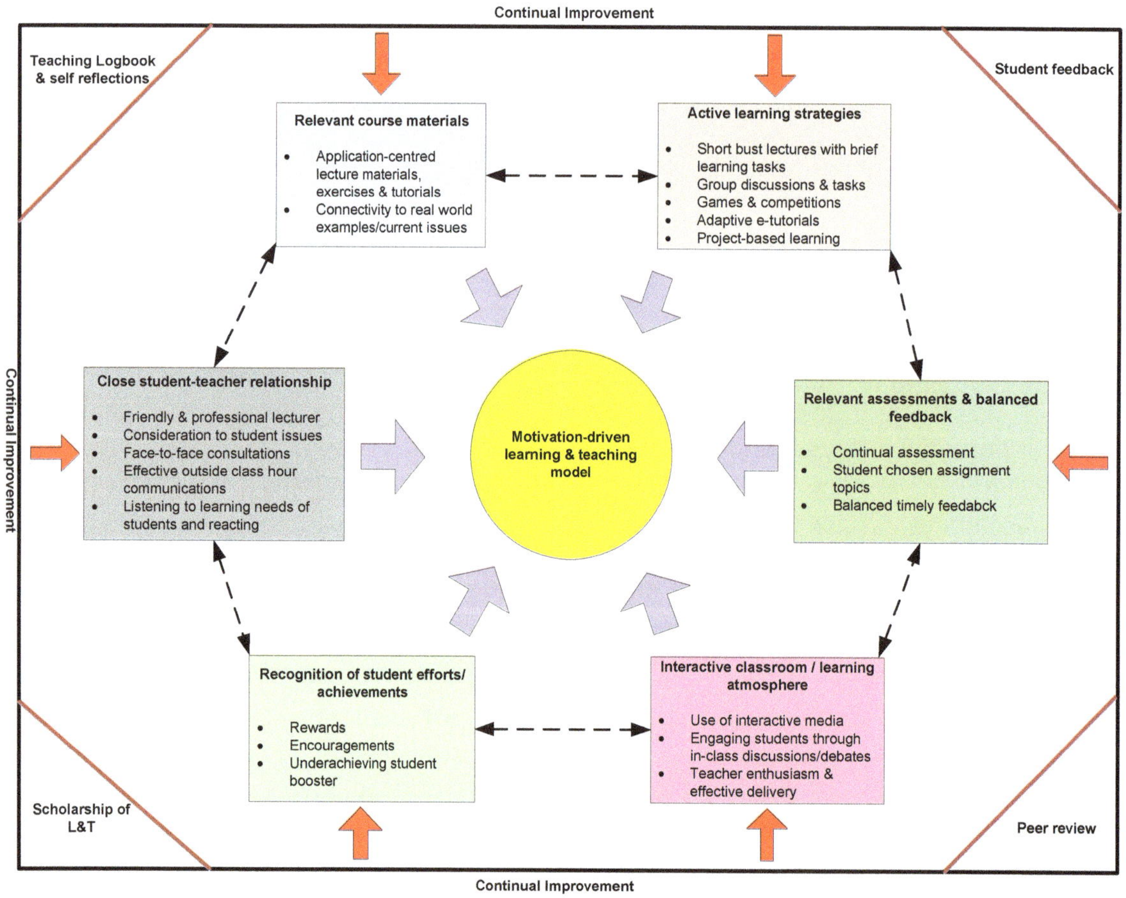

Figure 1 Motivation-driven learning & teaching model

Research Method

The study adopted an action research approach to validate the model. Action research is a process in which participants examine their own educational practice systematically and carefully, using the techniques of research (Ferrance 2000). Action research is a systematic process that allows educators to try out different ways of doing things in the classroom, until they find something that really works for them and students (Laycock & Long 2009). Mertler & Charles (2008) argued this approach is better for lecturers because it:

- deals with their own problems, not someone else's
- can start now—or whenever they are ready—providing immediate results
- provides them with opportunities to better understand, and therefore improve, their educational practices
- provides educators with alternative ways of viewing and approaching educational questions providing a new way of examining their own practices

A first year core course, Construction and Property Economics, from the Bachelor of Construction Management and Property degree program was chosen for the study. The selection was dictated by the nature of the course and the class, viz.:

- The class has always had student enrolments close to a hundred, causing challenges for the lecturer in terms of interaction, classroom management, assessment and feedback.
- Student feedback on the lecturer's teaching quality has been less-than-favourable in the past years.

Three stages were involved in undertaking the action research towards validating the model: (1) the model was operationalised and implemented in the course over a full teaching session, (2) student feedback on their learning experiences was received through an online questionnaire survey at the end of the session, and (3) survey data was analysed, findings were compared with previous years and reflections were made. Following sections elaborate on these stages vividly.

Operationalisation and Implementation of the Model

The author implemented the above learning and teaching model in the course, which covers such topics as design economics, project economic appraisal, cost benefit analysis for public projects and the construction industry and the national economy. The author had taught the course in the previous five years using the traditional model of delivery, involving lectures, tutorials, assignments and exams. The author's experience of teaching the same course over the past years suggested some insights/reflections that are also crucial for redesigning the delivery mode. These are:

- Students are more grade-focused than learning-oriented. They appear interested in learning tasks and active participation in lectures or tutorials if these are directly related to an assessment or grade
- Students favour the continual assessment system over the final exam system
- Students prefer hands-on tasks over formal written submissions
- Flexibility, informality and simplicity are three features that students want in their studies

In light of the proposed new model and insights from previous experience, the author implemented a motivation-driven teaching practice for the course, quite differently to past years. The account below explains how the six pillars of the proposed motivation-driven learning and teaching model were operationalised.

Designing Relevant Course Materials

In designing lecture materials, exercises and tutorial tasks, cases were drawn from real world situations. For example: when discussing economics of building designs interactive videos and actual house plans from the Australian home building industry were used; in teaching project economic appraisal examples from daily lives of students and local property development projects were utilised; the proposed East Coast High Speed Rail Network in Australia was chosen for exercising cost benefit analysis for public projects; and recent national and state policies, proposals and projects were weaved into discussions on the dynamic connections between the national economy and the construction industry.

Adopting Blended Active Learning Strategies

A blended delivery strategy was adopted, which contained lectures, tutorials and exercises, discussions and debates, competitions, projects, class presentations and adaptive e-tutorials.

- The lectures were broken down into short bursts and a variety of brief learning tasks were given to students in between the bursts to provoke interactions and discussions. The lecture notes were made available to students in advance of lectures, which enabled them to write additional self-notes during lectures. Also, students were given time or pauses to allow them jot down extra notes against points in the lecture slides.
- Tutorials and exercises were designed so that the classroom theories were applied to problem solving. Students were encouraged to work in small informal groups in the tutorial classes, discuss and debate solutions for problems given. An additional layer was added in that competitions between student groups were conducted and winning groups/students were awarded prizes.
- A project-based assignment was given to students in which students worked in small groups and each group explored a different subject matter. Continual consultations were in place to guide students in their self-directed learning. The submission consisted of a written report and a class presentation.
- An advanced form of eLearning, known as adaptive eLearning was also practised. An adaptive e-tutorial using a special software platform, called Smart Sparrow™, was made available to help students to prepare for a class test; this approach boasted with improved interactivity, real-time feedback and flexibility.

Applied Assessments and Prompt Feedback

A continual assessment system was used in the course with two class tests and a group assignment. Connections between lectures, tutorial contents and impending class tests were constantly made to spice up students' attention. An opportunity for practice tests was provided through the adaptive e-tutorial portal. The utilisation of the adaptive e-tutorial helped students tremendously to understand concepts and apply them to different situations, as indicated by students.

The assignment was entirely project-based, involved research and analysis, and learning was directed by students according to their learning need, pace and convenience. Clear assessment criteria were provided in advance to facilitate student learning. Students were expected to study and analyse real world cases/situations relevant to their topic for better grades. Intermediate consultations and feedback were held to check on the learning progress. Final submissions included a written report and a class presentation. Because the assignment covered a range of topics, students were exposed to twenty different discussions that were not covered in the scheduled lectures. This added more value to students' knowledge bases and eradicated monotony.

Assessment results and feedback were released within a few weeks after the assessment. This allowed students to understand their standing and areas that need improvements. Balanced and prompt feedback on students' efforts and submissions was provided to sustain student motivation. The continual assessment scheme was a good way for having multiple opportunities for formal feedback and drive students to improve on their performance.

Classroom Atmosphere and Interactive Learning Environment

Several tactics were used to make the classroom atmosphere and the learning interactive. First, the author showed enthusiasm and a passion for teaching the students by being active, walking around the class, talking vibrantly and enthusiastically engaging with students with a positive attitude. At times, entertaining students with jokes helped to eliminate boredom. During the delivery of lecturers, students were constantly engaged through short learning tasks and by both selectively and voluntarily asking students for their inputs. Interactive media such as graphics, animations, simulations and documentaries/digital stories were utilised to reinforce the lectures. In tutorial classes, students were invited to come to the board and explain their answers to fellow students, who were suggested to compare and contrast their answers. Whenever variations occurred, discussions emerged. These made collective learning possible and peer review took a strong place. The lecturer's evaluation and feedback at this juncture acted to reinforce their understanding and clarify grey issues. Moreover, three competitions were held in tutorial tasks, which helped to retain students' focus and engagement in learning. In the project-based learning task, dedicated time slots were allocated for face-to-face consultations between group members and with the lecturer. Moreover, discussion forums and file sharing arrangements were set up on the online learning platform, Blackboard. Apart from the usual online platform that is used at the university, the author utilised the adaptive eLearning platform, Smart Sparrow™, which improved flexibility, interactivity, real-time feedback and stimulating students' interest, as evidenced in qualitative feedback received on that particular tactic.

Recognising Student Efforts and Achievements

Several tactics were used to stimulate students' attention, confidence and motivation through recognition of students' efforts and achievements. Very basically, students who voluntarily contributed to learning tasks during lectures, exercises and tutorial classes were openly encouraged with positive remarks. Prizes were awarded to winners of the competitions in three occasions. And, the winners were publicised and encouraged before the class. Bonus points were made available for completing certain difficult exercises during tutorial classes. These were separate from the prizes for competitions. Individual emails were sent to high performing students in class tests, encouraging them to continue the good work. And, comforting emails were sent to underachieving students and additional support was also provided to them. Special consultation hours were dedicated every week for additional support.

Student-Lecturer Relationship

In order for students to be interested in a course, they should like it. For students to like a course, the lecturer must be likeable to them first. Strengthening the lecturer-student relationship in a professional manner is therefore paramount. The author practised the following habits in that venture:

- The author put an effort in knowing most students by their names and their study circumstances individually. This enabled close interactions throughout the course and students liked it when they are known to the lecturer personally.
- The author was friendly and kind, yet professional with every student. Even some disruptive and rude students were calmed down by a soft approach.

- Paying good attention to individual circumstances of students is crucial. The author took every effort to make students who had study disabilities or personal issues feel that their problems were considered promptly and appropriately.
- Swiftly responding to students' emails and enquiries increased students' confidence in the lecturer
- Informal conversations about daily issues during breaks were helpful to get to know students better.

Effectiveness of the Model

The effectiveness of the proposed new motivation-driven learning and teaching model, which was mirrored in the above teaching practice, was evaluated through a formal feedback system used by the author's university. At the conclusion of the teaching session, the university conducted an anonymous online questionnaire survey amongst students to receive their feedback on the quality of teaching of lecturers. The survey evaluated the following nine quality attributes:

- Effectiveness of communication
- Stimulating students' interest in the subject
- Encouraging critical thinking
- Providing feedback on students' progress
- Interactivity in delivery
- Helpfulness and support to students' learning
- Course structure
- Lecturer's enthusiasm
- Encouraging students to perform

The questionnaire contained ten questions related to the attributes above and student responses to the questions were collected on a 6-point Likert scale. Qualitative comments were also received for the following two questions:

- What were the best features of this lecturer's teaching?
- How can this lecturer's teaching be improved?

All students who enrolled in the course (total of 85 students) were invited to participate in the online survey but only 40 students responded, making a response rate of 47%. Table 1 shows the summary of the survey results. Furthermore, these survey results were compared with the corresponding results of previous years' surveys for the same subject when the motivation-driven teaching practice was not in place but the topic coverage and the expected learning outcomes were the same. The mean response for each question for this year was higher than the values for previous years. The comparison chart in Figure 2 better illustrates this. The X-axis represents the survey questions and the Y-axis represents the corresponding mean student rating for each question for the years considered. The survey results were further analysed in conjunction with the qualitative comments provided by students. Given the nature of qualitative comments, the findings are discussed under five themes in the sections below.

Survey question	Mean rating	Standard deviation
1. This lecturer communicated effectively with students (e.g. He / She explained things clearly).	4.9	1.2
2. This lecturer stimulated my interest in the subject matter he/she was teaching.	4.9	1.1
3. This lecturer encouraged me to think critically.	5.0	1.1
4. This lecturer provided feedback to help me learn.	4.8	1.2
5. This lecturer encouraged student input and participation during classes.	5.2	1.0
6. This lecturer was generally helpful to students.	5.1	1.0
7. The lecturer was well prepared and structured the subject matter well.	5.2	1.0
8. The lecturer was enthusiastic about the material he/she was teaching.	5.3	1.0
9. The lecturer encouraged students to do their best work.	5.3	1.0
10. Overall, I was satisfied with the quality of this lecturer's teaching.	5.1	1.1

Table 1 Summary of survey results

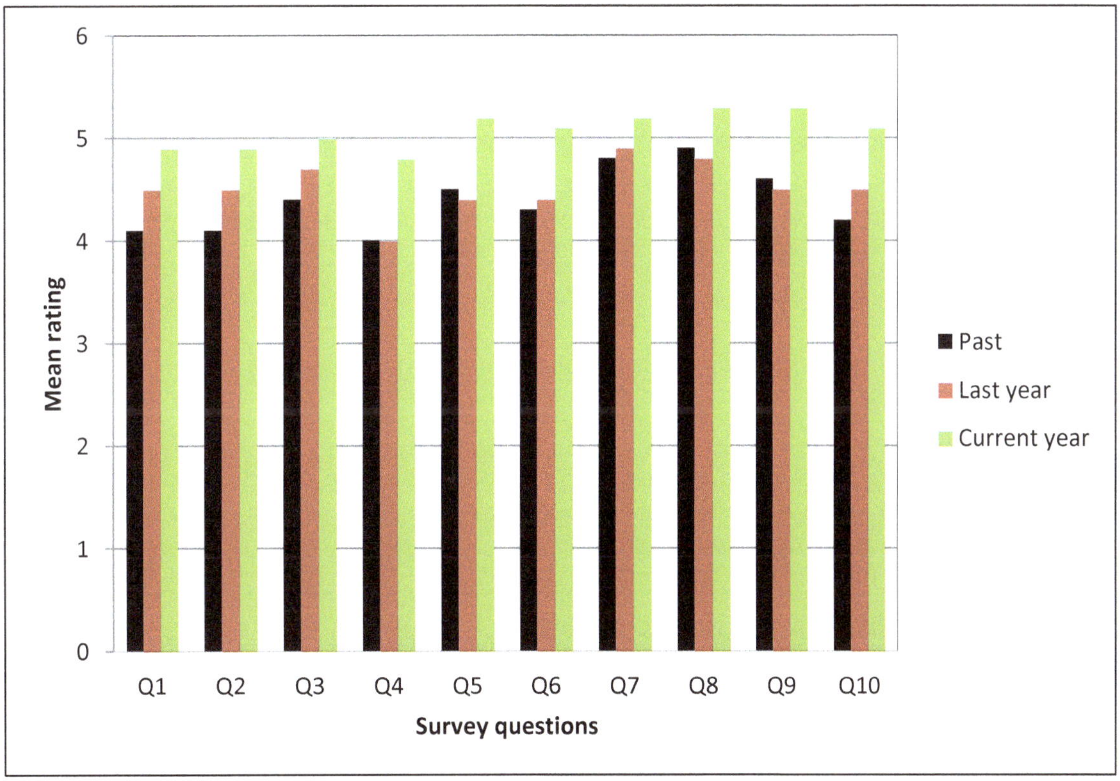

Figure 2 Comparison chart

Effectiveness of Course Delivery

The effectiveness of course delivery was measured by four attributes, including: lecturer's communication effectiveness, interactivity in delivery, course structure and lecturer's enthusiasm. The individual mean ratings received for these attributes were accumulated as shown below and an overall effectiveness indicator of 5.15 (out of 6.0) was yielded.

Q1: This lecturer communicated effectively with students (e.g: he/she explained things clearly) = 4.9

Q5: This lecturer encouraged student input and participation during classes. = 5.2

Q7: The lecturer was well prepared and structured the subject matter well. = 5.2

Q8: The lecturer was enthusiastic about the material he/she was teaching. = 5.3

Cumulative mean rating for 'effectiveness of course delivery' = 5.15

Furthermore, the qualitative comments were analysed to identify what aspects of course delivery resulted in obtaining the high overall rating. It is evident that students look for five elements in appreciating effectiveness of course delivery, which are:

- lecturer's knowledge of and experience in the subject matter
- Well organised and clear lecture notes and tutorial exercises
- Variety and innovation in teaching style
- Interactions and student engagement in classes
- Enthusiastic, cheerful and easy-going nature of the lecturer

Below are some of the direct student quotes, relating to this indicator, for the question "what were the best features of this lecturer's teaching".

Very good academic knowledge of the subject; helpful when asked questions

Very well organised with lecture notes and tutorial exercises

Lecture notes were very good and teaching style was awesome

Good, clear teaching skills, able to deliver the course outline well and keep the lectures productive

The lecturer used the most modern methods of teaching and interactive learning rather than sitting on a stool and reading through boring lecture slides

The variety of teaching styles used was excellent; engaging and tried to get everyone involved

Always taught with a smile; interested in the education of his classes; innovative in his educational methods

This lecturer made an effort to spend extra time during the lecture to focus on aspects of the course which were either important or difficult to understand in order to ensure that students did not get lost or confused. This lecturer was also very good at minimising disruptions caused by talkative students during the lecture

Stimulating Students' Motivation to Learn

The lecturer's ability to stimulate and improve students' motivation to learn was measured by three attributes, including: stimulating interest in the subject, encouraging critical thinking and encouraging best performance. The ratings received for the corresponding questions in the survey were aggregated, which resulted in an overall mean rating of 5.07 out of 6.0.

Q2: This lecturer stimulated my interest in the subject matter he/she was teaching = 4.9

Q3: This lecturer encouraged me to think critically = 5.0

Q9: The lecturer encouraged students to do their best work = 5.3

Cumulative mean rating for 'stimulating students' motivation' = 5.07

As in the previous indicator, the qualitative comments were studied to identify what students perceive as motivating factors. Four aspects in the course seem to have motivated student learning:

- Nature of tutorials and exercises (real-world oriented; designed as competitions and games)
- Use of adaptive eLearning tasks, which provided improved flexibility, interactivity and reflective learning opportunities
- Incentives/rewards for better performance
- Personal encouragements given by the lecturer

Some direct student quotes on these are as follows (these were given to the question "what are the best features of this lecturer's teaching"):

Tutorial exercise games were best

e-tutorials really helped me to prepare for exams

Giving incentives to students is very encouraging as it makes the subject more fun and enjoyable

Encouraged student learning and participation with bonus points

Friendly approach and encouraging attitude; very involved with the class and is enthusiastic about the material taught

Feedback and Learning Support

The level of learning support offered to students was measured by two attributes, including: feedback and helpfulness. The aggregation of the ratings received for these individual attributes resulted in 4.95 out of 6.0.

Q4: This lecturer provided feedback to help me learn = 4.8

Q6: This lecturer was generally helpful to students = 5.1

Cumulative mean rating for 'feedback and learning support' = 4.95

Being helpful, approachable and prompt in replying enquires/questions have supported student learning in the course, as evidenced by the following direct quotes of students (given as answers to the question "what are the best features of this lecturer's teaching"):

Enthusiastic, cheerful and always happy to help

Enthusiastic attitude and willingness to help

Helpful and interactive

Very personal, and provided a lot of extra assistance for students

The lecturer seemed eager to help students and he put together slides that were helpful and easy to understand

The lecturer was a great teacher and assisted in my learning wherever possible. He was also very prompt in responding to email enquiries regarding course work

The author believes that the slightly low overall rating (4.95) for this indicator, as opposed to other indicators, maybe due the less-adequacy of formal written feedback given to students.

Overall Student Learning Experience

The overall student learning experience was computed by aggregating the mean ratings for all the nine attributes (Q1-Q9) that collectively characterise teaching quality. Hence, the overall mean rating for Q1-Q9 was computed to be 5.08, which is almost the same as the rating for Q10. On the whole, it is evident from the survey findings that the proposed motivation-driven learning and teaching model has been effective in improving teaching quality and student learning experience. An all-encompassing comment on this by a student reads as follows:

"The lecturer has clearly defined the course to make it as interesting, engaging and useful as possible. He is happy to help students out and he explicitly defined what was required for assessment tasks to help students prepare. He provided us with plenty of materials of revision, which not many lecturers do. Moreover, he is very knowledgeable and I can walk out of every one of his lecture knowing something new"

Potential Improvements

Continual improvement of tactics is a key element in the proposed motivation-drivel learning and teaching model. Hence, students' suggestions for potential improvements were looked at attentively. The comments on this account show three patterns. Thirty-two per cent of the respondents felt that there is no need for further improvements of the methods used. Responses from another twenty-two per cent of participants suggest that more case studies and practical exercises would be better. The rest of the comments were related to communication. The author is a non-native English speaker and this has inevitably resulted in his accent being different, which attracted some comments. On a positive outlook, the author has set out to revamp the course and delivery in the future on account of all these comments.

Challenges for Implementation

The implementation of the motivation-driven learning and teaching model has resulted in a good learning experience for students and the student feedback is promising for the lecturer. Nonetheless, the lecturer had to overcome a few challenges in successfully implementing it, as discussed below.

Pedagogical Inputs

The lecturer's pedagogical inputs, time and effort invested were more in this approach than that of in the traditional approach. While this was a challenging part, it has been a core element in attracting better reviews from students. The experience of implementing the new learning and teaching model discussed in the paper endorses that the extra effort will pay-off and is worth the investment.

Students' Negative Perceptions

Students' perceptions and feelings about the benefits of various learning tasks used in a course will influence their level of motivation, participation, learning process and thereby their feedback on teaching quality. Hence, for a successful implementation of a motivational strategy, the lecturer should convince students/learners that participation will provide benefits worth the time and effort. Assessments can be used as the point of reference; doing so will stimulate motivation to participate. In this particular case, bonus marks were provided for some difficult tutorial tasks. Also, the adaptive e-tutorial was linked to a class test in that the e-tutorial was provided to students as a supplementary resource to prepare for the test. These tactics encouraged students to actively participate in practising and learning.

Technology

As indicated above, an adaptive-eLearning task was utilised in the course to enhance prompt feedback and student learning. Most of the widespread eLearning systems that are used in universities do not have adaptive eLearning capabilities. It was therefore necessary to use a suitable adaptive eLearning system. This was additional to the eLearning platform that was used in the course. The lecturer had to climb a learning curve and dedicate extra time and effort to familiarise himself with the usage of the new technology.

Conclusions

Traditional instructional models regard learning and teaching as a process-oriented, mechanical practice. However, effective student learning occurs when instructional models incorporate social and psychological dimensions in designing curriculums. Motivation of students to actively engage in learning is a strong psychological force that can deliver desired outcomes for students and lecturers. It is the responsibility of lecturers to motivate their students to achieve the high expectations they have of them. The presence of seven elements in course designs will help lecturers to exercise a motivation-driven learning and teaching practice, which include:

- Relevant course materials that demonstrate applications of theories to real world issues
- A blended delivery strategy with a variety of active learning tactics
- Relevant and application-focused assessments, and prompt and balanced feedback on student performance
- Recognition of student performance through rewards and encouragements
- Interactive classroom atmosphere, filled with intellectual engagement of students and enthusiasm of the lecturer
- Close yet professional relationships between students and the lecturer
- Continual improvement of the tactics used to achieve the above six elements on account of student feedback, peer review, self reflections and scholarships of learning and teaching

The operationalisation of this model of learning and teaching is highly context specific. Lecturers should consider graduate attributes, course aims and expected learning outcomes in developing an effective teaching plan for a given course. Moreover, advanced planning is essential for an effective operationalisation of the motivation-driven learning and teaching strategy and the implementation could demand extra efforts from lecturers. However, the extra efforts would be rewarded pedagogically and in the form of improved student learning and favourable feedback on teaching quality.

References

Beldagli, B. & Adiguzel, T. (2010) 'Illustrating an ideal adaptive e-learning: a conceptual framework', *Procedia-Social and Behavioural Sciences*, **2** (2), 5755-61

Edelson, D.C. & Joseph, D.M. (2004) 'The interest-driven learning design framework: motivating learning through usefulness', *Proceedings of the 6th international conference on Learning sciences*, 166 – 7, Retrieved from http://dl.acm.org/citation.cfm?id=1149145. [14 Aug 2012]

Ferrance, E. (2000) Action research. Brown University, Retrieved from http://www.alliance.brown.edu/pubs/themes_ed/act_research.pdf. [2 Dec 2012]

Freeman, M. & Blayney, P. (2005) 'Promoting interactive in-class learning environments: a comparison of an electronic response system with a traditional alternative', *Proceedings of the 11th Australasian Teaching Economics Conference*, 11-12 July, University of Sydney, 23-34. Retrieved from http://ses.library.usyd.edu.au/bitstream/2123/199/1/03%20Freeman%20Blayney.pdf. [29 Sept 2012]

Gavrin, A.D. & Novak, G.M. (1999) 'What is physics good for? motivating students with online materials', Retrieved from http://serc.carleton.edu/resources/14235.html. [30 Sept 2012]

Jenkins, A. (1992) 'Encouraging active learning in structured lectures', Gibbs, G. (ed.), *Improving the Quality of Student Learning*, Technical and Education Services Ltd, Bristol, 46-58

Keller, J. & Suzuki, K. (2004) 'Learner motivation and e-learning design: a multi-nationally validated process', *Journal of Educational Media*, **29** (3), 229-39.

Kember, D. & McNaught, C. (2007) *Enhancing University Teaching*, Routledge, London

Kember, D., Ho, A. & Hong, C. (2008) 'The importance of establishing relevance in motivating student learning', *Active Learning in Higher Education,* **9** (3), 249-63

Laycock, D. & Long, M. (2009) Action Research? Anyone can! IBSC Global Action Research Project. Retrieved from http://drjj.uitm.edu.my/DRJJ/MATRIC2010/5.%20Anyone_can_Action_Research-DRJJ-02022010.pdf. [2 Dec 2012]

Liu, E. Z., Lin, C., Jian, P. & Liou, P. (2012) 'The dynamics of motivation and learning strategy in a creativity-supporting learning environment in higher education', *The Turkish Online Journal of Educational Technology*, **11** (1), 172 -80

Mertler, C.A. & Charles, C.M. (2008) *'Introduction to education research*, 6th Ed., Allyn & Bacon Boston, Mass

Ramsden, P. 2003, *Learning to teach in higher education*, 2nd Ed., Routledge, London

Underlying Causal Factors Associated with Construction Worker Fatalities Involving Stepladders

3

Aneurin Grant, (University of West Florida, USA)
Jimmie Hinze, (University of Florida, USA)

Abstract

Stepladders are frequently utilized on construction projects as a means of access to elevation. Stepladder heights commonly range from 4 feet (1.2 m) to 14 feet (4.3 m). Since these heights are not extreme, there is a common misperception that stepladder use presents a low risk. On the contrary, extreme care must be exercised to ensure that work on stepladders is performed safely, as described in the conspicuously located recommendations and brightly-colored warnings that adorn virtually all newly-purchased equipment. Despite this, accidents involving stepladders occur on a regular basis. This study was conducted to better understand the underlying causes of these accidents. The narrative descriptions of 180 stepladder-related fatalities were obtained from the U.S. Occupational Safety and Health Administration and were analyzed to identify the at-risk behaviors that preceded the fatality incidents. The results showed that most of the incidents should have been anticipated and could have been avoided. Unsafe practices such as improper lockout-tagout of electrical equipment, loss of balance, working on a folded stepladder, over-reaching, straddling the ladder, "walking" the ladder, poor footing, and unstable/shifting ladders, among others were identified. Virtually all of these fatalities could have been avoided by adhering to the guidelines posted on the stepladders and by complying with basic safe construction practices.

Keywords: Accident causation, Construction worker fatalities, Safe work practices, Stepladders

Introduction

On building construction projects many different crafts are commonly involved and they perform many different tasks. Much of the work is performed at slight elevations and over a relatively short period of time. These tasks are often performed above the finished ceiling height, and include the installation of electrical lines and fixtures, plumbing lines, ductwork, insulation, sprinkler systems, and dropped ceilings. In many instances, the work performed cannot be completed at a single point in time, i.e., some workers may return to the same elevated location on a number of occasions before the installation is complete. When these individual tasks are performed in a short period of time, it is common for workers to perform the work with the use of stepladders, a temporary support system that can be set up and transported quickly.

Stepladders on construction projects have broad utility. Depending on their height, stepladders make it possible for work to be performed at elevations of a few feet to about 14 feet (4.3 m) or more. It is common for stepladders to be used by electricians, sheetmetal workers, carpenters, sheetrock installers, painters, plasterers, wallpaper hangers, plumbers, sprinkler installers, and many other trades.

Stepladders have been in wide usage on construction projects for many years; however, the use of stepladders is inconsistent in the industry. Accidents involving stepladders still occur and some of these result in fatalities.

This study was undertaken to identify the causal factors of stepladder fatalities, and more broadly, the initiating factors in stepladder accidents. This research objective was pursued through the examination of narrative descriptions of fatality cases, considering both the physical conditions that might have contributed to the incident, and the procedures taking place at the time of the fatality incident. An additional objective of this study was to determine if typical manufacturers' warnings and instructions concerning the safe use of stepladders were adequate, or if worker error and improper use were more likely to explain the causes of stepladder fatality incidents.

Literature Review

Falls from elevation are the most common cause of construction worker fatalities. A study of U.S. construction worker fatalities that occurred from 2007 through 2009 revealed that 41% of those fatalities were due to falls (Hinze and Coates 2011). Of these fall fatalities, 13 percent involved ladders. In addition, more than one of every seven ladder-related fatality cases involved stepladders. It should be noted that most of the ladder-related fatalities involved extension ladders. Stepladders are generally of a lower height than typical extension ladders, and would be expected to be involved in fewer fatalities. In an earlier study of the causes of construction worker fall deaths it was found that 11.33% of the fatalities were the result of falls from ladders (Huang and Hinze 2003). Any differences between these statistics have not been determined to be significant.

The dangers associated with the use of ladders have been examined in prior research and are well-documented (Sharp, 2004). Some have attempted to quantify the epidemiological aspects of ladder injuries. Fractures were the most likely injuries to occur from ladder use, and the body parts most likely to be injured were the extremities. Male workers had a higher incidence of injuries associated with ladder use than women, and hospital admission rates increased as the elevation of the falls increased (D'Souza et al, 2007). This latter finding seems to corroborate the findings of Hinze and Coates (2011), who stated that the use of extension ladders is more likely to cause a fatal accident due to the height of the equipment. A separate study observed a relationship between the injury pattern and the elevation of the fall, and that upper extremity fractures are the most likely type of injury (O'Sullivan et al, 2004; Mitra et al, 2007). Fractures are also cited as the most frequent type of injury by other researchers (Partridge et al, 1998; Smith et al 2006), while others found that sprains and contusions occur more frequently (Faegermann and Larsen, 2000; Faegermann and Larsen, 2001).

Some researchers have concluded that accidents are more likely to occur in certain demographics, specifically men (D'Souza et al 2007; Harvinder and Goldbloom, 2008), and older men (Camino et al, 2011; Mitra el al, 2007). Mitra et al. have suggested that intrinsic and extrinsic variables should be examined, which may identify a user's propensity to injury. Intrinsic factors include the individual worker characteristics that might influence the probability of being injured, such as the user's age, weight, and medical condition. Extrinsic conditions pertain more to the physical conditions existing when a task is being performed by the worker, including activity-related and environmental factors, such the task being performed and surface and other localized conditions. Cohen and Lin (1991) examined various extrinsic factors associated with falls from ladders, including overreaching (19%). Faegermann and Larsen (2000; 2001) also concluded that the majority of accidents occur when users reach too far sideways, causing the ladder to tip or slip. Partridge et al. (1998) determined that 79% of the injuries resulting from falls were caused by over-reaching or incorrect ladder placement.

Some researchers have examined ladder mechanics in an attempt to improve safety. Juptner (1976) studied the angle and slip coefficient of the rungs, in addition to the ladder's lateral stability. The "slip coefficient" for the anti-skid mechanisms has been studied in conjunction with several types of ladders of varying lengths (Hakkinen et al 1998).

Countermeasures have also been proposed by Harvinder and Goldbloom (2008), who focused on user competency, personal protective equipment, as well as appropriate conditions, set-up and ladder design. Ladder mechanics have also been analyzed with a focus on the users center of mass or center of gravity, where it has been found that lateral stability is dependent not only on the users movements from side-to-side, but also on the height of the rung above the ground, a finding that suggests that the design of ladders might best be re-examined (Yang and Ashton-Miller, 2005). It should be noted that these prior studies have been focused primarily on the use of extension ladders.

Safety is also a serious concern as it pertains to the use of stepladders. Many construction workers are injured and even killed in incidents in which stepladders are being used at relatively low elevations. These fatalities are especially unfortunate as it is widely felt that the safe use of stepladders is well-publicized. When a new stepladder is purchased, it is common for the stepladder to have several safety stickers attached to it. These stickers or signs on the stepladders are inscribed with safety rules from top to bottom, including instructions for setting up and using the stepladder safely. Examples of such rules include the following:

- Do not over-reach
- Do not straddle front and back
- Keep the body centered between the side-rails
- Do not step on the top two steps
- Do not walk or shift the stepladder while on it
- Set all four feet on a firm level surface
- Lock the spreaders when using the stepladder
- Wear slip-resistant shoes
- Abide by the weight limit
- Do not use folded stepladders as a single ladder
- Always face the stepladder when ascending or descending

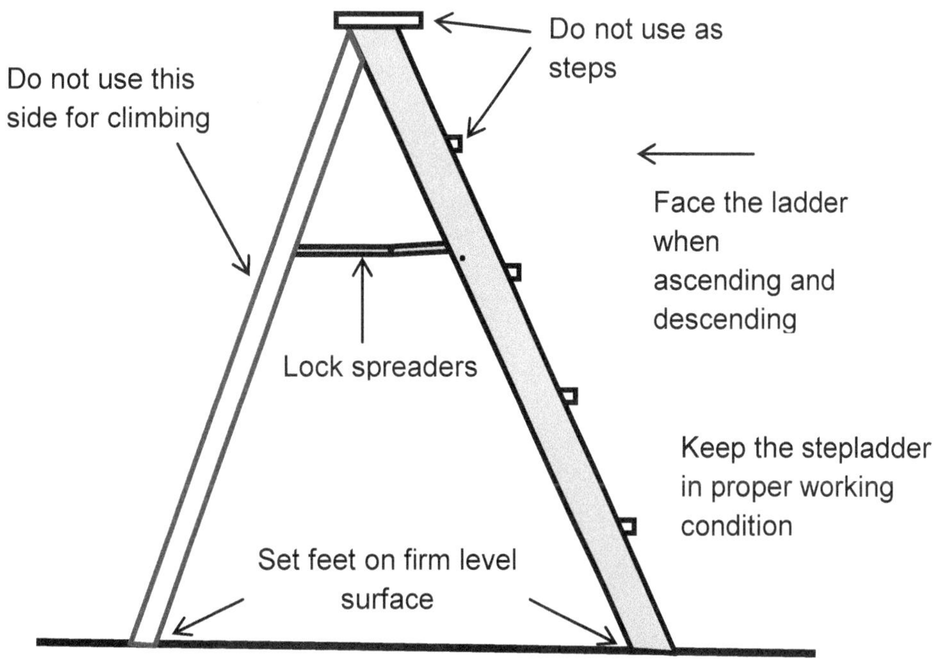

Figure 1 Safety guidelines for typical stepladders

These instructions for the safe use of stepladders are clear and appear to be self-explanatory, however, additional safe work practices involving stepladders can be identified, as not leaving tools or equipment on the top step or tray if the ladder is vacated for any period of time, cordoning off high-traffic areas such as doorways where stepladders must be positioned, and keeping the work area around the stepladder clear. Some of the guidance tips that are common on stepladder stickers are shown in Figure 1. Note that these instructions are valid for stepladders of any size.

Guidance is commonly provided on the inspection of stepladders, the selection of stepladders, and the proper care/storage of stepladders. Furthermore, the safe use of stepladders has been investigated at length and is well documented. Yet, it is not clear that established safety protocol is consistently being followed. On the contrary, it is believed that convenience and the rewards of expedient completion often influence the execution of construction work, including the use of stepladders. As such, one of the main research objectives of this study concerned the identification of workers' disregard for the recommended safety procedures.

Research Method

The data for this research were provided by the U.S. Occupational Safety and Health Administration (OSHA) Office of Statistical Analysis. The database of these fatality cases consisted of the inclusive years of 1990-2009. The database contained narrative descriptions of over 15,000 construction worker fatality incidents that occurred in the twenty-year period.

While the data for some years were coded, this was not done consistently for the various years. To ensure a consistent examination of the data, the causal factors were categorized according to the narrative descriptions of each fatality case. While the descriptions were not prepared in any particular format, it was felt that the most relevant information about the incidents could be obtained in this manner.

First, the incidents involving stepladders were indentified. Specifically, incidents in which stepladders were deemed to be part of the chain of events contributing to the incidents were extracted from the larger database. The causal factors for the stepladder incidents were coded so that the information could be examined through statistical analysis, in this case, with the use of the Statistical Package for the Social Sciences (SPSS). Information about the incidents extracted from the database included the heights of the stepladders, the tasks being performed on the stepladders, the material makeup of the stepladders, the causes of the incident, the types of projects being constructed, the heights of the falls (fall incidents only), and whether the incident was foreseeable and thereby preventable. There was considerable variation in the level of detail provided in the narrative descriptions. As a result, some incidents lacked sufficient information for the purposes of categorization. For example, not every description provided the material makeup of the stepladder, or the height of the ladder.

Results

A total of 180 fatality incidents were identified in which stepladders were involved. One of these incidents involved two fatalities, while all others consisted of a single fatality. In some cases, the stepladders were used in an inappropriate manner. Other cases, especially those involving electrical work, involved poor safety protocol when working on the stepladders. For example, some workers elected not to employ proper lockout/tagout procedures when working with electrical components while on a stepladder. While some workers died from heart attacks or heat stroke, these were not assumed to be closely-associated with the actual use of the stepladders and were not included among the final 180 cases, as the stepladders did not appear to be primary contributing factors in the incidents.

The material makeup of stepladders might be an important selection criterion for some tasks, i.e., aluminum or metal stepladders are not advised for working in locations where electrical contact might be made. The material makeup of the stepladders was identified in 40 incident descriptions. If these incidents are considered to be representative, then most (57%) stepladders are made of fiberglass, with the remaining stepladders being equally divided between wood and aluminum.

As previously noted, many different trades rely on the use of stepladders to perform at least some of their tasks at elevation. Electricians were most numerous among the victims. More than one-third of the victims were electricians, with nearly half dying from electrical shock, while most others died from falls, often caused by an initial electrical shock. Other trades that were involved in stepladder incidents included carpenters, demolition workers, heating ventilation and air conditioning (HVAC) workers, steel erectors, painters, insulation workers, window installers, drywallers, and others.

While stepladders are used most extensively on building projects, the stepladder incidents occurred in a variety of locations, including industrial, commercial and residential projects. Since stepladders are best suited for use on flat surfaces, it is logical that 77% of the incidents occurred indoors.

Workers typically work on stepladders to gain access to a work area that is at a slight elevation, often carrying their tools and materials with them as they ascend, as opposed to ascending the ladders and hoisting the tools and materials with the assistance of a co-worker. In over half of the incidents, it was evident that the victims were carrying either tools or materials that compromised their ability to work safely on the stepladders.

The type of movement on the stepladders (ascending, descending or being stationary on the stepladders) by the workers who were fatality victims was provided for 140 incidents (there were no witnesses for 40 incidents). Of the 140 incidents, one worker lost his balance as he was descending the stepladder while facing away from the ladder, a practice that is clearly contrary to safe work procedures. In most of the remaining incidents workers were stationary, that is, not ascending, descending, nor reaching. When workers were stationery on the stepladders, they were often engaged in some type of work activity For example, in 31 instances the workers were engaged in cutting or drilling tasks. These activities often resulted in unexpected movement, which in turn precipitated fatal fall. These data are summarized in Figure 2.

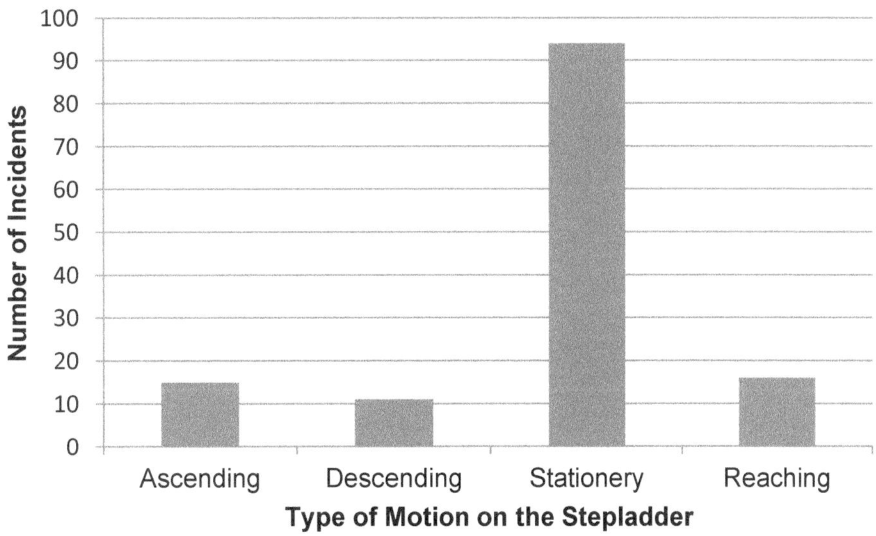

Figure 2 Type of worker motions at the time of the fatal incidents (N = 136)

The heights of the stepladders should be determined by the specific needs of workers. At times, the size of stepladder is determined by convenience, with workers choosing ladders based on simple availability. While nearly half of the stepladders were 8 feet (2.4 m) in length, they varied considerably, ranging from 3 feet (0.9 m) to 14 feet (4.3 m) in length (see Figure 3). The length of the stepladder is measured along the side rails. Thus, the height of a stepladder with 8 feet (2.4 m) long side rails will be slightly less than 8 feet tall when properly set up. Also, since the top two steps are not to be utilized for foot support, the highest standing point on an 8-foot ladder is about 5 feet-9 inches (1.75 m). Despite the relatively shorter heights involved with stepladders, 80 percent of the fatalities analyzed in this study were the result of falls.

Of the fall incidents, it was noted that in 21 incidents the fall height exceeded the height of the stepladder. This occurred when stepladders were placed next to open balconies, windows, elevator shafts, or some other opening. These falls exceeded the stepladder height from 2 (0.6 m) to 200 feet (61 m).

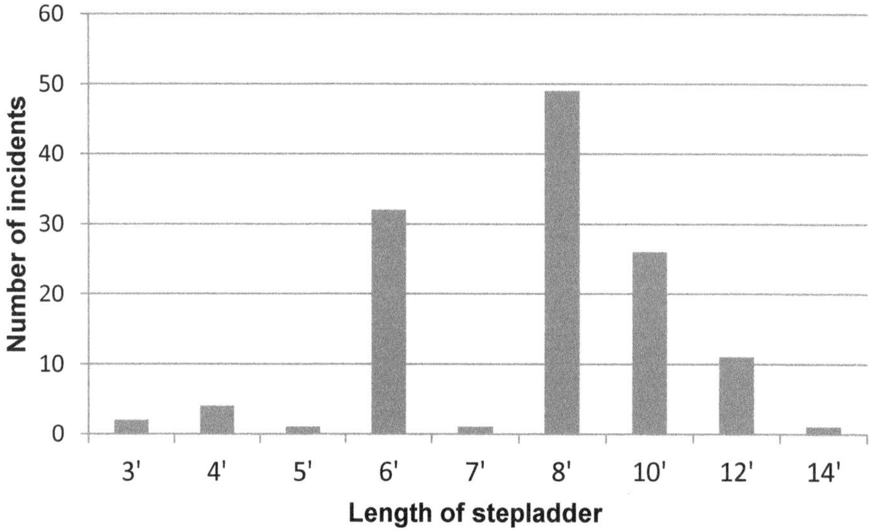

Figure 3 Length of stepladders involved in fatal incidents (N = 127)

The performance of many different tasks was described in the narrative descriptions of stepladder incidents. Many fatalities were not a direct result of stepladder misuse. Rather, many fatalities occurred due to poor electrical safety and lockout/tagout procedures, whereby workers were electrocuted or shocked at elevation and subsequently fell. It is unclear whether these electrocutions and electrical shocks occurred due to incidental contact of energized circuits or materials, or if deliberate work was being performed on hot circuits. Similarly, the narrative descriptions of these incidents did not indicate whether proper lockout/tagout procedures were ignored as a matter of "convenience". As such, the findings of this study do not identify specific frequencies of chance versus overt negligence.

Several specific uses of stepladders are known to place workers at risk. Nonetheless, these unsafe uses continue to be commonplace on many construction sites. Some of these practices will be described more fully, as follows.

Electrocution: A total of 31 electrocutions (17.2%) occurred while workers were working on stepladders. It was also evident that in many of these incidents the workers recognized the electrical hazards, but they ignored them for some reason. It is common practice to lock out electrical current before performing work on a "live" wire or fixture. While aluminum stepladders would be suspected of presenting a greater electrical hazard than the use of wood or fiberglass stepladders, this did not appear to be a factor in the electrocution incidents. In many instances, the electrical current did not pass through the stepladder, but to another grounded source that the worker happened to touch. A typical case is the following:

> An electrician was working from a 6-foot (1.8 m) stepladder installing a lighting fixture. The lighting circuit was energized while he was working on it. The electrician contacted an energized part of the circuit and was electrocuted.

Electric Shock/Fall: There were 30 incidents (16.7%) in which a worker received an electric shock, but was not electrocuted while working on the stepladder. In these cases the electrical shock was followed by a fatal fall. The following is a typical case:

> An electrician was working from a 10-foot (3 m) high stepladder, installing overhead lighting fixtures in a new department store under construction. The electrician was working with the circuit energized, in violation of company policy. He apparently received an electric shock, which caused him to fall from the stepladder to the smooth concrete floor several feet below. (The ladder was not defective or damaged in any way.) The electrician struck his head on the concrete floor and died as a result of the fall.

Loss of Balance: Several of the incident descriptions mentioned that the workers fell from stepladders when they lost their balance. For 24 worker deaths, the cause was noted as loss of balance, without describing the specific cause of the loss of balance. In many of these instances, there were no witnesses. The following is a typical description:

> Worker was standing on a stepladder preparing to use a cutting torch. He lost his balance, fell onto the concrete floor, and was killed.

Leaning a Folded Stepladder: Because of their basic configuration and the locking of the spreaders (see Figure 1) stepladders may be separated from a facing wall. In some circumstances, this distance can be problematic. Space permitting, this problem can be addressed by turning the stepladder 90 degrees so that work can be performed directly to the side of the stepladder. In no circumstance is it acceptable to fold the stepladder and lean it against the wall so that the work can be more easily accessed. The feet of stepladders are not designed for this type of use. Such a use can result in the stepladder slipping on the floor if it is placed at a flat angle, or if the angle is too sharp, the worker's center of gravity might cause the worker to fall backwards. This is similar to what happened to 10 construction workers, as described in the following example:

> Worker was working from a folded 6-foot (1.8 m) stepladder on top of a dust collector system. He lost his balance as the ladder tipped. He fell approximately 14 feet (4.3m) to the concrete pad and died.

Overreaching on the Stepladder: In many ways, the stability of a stepladder is dependent on the user's center of gravity. There is a common "belt buckle rule" that states that the belt buckle of the worker is to always remain between the side rails. In other words, once the belt buckle is beyond the side rail, the worker's position will become unstable. This may occur when work is performed in a linear fashion and the unfinished work extends beyond the side rails. It might be tempting for the worker to stretch just a little bit in order to get a little extra work performed before moving the ladder, but this can be dangerous. The worker should descend the stepladder, realign the stepladder and then ascend the stepladder to perform the next task. In many cases, workers attempt to gain efficiency by minimizing the movement of stepladders. Of course, these attempts to gain efficiency imply the unsafe practice of over-reaching, which makes a fall from elevation more likely. This occurred nine times in the database. A particularly egregious example is as follows:

> A worker was installing a crown molding piece from a tubular welded scaffold that was approximately 16-feet (4.9 m) high and placed around a corner. He placed a stepladder on a balcony bench to extend his reach. Other workers in the area felt that the worker had lost his balance when the ladder moved as he was reaching. He fell approximately 15 feet (4.6 m) to the ground and was killed.

Straddling the Ladder: There were seven instances in which workers used the top step of a stepladder as a seat, generally with one leg straddling each side of the stepladder. This is of course a noncompliant use of the equipment. It is not clear how these falls were initiated in these seven incidents, but it is possible that the workers shifted their weight in such a manner that they initiated a fall.

Standing on the Top Step: The top two rungs are specifically not to be used as steps (see note on Figure 1), as clearly indicated on the safety signs on the side rails of newly purchased stepladders. This rule was violated by seven workers who died when they fell from the stepladders. A typical case is as follows:

> The worker was installing insulation in the garage of a single family dwelling. He was standing on the top step of a 10-foot (3.0 m) stepladder when he fell to the concrete slab floor striking his head on the floor and succumbing to head trauma.

Walking the Stepladder: For some types of work, brief tasks are performed while working on a stepladder. As the next task may be a short distance away, workers may be reluctant to descend the stepladder, reposition it and ascend the stepladder to continue the task. Some workers find the constant descending and ascending effort to be laborious and time-consuming. As such, some have modified the stepladders so that the locking device is inoperative. They can then shift their weight on the stepladder and "walk" the stepladder to the next location. There are different ways of "walking a ladder", including raising the stepladder so all the support is on two legs and then shifting the weight from one side to the other to advance the stepladder. Some workers "walk" with the stepladders by straddling the top of the stepladder and shifting the weight from one pair of legs to the other. Both of these practices are considered to be unsafe. This practice was found to result in the death of four construction workers.

Defective Stepladder: Before each use, stepladders should be inspected for their integrity. The rungs should be tightly fixed, the locking device must be operational, and there should be no broken components. The consequences of defective components are self-evident, yet this was noted as the cause of death for one construction worker.

Lock Spreader not Locked: The lock spreader has the function of keeping the rails of the stepladder at a fixed distance. If the lock spreader is not fully locked, the rails of the stepladder are closer together and this reduces the stability of the stepladder as the rails will not be fixed in place. There was one instance, resulting in a worker death, where the lock spreader was not properly positioned.

Other Causes: There were many other causes of the incidents involving stepladders. In some instances, the cause could not be ascertained due to the lack of witnesses. In some instances, the causes were apparent and should have been anticipated by the workers. For example, the following case involves wind which might have been predicted:

> Two workers were working from stepladders moving sheets of siding. A strong gust of wind apparently caused one of the ladders to tip. The worker jumped free of the stepladder and landed on his feet after dropping 3 to 4 feet. His forward momentum caused him to continue pitching forward striking the top of his head on the asphalt paving knocking him unconscious. He was pronounced dead 2 hours later.

Conclusions

The examination of the narrative descriptions of the fatalities involving stepladders revealed that almost all (95%) of the incidents could have been anticipated, and consequently avoided. Workers often took short-cuts and ignored safe work practices. This may be done in an effort to save time, but ignores the risks to workers' lives. Workers may be at least partially responsible for their own deaths by employing poor work practices. The manufacturers' instructions and warnings commonly affixed to stepladders appear to be more than adequate. Rather, compliance is the problem. In fact, many of the causes of the fatalities involving stepladders were in direct contradiction of the stated instructions and guidance commonly affixed to the stepladders. The question arises as to why these instructions were not followed. Perhaps workers become accustomed to performing work while employing unsafe procedures. Employers cannot let this happen. There is no indication that there is any fault in the design of stepladders.

It may also be concluded that exposure to energized electrical components poses a tremendous risk to workers on step ladders, as is evident in 61 of the fatality incidents discussed in this study. Logically, this type of exposure is germane to electrical workers. However, many other trades are involved in tasks that expose the workers to energized electrical components. For example, workers installing sheet metal ductwork, fire sprinklers, ceiling tile and audio visual equipment may also be exposed to energized components, often through inadvertent contact.

Construction firms should carefully review their internal work rules and procedures. Perhaps the first serious consideration should be whether alternative means are available for workers to work at elevation. Consideration should be given to the use of platform scaffolds or movable stair platforms. Some construction firms have been known to ban the use of stepladders on their projects.

Additional consideration should be given to lockout/tagout procedures. The results of this study suggest that contact with energized components is a major risk associated with the use of step ladders. A re-examination of lockout/tagout policies and procedures may identify unknown, hazardous conditions. Moreover, the identification of such hazards may help to clarify where and when it is appropriate to turn the power off, without impeding the progress of work.

When stepladders are to be employed on a project site, it is imperative that workers are required to follow good safe work procedures. This begins with a clear orientation on the proper use of stepladders and strict enforcement of the safe work procedures.

References

Huang, X. and Hinze, J. (2003) 'Analysis of Construction Worker Fall Accidents', *Journal of Construction Engineering and Management*, ASCE, **129** (3)

Hinze, J. and Coates, W. (2011) 'Trends in Construction Work Fatalities', *Proceedings of CIB W099 Prevention: Means to the End of Injuries, Illnesses, and Fatalities*, Washington, DC, August 24, 2011

Camino Lopez MA, Ritzel DO, Fontaneda Gonzalez I, Gonzalez Alcatara OJ (2011) 'Occupational accidents with ladders in Spain: Risk factors', *Journal of Safety Research*, **42**, 391-398

Cohen, HH and Lin, L. (1991) 'A Scenario Analysis of Ladder Fall Accidents', *Journal of Safety Research*, **22** (1), 31-39

D'Souza AL, Smith GA, Trifiletti, LB (2007) 'Ladder-Related Injuries Treated in Emergency Departments in the United States, 1990–2005', *American Journal of Preventative Medicine*, **32** (5)

Faegermann C, Larsen LB (2001) 'The mechanism and severity of nonoccupational ladder fall injuries', *Journal of Safety Research*, **32**, 333-343

Faegermann C, Larsen LB (2000) 'Non-occupational ladder and scaffold fall injuries' *Accident Analysis and Prevention*, **32**,745-750

Hakkinen KK, Pesonen J, Rajamaki E (1988) 'Experiments on safety in the use of portable ladders', *Journal of Occupational Accidents*, **10**, 1-19

Harvinder SB, Goldbloom D (2008) The Journal of Trauma, Injury, Infection and Critical Care, June 2008

Juptner H (1976) 'Safety on ladders: an ergonomic design approach', *Applied Ergonomics*, **7** (4), 221-223

Mitra B, Cameron PA, Gabbe BJ (2007) 'Ladders Revisited', *Medical Journal of Australia*, **186** (1)

O'Sullivan J, Wakai A, O'Sullivan R, Luke C, Cusack S (2004) 'Ladder fall injuries: patterns and cost of morbidity', *Injury: International Journal of the Care of the Injured*, **35**, 429-431

Smith GS, Timmons RA, Lombardi DA, Mamidi DK, Matz S, Courtney TK, Perry MJ (2006) 'Work-related ladder fall fractures: Identification and diagnosis validation using narrative text', *Accident Analysis and Prevention*, **38**, 973-980

Partridge RA, Virk AS, Antosia RE (1998) 'Causes and Patterns of Injury from Ladder Falls', *Academic Emergency Medicine*, **5**, 31-34

Sharp D (2004) 'Four Up, One Back – And No One Off', *The Lancet*, **363**, 1252

Yang BS, Ashton-Miller JA (2005) 'Factors affecting stepladder stability during a lateral weight transfer: A study in healthy young adults', *Applied Ergonomics*, **36**, 601-607

Comparison of Time Adjustment Clauses between DZ3910, AS4000 and STCC

4

David Finnie, (Otago Polytechnic, New Zealand)

Abstract

This article examines time adjustment clauses, as they relate to time adjustment between standard terms of construction contracts. DZ3910, AS4000 and STCC were compared on the basis of how risks are allocated, how this may impact on the contractor's pricing, and ease of understanding for each clause. ASTCC was found to be the most easily interpreted contract, followed by AS4000 and then NZS3910. These assessments were based on the following: a) whether each contract contains words with multiple meanings, b) the number of words used per sentence, c) the amount of internal cross-referencing, and d) the clarity of the contract structure. The allowable pre-conditions for the contractor to claim a time adjustment are similar for all three contracts, and none of them expressly state which party is to bear the risk of buildability, or address the risk of a designer's disclaimer clause. All of the contracts adopt the principle of contra preferentum which means that the employer bares the risk of variance if there are any ambiguities in the design documentation. Due to their similarities of risk allocation, all of the contracts provide the employer with a similar amount of price surety. AS4000 is the only contract to contain a stringent time-bar clause, limiting a contractor's time adjustment claim. ASTCC requires the contractor to apply 'immediately' and DZ3910 provides a time-bar of 20 working days or as soon as practicable. None of the contracts clarify whether their timing requirements take precedence over the prevention principle, or over any other ground for claiming a time adjustment. The effect of DZ3910's pre-notification clause 5.19.3 is discussed, and an alternative contents structure is recommended for DZ3910, using a project management method.

Introduction

An important aspect of any construction contract is the legal machinery to deal with time adjustments throughout the course of the project. In standard terms of contract, these provisions are commonly called extensions of time (EOT).

This report will compare and critique three conditions of contract:

1. Australian Standard AS 4000 – 1997 General Conditions of Contract
2. Standard Terms of Contract for Small Projects 2012: This is a contract recently drafted by Naseem Ameer Ali, lecturer at Massey University
3. New Zealand Standard NZ3910: 2003 Conditions of Contract for Building and Civil Engineering Construction which is currently being revised to DZS3910. Opportunity for public comment has recently closed, through Standards New Zealand, for its revision

For the purpose of this report, these standard terms of contract will be abbreviated to, AS4000, ASTCC and DZS3910 respectively. Each of these contract clauses will be compared and critically reviewed in relation to the way that they allocate risks between parties; how this may affect the contractor's pricing; and the ease with which the clause may be readily understood. Recommendations are made for improvements to each clause.

While contracts are not necessarily instruments designed to flag the legal issues that lie beneath their expressed terms, they may usefully highlight significant legal principles to the user. Users of construction contracts are likely to be professionals such as contractors, architects and engineers rather than lawyers. A clause may be perfectly comprehensible to read, but leave the reader unaware of its significance within the law of contract. This paper critiques each contract's time adjustment clause for its readability and the ease with which each clause can be understood. Further conclusions have also been drawn concerning the degree to which each clause reveals the related underlying legal principles.

Terms used in each contract also vary in important ways, which can impede understanding. To clarify: the term client means the same as employer or principle, and the term administrator means the same as engineer or superintendent. Copies of each of the clauses are contained in Appendices 2 – 4.

Risk Allocation for Time Adjustment

This paper will first determine what the general legal position is, where there are no expressed terms. Each contract is then compared to this base position in order to determine the impact of its clauses.

There are two main expressed contractual conditions which impact on the allocation of risks between parties for EOT claims. These are the listed pre-conditions by which the contractor may claim an EOT, and any time-bar clauses which limit the period in which they may do so. A third associated condition would be the provision of any liquidated damages. These are a genuine pre-estimate of the damages incurred by the client if the project is completed late (Murdoch & Hughes, 2008). These damages would usually be listed in the specific conditions of contract. In the absence of any specified liquidated damages the client still has the right to claim for general damages. Therefore, the provision of damages has not been critiqued in this paper. In addition to these expressed provisions, there are some other matters relating to general law which also need to be considered. These are the risk of buildability, designer's disclaimer clauses, and the contra-preferentum principle.

The Prevention Principle

A list of pre-conditions may be included in a contract to allow the contractor a means to claim an EOT for matters which are largely out of their control. If there were no listed pre-conditions for which the contractor may claim for delays, there remains an implied legal principle. The prevention principle provides that if the employer delays the contractor, the contractor may claim for their contractual completion date to be extended in compensation for the delay. In fact, if there were no legal mechanism to extent time at all, then the due date could not be extended and time would be at large. This would weaken the employer's position because they would no longer have a date from which to enforce liquidated damages, and the contractor would only need to complete the work in reasonable time as deemed by the courts. Reasonable time is likely to be a less demanding requirement to the contractor than the specified completion date (Murdoch & Hughes, 2008). Therefore, EOT clauses are necessary to provide protection to both the contractor and the employer.

Time-bar Clauses

A strict time-bar clause may be included in a contract to limit the timeframe by which the contractor can claim for an EOT after a delay has occurred. There has been some debate about whether a time-bar clause should take precedence over the prevention principle. One approach is to conclude that a contractor's failure to comply with a time-bar would constitute a breach of contract. Another is that the prevention principle should only be used to reduce the damages for periods of delay when caused by the principle (Finnie, 2011). However, Lim (2009) has asserted that 'by asking for performance when the promise is guilty of delay is to ask the promisor to do something that he or she never agreed to.'

Some guidance has been provided from recent New Zealand and Australian case law. Regarding Andrews v Australia and New Zealand Banking Group Ltd [2012] HCA 30, Heintzman (2013) stated:

> The central conclusion in Andrews v. ANZ is, of itself, extremely important for the law relating to building contracts. All the clauses in a building contract which provide for monetary payments should be closely examined to see if they impose unlawful penalties. Even those clauses which do not give rise to a breach of contract should be examined. If those payments are out of line with the actual detriment arising from the relevant conduct, potentially they are penalties and unlawful, whether or not they are in relation to a breach of contract.

This would indicate that a more liberal approach may be adopted by the courts with regard to the application of time bar clauses.

Buildability Risk

Contractors may be interested to read about the allocation of design risk under traditional building contracts. In a recent paper published by the Society of Construction Law UK 'Design risk allocation under a traditional build contract', (Rosenberg, 2012) highlights that contractors take on the risk of design where it relates to buildability in all but a narrow range of circumstances. Case law examples from within both the New Zealand and the UK legal systems have favoured the employer and engineer over the contractor. For example in the case of Wilkins and Davies Construction Co Ltd v Geraldine Borough [1958], in the Supreme Court, Timaru, New Zealand, the contractor was to sink a sewage tank and pump chamber below ground level, the floor of the tank to be kept dry by pumping from a 44-gallon drum.

However, un-expected ground conditions prevented the sinking of the tank and pump chamber. This resulted in the whole design being abandoned. The contractor claimed for the additional costs associated with the new design. However, the court ruled against the contractor on the grounds that the new work had not been frustrated, nor was it outside the original contract scope (Rosenberg, 2012). Another case in New Zealand involved R M Turton & Co v Kerslake & Partners [2000]. In this case the engineer incorrectly specified heat pumps to achieve a certain level of performance. The judge ruled against the contractor's claim for compensation to correct the heat pumps, on the basis that the engineer did not have a duty of care over the contractor, and because the engineer had defined their liability by way of a disclaimer (Rosenberg, 2012). The disclaimer stated: 'By submitting a tender, the tenderer guarantees that the equipment installed will perform as described in the specification/s' (Kennedy-Grant, 2008).

Bailey (2007) also provided that 'Likewise a proprietor who supplies a design to a contractor, which the contractor promises to implement, does not warrant that the design is "buildable".' In fact, where the contract contains no express terms to the contrary, any increased work resulting from unexpectedly difficult ground conditions becomes the contractor's responsibility. This is further reinforced by Hibberd (2006) who, when referring to JCT's Major Project Construction Contract (MP 05) states:

> In terms of ground conditions (not expressly covered in DB 05) the default position under MP 05 is that the contractor takes the risk, so any necessary changes do not give rise to a Change formally (i.e. no additional cost recovery or extension of time).

Again, Murdoch & Hughes (2008, p177) state that 'As a basic principle, it is clear that the risk of adverse site conditions rests with the contractors.'

Bailey (2009) highlighted the disadvantage of this general principle, which is that if a party (e.g. the contractor) encounters severe unexpected difficulties and cannot recoup the additional cost involved, they may prefer to default rather than incur the associated additional expense. However he went on to conclude that while a legal doctrine based on good faith has its benefits, in reality it would be difficult to implement in the face of contract law. For example a party claiming a breach of good faith would, in effect, be claiming against a breach of contract.

An exception to the contractor's liability regarding site conditions may be established where documents are provided by the principal at the time of tender, without any disclaimer (Bailey, 2007). An expression of this principle can be found in Standard's Australia's General Conditions of Contract AS4000 – 1997 which provides (Standards Australia):

25.1 Scope
Latent conditions are physical conditions on the site and its near surrounds, including artificial things but excluding weather conditions, which differ materially from the physical conditions which should reasonably have been anticipated by a competent Contractor at the time of the Contractor's tender if the Contractor had inspected:

(a) All written information made available by the Principal to the Contractor for the purpose of tendering;
(b) All information influencing the risk allocation in the Contractor's tender and reasonably obtainable by the making of reasonable enquires; and
(c) The site and its near surrounds.

This principle can also be seen within the JCT editions. When discussing a clause in which the contractor is not responsible for the employer's requirements, or for verifying the adequacy of any design contained within them, Hibberd (2005) states:

> The other view of this exclusion is that it only provides the contractor with a limited amount of protection. This is because he has a contractual obligation to give notice if he finds an error, omission or inadequacy in, amongst other things, the Employer's Requirements, which taken together with an implied duty to warn of defects that should be apparent to a competent contractor and an obligation to comply with statutory requirements means a significant level of responsibility is still with the contractor.

Disclaimer Clauses

Importantly, engineers and architects have a duty to take all necessary steps to correct faults in their design which have been brought to their attention. This is dependent however, on their contractual duty, and whether their employment is to produce an initial design only or includes supervision of the works. The engineer/ architect may issue instructions to the contractor via variations, to remedy such design deficiencies (Nissen, 2008), but this obligation nevertheless relies on the contractor to notify them of errors in the first place. As pointed out by Rosenberg (2012), the engineer/ architect may further protect their position by obtaining a proposal from the contractor before issuing a variation, to ensure that the contractor is responsible for any associated costs and/ or time.

The Contra Preferentum Principle

A further point to consider is the contra preferentum principle, whereby any ambiguities within contract documents err against the party who wrote them. However this principle only applies where the document was drafted by only one party to the contract. Where documents have been written by committee, with parties representative of both contractual sides (such as contractors and engineers) contra preferentum would not apply. Rather, any additional cost and/ or time resulting from the ambiguity could be shared between the parties (Murdoch & Hughes, 2008).

Drawings and specifications would not usually be prepared by a committee, but rather by the design team employed by the principal. Therefore the contractor would expect to be reimbursed for any increased cost and/ or time relating to design ambiguities.

An example of expression of this principle can be seen in the following clause from AS4000 – 1997 (Standards Australia) for dealing with document discrepancies:

8.1 Discrepancies
Figured shall prevail over scaled dimensions in a discrepancy. Otherwise, if either party discovers any inconsistency, ambiguity or discrepancy in any document prepared for the purpose of carrying out WUC, that party shall give the Superintendent written notice of it. The Superintendent, thereupon, and upon otherwise becoming aware, shall direct the Contractor as to the interpretation and construction to be followed.

If compliance with any such direction under this subclause causes the Contractor to incur more or less cost than otherwise would have been incurred had the direction not been given, the difference shall be assessed by the Superintendent and added to or deducted from the contract sum.

Who Should Bear Design Risk?

A common principle for allocating contractual risk is to allocate it to the party most capable of managing the risk (Bailey, 2007; Rosenberg, 2012). The client's preference for risk allocation may also be considered.

The client's preference for price surety will influence which type of procurement strategy and form of contract they adopt. Contracts may be categorised as design and build, pure lump sum, lump sum with quantities, re-measurement, and cost reimbursement. A pure lump sum contract is where the contractor is responsible for both the quantum and the rates. In a re-measurement contract the contractor is responsible for the rates, but the client is responsible for the accuracy of the measured quantities. Under a cost reimbursement contract, the contractor is reimbursed for the actual work carried out, so the client adopts the risk of the quantities, the costs and the rates (Kirkham, 2009).

If price surety is a high priority, transferring risk to the contractor will be preferable. However this is likely to increase the contract price, because the contractor may increase their price to account for the risk. For example, if the contractor is to bear the risk of difficult ground conditions, they may base their price for excavation and foundation work on the likelihood of striking poor ground. This will provide the client with price surety, but means that they are likely to pay for the worst case, regardless of whether it eventuates. This could equally apply to the risk of document ambiguities.

Under turnkey contracts (design, build and finance) for public works, the client is more likely to be motivated by price surety over lowest cost (Black, 2005). On the other hand, a private developer with a good understanding of construction may be prepared to accept some of the risk in order to gamble on achieving a lower price. Likewise, a private hospital constructing a series of highly technical extensions may opt for a target cost contract where risks are shared between the parties.

This presents a problem for DZ3910. Unlike the collection of contracts within the FIDIC or JCT series, DZ3910 is a contract which can be either a pure lump sum, re-measure or cost reimbursement contract.

Risk of Ambiguities within Design Documents

The professional designer (client's employee), would be the expert most able to manage the risk of design ambiguities. The contractor is expert at pricing and constructing the works, but not design. On this basis, the risk of ambiguities should be allocated to the client by way of their agent (assuming a traditional contract arrangement).

Under a typical design and build contract, the contractor should be responsible for the design (or design elements that the contractor has produced).

Risk of Design Relating to Buildability

In this case, the contractor should have expert knowledge regarding the buildability of the proposed design. Generally, under any type of contract arrangement (such as lump sum, cost reimbursement or design and build), the contractor would be the most competent to determine whether the design can actually be built.

Comparison of Risk Allocation between Clauses

In section 2 above we have defined the contractual risks, and where (in theory) the various risks should be allocated. We can now interpret each contract type, to determine how they allocate risks between parties and how this may affect the contractor's pricing.

AS4000

- Allowable pre-conditions are called qualifying events. These include:

Qualifying cause of delay Means;

(a) Any act, default or omission of the Superintendent, the Principal or its consultant, agents or other contractors (not being employed by the Contractor); or

(b) Other than:

i) A breach or omission by the Contractor;
ii) Industrial conditions or inclement weather occurring after the date for practical completion; and
iii) Stated in Item 23;

Part (a) and (b) (i) are nothing other than a clarification of the prevention principle. That only leaves an industrial condition or inclement weather as qualifying events. The reader is left to interpret the definition and extent of an industrial condition. I was left wondering whether this means something such as a strike or lock-out as listed in DZ3910, or something more general relating to the construction industry. Importantly, the contractor is left accepting the risk of delay caused by anything else. This would include buildability issues or designer's disclaimers. The contractor needs to be fully aware of their exposure, and either exclude risks in the tender offer or accept it by perhaps increasing their price to allow an amount for acceleration costs that may be required.

- Contains a time-bar clause of 28 days for the contractor to claim for an EOT in Clause 34.3 (b). This is followed by Clause 34.5 which allows the superintendent 28 days to respond. This increases the contractor's risk, in that if they have poor time-management procedures they may lose the opportunity to claim an EOT for an otherwise allowable circumstance (subject to the prevention principle vs. time-bar clause debate mentioned previously). Likewise, if the superintendent fails to respond on time, the contractor's EOT claim will stand, regardless of whether it is valid.

- Clause 34.4 states that when assessing an EOT claim:

34.4 Assessment when both non-qualifying and qualifying causes of delay overlap, the Superintendent shall apportion the resulting delay to WUC according to the respective causes' contribution.

In assessing each the Superintendent shall disregard the question of whether:

(a) WUC can nevertheless reach practical completion without an EOT; or

(b) The Contractor can accelerate, but shall have regard to what prevention and mitigation of the delay has not been effected by the Contractor.

This clause is perhaps unnecessary, because the evaluation of which delaying events impacted on the programme, and granting an EOT for the qualifying events only, would in-effect be the process in the absence of this expressed clause. Perhaps it provides some useful guidance.

- The risk of ambiguities within design documents remains with the employer under the contra-preferentum principle.

ASTCC

- Allowable pre-conditions are listed under Clause 2.5.3 as follows:

2.5.3

(i) the act or failure to act by the Client, the Contract Administrator, or others engaged by the Client or anyone within any of their control;

(ii) exceptionally bad natural environmental event (including exceptionally bad weather) beyond the parties' control occurring at the project site;

(iii) a government authority action;

(iv) the Contractor suspends the work following an instruction from the Contract Administrator to suspend the work; or

(v) the Contractor suspends the work following non-payment as provided under the termination provision in this contract.

Items (i) and (iv) simply confirm the prevention principle. Item (v) is confirming what is covered by statute under New Zealand's Construction Contracts 2002 regarding consequences for non-payment. This leaves natural disasters and government authority actions. Once again, the contractor accepts the risk for any buildability issues and designer's disclaimers.

- The risk of ambiguities within design documents remains with the employer under the contra-preferentum principle.

DZ3910

- The acceptable pre-conditions for claiming an EOT are listed under Clause 10.3.1 as follows:

  (a) The net effect of any Variation; or

  (b) Weather sufficiently inclement to interfere with the progress of the works; or

(c) Any strike, lockout, or other industrial action; or

(d) Loss or damage to the Contract Works or Materials; or

(e) Flood, volcanic, or seismic events; or

(f) Any circumstances not reasonably foreseeable by an experienced Contractor at the time of tendering and not due to the fault of the Contractor.

These seem to be a reasonable list of pre-conditions. Note that the contract does not state any delay caused by the principle or their engineer. This would still be a claimable delay under the prevention principle. In the absence of any expressed term, the contract relies on users having knowledge of this legal principle. This is an important point, because one of the principles of modern plain English drafting is that the contract wording should be appropriate for the parties using it. Users of a construction contract are likely to be professionals, but without specialist legal knowledge (Ameer Ali, 2008). However in this instance, perhaps it could be deemed as covered under Clause (f).

A further important question here is whether Clause (f) covers matters relating to buildability. A good example of this issue is where the contractor encounters unexpected ground conditions which affect buildability. Guidance is provided by Bailey (2007), who explains what the general legal position is:

> If a contractor has promised – in unqualified terms – to construct a building or other structure for a lump sum price, that is what the contractor must do: the agreed price is all that the contractor is entitled to be paid for doing so. It is no excuse for late performance or non-performance that the site conditions which the contractor encountered were worse than it anticipated. In short, the risk of adverse site conditions, whether they be known or latent, is with the contractor. It is therefore only if a contract makes express provision for the payment of additional money in the event of bad ground being encountered that the contractor will be entitled to additional payment (and usually also an extension of time).

The question then is whether Clause 10.3.1 (f) specifically addresses the issue of buildability. This is further clarified again by Bailey (2007):

> The encountering of unforeseen site conditions affords no excuse for not meeting these contractual commitments. It follows that unless a contract makes specific provision for a contractor to be entitled to an extension of time for completion, additional compensation, or both, in the event of adverse site conditions being encountered, the risk of those conditions will remain with the contractor. This is essentially the position adopted by the JCT 05 form, whose clause 2.29 lists exceptionally adverse weather conditions and force majeure as "relevant events", potentially entitling the contractor to an extension of time. But the contract makes no provision for any extension of time or for that matter any increased compensation, in the event of unforeseen bad ground being encountered. Hence the general position applies, i.e. the contractor bears this risk.

DZ3910 Clause 10.3.1 (f) also does not specifically state unforeseen ground conditions, or any other matter relating to buildability. Moreover, the general legal position is that a reasonably competent contractor should be able to peruse the documents and investigate the site, and foresee any issues relating to buildability. Consequently, as under JCT 05, the general legal principle applies here too, that the contractor accepts the risk for any buildability issues and designer's disclaimers.

A further point to consider is that DZ3910 contains a clause 5.19.3 which states:

> If the Contractor does not notify a matter which it reasonably ought to have notified, any Variation arising out of the matter shall be valued by the Engineer as if notice had been given and that notice might reasonably have resulted in the effect of the matter being avoided or minimised.

This Clause expressly shifts the risk to the contractor, unless they notify the engineer of such matters (within a timeframe not yet stipulated).

The risk of ambiguities has been allocated to the contractor for any matters of which they fail to advise the engineer. This increases the contractor's liability from the implied law stance pertaining to the contra preferentum principle. However, this does provide the principal with greater price surety, especially if a time-bar clause requires the contractor to notify any matters within say, 1 month of being awarded the contract. Would this be such a difficult and unusual undertaking for the contractor? In times of a competitive market, contractors may choose to submit bids at or below costs, which they believe to be the likely market price. They then generate claims and variations post-contract in order to recover the shortfall (Heaphy, 2011). This would imply that a contractor's procedure would be to immediately scan the documents for any ambiguities upon being awarded the contract. Furthermore, provided they do carry out this procedure, they are still entitled to any resultant cost and/ or time adjustments, and the principal gains their surety.

An interesting question here is whether the contractor's notification to the principle of any matter 'it ought to' would imply that any associated compensation is then claimable through variations? When it comes to design deficiencies, would this take precedence over the general legal position about buildability? If so, this would certainly benefit the contractor. For example in the case of Wilkins and Davies Construction Co Ltd v Geraldine Borough [1958], relating to the concrete pump dispute discussed earlier, had the contractor notified the engineer that they had a concern about the design, would they have been awarded compensation?

Likewise, in the case of R M Turton & Co v Kerslake & Partners [2000] relating to the under specified heat pumps, would Clause 5.19.3 override the engineer's disclaimer if the contractor raised concerns with the engineer about the capacity of the specified heat pumps?

This does however, have the potential to lead to a ridiculous situation whereby the contractor limits their design risk liability by notifying about all sorts of design matters wherever there could be any possible doubt about its buildability. In order to remove any ambiguity, such clarification should be incorporated into the wording of the clause

.

- Includes time-bar clauses of 20 Working Days for the contractor to claim an EOT and 20 Working Days for the engineer to respond. However, these are softened with the addition of 'or as soon as practicable thereafter.'

- The risk of ambiguities within design documents remains with the employer under the contra-preferentum principle, subject to clause 5.19.3.

Ease of Understanding

Before each contract's ease of understanding is evaluated, it is first necessary to understand what it is, in fact, that makes a contract easy to understand. There has been an international movement towards modern plain English drafting. This aims at writing contracts in a manner which expresses rather than impresses. The focus is on removing long-winded complicated legalese and replacing it with clauses which are as simple as possible without losing their legal intent (Ameer Ali, 2008). Some of these principles include:

1. Having a clear structure which is easy to follow
2. Avoiding the use of words which have multiple meanings such as shall or and/or
3. Words being used in their ordinary sense
4. Using short simple sentences (average of 20 words per sentence is appropriate).

Each clause should be as simple as possible without losing its legal intent.

With this in mind, we can investigate the clauses in each contract:

AS4000

- The contents page is not set out in a project management-type structure, nor does it follow a chronological order. For example, out of 43 items, item 26 is 'Setting out the works'. This is followed by item 27 'Cleaning up'.

- Uses the word shall which has multiple meanings. This is interesting because the Australian and New Zealand governments seem to have moved away from using such words. For example, Ameer Ali (2008) noted:

> there are no "shall" or "and/or" in the New South Wales, Queensland and Victorian Acts. Nor is there any in the New Zealand Construction Contracts Act 2002. And instead of "null and void" the Queensland Act, for example, uses just "void".

Interestingly, AS4000 Clause 34 contains shall ten times!

- Clauses 34.1 – 34.5 contain twelve sentences with an average word count of 25 words per sentence and a maximum of 42 words. This seems reasonable.

- Abbreviations are used, which forces the reader to look up meanings in the glossary. For example WUC means 'work under the contract'. This does not seem excessive. But to find the acceptable pre-conditions for claiming an EOT, I had to look up qualifying event in the glossary. There it lists all the acceptable events, including 'Stated in Item 23;' So next I had to look up Item in the glossary, which for some reason is not listed in alphabetical order like the other terms, which provided 'Means an Item in Annexure Park A;'. Besides the fact that there is a spelling mistake (park should be part), I found it a long and cumbersome process to find something as vitally important as the pre-conditions for claiming EOT.

ASTCC

- ASTCC adopts a project management structure for its contents. Its author, Ameer Ali (2008) promotes this by stating that it 'will lead to greater efficiency during contract administration.' This very quickly becomes evident as I turn to the contents page, see a sub-heading for time and am immediately directed to page 9. Easy!

- There is no internal cross-referencing.

- Words have been used in their ordinary sense without legal jargon. An example is the heading of Clause 2.5 'Adjusting the date for completing the work'. Not only is this title clear to any reader, but it also infers that time may be adjusted both positively and negatively, rather than just being 'extended'.

- There are no words with multiple meanings.

- Clause 2.5.2 states that the 'Contract Administrator must then assess and decide if the date for completing the work is affected.' This means the same as AS4000's clause 34.4. However it uses sixteen words instead of seventy.

- Clauses 2.6, contains ten sentences with an average word count of 18 words per sentence, and a maximum of 25 words. This is less than the recommended average word length of 20 discussed earlier, and provides for easy reading.

DZ3910

- The contents page is not set out in a project management-type structure, although it does follow a roughly chronological order. It has the longest contents page, at three pages. This perhaps reflects the size of the contract, which is 148 pages long. (AS4000 is 39 pages and STCCC is 19 pages).

- Uses the word shall which has multiple meanings. Clause 10.3.contains shall nine times. Clause 10.3.4 alone contains three:

10.3.4
Upon receipt of notice of a claim for extension of time the Engineer shall investigate the claim. The Engineer shall within 20 Working Days or as soon as practicable thereafter determine whether or not the Contractor is fairly entitled to an extension and shall notify the Contractor of his or her decision.

- Clauses 10.3 contains 20 sentences with an average word count of 24 words per sentence, and a maximum of 58 words. While the average word count seems reasonable, there are two sentences with over 50 words, which perhaps could be better structured.

- There are internal references for Clauses 10.3.1 and 10.3.2 to the Guidelines section. Here there is another whole page of writing in relation to Clause 10.3 for Extension of Time. The introduction to the Guidelines state that they 'do not have contractual status' so they have not been evaluated in this paper. However, it is worth noting that this is the only contract which includes a guidelines section. Surely any contract should be drafted in a manner which is clear from the onset, and does not require a further document to explain how to read it.

- Clause 10.3.6 (below) is an example of a clause with excessively long sentences. It describes information about compensation for the contractor's acceleration costs, which is not expressed in either of the other contracts.

10.3.6
In any circumstances in which the Engineer determines that the Contractor, is entitled to an extension of time the Engineer may, subject to the agreement of the Principal and the Contractor, approve the taking by the Contractor of steps to accelerate the work. In such case the Engineer shall grant either no extension of time or such lesser extension as may be agreed. The Contractor shall be entitled to be paid such sum as may be agreed or failing agreement such sum as is reasonable to compensate the Contractor for the additional Cost of the agreed steps taken to accelerate the work and of the effect of those steps on other aspects of the work together with an allowance for profit. The sum payable shall be calculated in accordance with 9.3.6, 9.3.8, 9.3.9, and 9.3.14.

Perhaps the other contracts do not need to address this, because parties to a contract are free to agree to anything lawful. It seems fairly obvious that any agreed costs for valuing a variation resulting from acceleration costs would be calculated in accordance with the clauses for dealing with valuing variations. An alternative wording to this clause is provided below:

10.3.6
Where all parties mutually agree, the Engineer may instruct the Contractor to accelerate their work. Where any change in cost cannot be agreed they will be calculated in accordance with 9.3.6, 9.3.8, 9.3.9, and 9.3.14. Any associated extension time may be adjusted or deleted by the Engineer in reflection of the acceleration.

Recommendations

AS4000

- Adopt a project management-type structure for the contents.
- Replace any words which have multiple meanings. For example replace shall with must, may, or will on every occasion.
- List the qualifying events for claiming an EOT under clause 34 (which deals with time), with reference to any additional events directly to the Annexure Part A.
- Include a clarification explaining how the risks of buildability issues are to be allocated. Also provide clarification that any clauses are subject to designer's disclaimers.
- Clarify what is meant by 'industrial condition' as a qualifying event.

ASTCC

- Include a clarification explaining how the risks of buildability issues are to be allocated. Also provide clarification to explain that all clauses are subject to designer's disclaimers.

DZ3910

- Adopt a project management-type structure for the contents. An example of an alternative structure is provided in Appendix 1.
- Replace any words which have multiple meanings. For example replace shall with must, may, or will on every occasion.
- Include a clarification explaining how the risks of buildability issues are to be allocated. Also provide clarification that any clauses are subject to designer's disclaimers.
- Re-structure some of the longer sentences (e.g. over 40 words long) to make them more readable.
- Where possible, remove the Guidelines by re-writing contract clauses in such a way that no further guidance is required.

Opportunity for Further Research

There seems to be a general misunderstanding in industry about the risk allocation of buildability under a construction contract. When writing this paper, I was surprised to learn about this matter. I asked some of my peers, who included architects, contractors, engineers and quantity surveyors the following question 'If a contractor submits a lump sum price to construct a building, based on drawings and specifications, and then encounters unexpected ground conditions which increase the cost of the work, who pays?' They all answered that the client would pay, because the unexpected ground would be treated as an unforeseen circumstance. I was not surprised by the responses, because that is how I would have answered the same question a few weeks earlier. In fact, a professional quantity surveyor said he had just authorised a sizable sum of money to a contractor for that exact reason.

It would be interesting to carry out a survey amongst industry practitioners to gauge their interpretation of this issue. The amount of money paid out to contractors for unexpected ground conditions could also be researched. Importantly, the issue is not that the client is being 'ripped off' by paying more than they should, because if none of the parties have an understanding of the legal position, the contractors have not priced for the risk in the first place. Instead it seems to be a matter of awareness. Once contractors become aware of the risk, they are likely to exclude it from their price anyway. It is still an important point however, because if the status quo is to remain, contractors will continue to expose themselves to the risk of paying additional costs for which they have not allowed, such as in the case of R M Turton & Co v Kerslake & Partners [2000].

Conclusions

ASTCC is the easiest contract to interpret, followed by AS4000 and then NZS3910. This assessment was made by comparing the numbers of words with double meanings in each contract; the number of words used per sentence; the amount of internal cross-referencing; and the clarity of the contract structure.

There are similarities between all three contracts in the allowable pre-conditions for a contractor to claim a time adjustment, and none of the contracts expressly state which party is to bare the risk of buildability, or address the risk of a designer's disclaimer clause. All three contracts adopt the principle of contra preferentum, which means that the employer carries the risk of variance due to ambiguities in design documentation. Due to their similarities of risk allocation, all of the contracts provide the employer with a similar amount of price surety.

AS4000 is the only contract to contain a stringent time-bar clause for the contractor to claim a time adjustment. ASTCC requires the contractor to apply 'immediately' and DZ3910 provides a time-bar of 20 working days or as soon as practicable. None of the contracts provide whether their timing requirements take precedence of the prevention principle or any other ground for claiming a time adjustment.
An alternative contents structure for DZ3910, using a project management method, is recommended.

References

Ameer Ali, N. (2008) 'Modern plain English drafting and construction: the Malaysian subcontract model terms'. Society of Construction Law UK. D90. http://www.scl.org.uk/node/828 October 2012

Bailey, E. (2009) A doctrine of good faith in New Zealand contractual relationships. Master's thesis, University of Canterbury, Canterbury, New Zealand. http://ir.canterbury.ac.nz/bitstream/10092/2862/1/Thesis_fulltext.pdf October 2012.

Bailey, J. (2007) 'What lies beneath: site conditions and contract risk', Society of Construction Law UK. 137. http://www.scl.org.uk/node/608 October 2012

Black, M. (2005) 'Design risk in FIDIC contracts', Society of Construction Law UK. http://www.scl.org.uk/node/579 October 2012

Finnie, D. (2012) 'Contract Delay what is it and how are we Performing?', *Australasian Journal of Construction Economics and Building*, **12** (1)

Heaphy, I. (2011) 'Do target cost contracts deliver value for money?' Society of Construction Law UK. D126. http://www.scl.org.uk/do-target-cost-contracts-deliver-value-money-1 October 2012

Heintzman, Thomas G. (2013) 'Andrews v ANZ: what are the consequences for building contracts?' Association of Corporate Council. http://www.lexology.com/library/detail.aspx?g=5c212e12-b283-49d2-a872-1e2544854850 October 2012

Hibberd, P. (2005) 'Is single-point design responsibility under JCT 05 illusory?' Society of Construction Law UK. D73. http://www.scl.org.uk/node/806 October 2012

Kennedy-Grant, T. (2008) 'Cross fertilisation – New Zealand developments in construction law and adjudication'. Society of Construction Law Singapore. http://www.kennedygrant.com/docs/sing_oct_08.pdf October 2012

Kirkham, R. (2009) *Ferry and brandon's cost planning of buildings*, 8th ed, Blackwell

Lim, T. (2009) 'Essence of Time in Construction Contracts', The Australasian Journal of Construction Economics and Building, **9** (2)

Murdoch, J. and Hughes, W. (2008) *Construction contracts law and management*, 4th ed, Taylor & Francis

Nissen, A. (2008) 'The designer's duty – time for review'. Society of Construction Law UK. 151. http://www.scl.org.uk/designers-duty-time-review-0 October 2012

Rosenberg, K. (2012) 'Design risk allocation under a traditional build contract'. Society of Construction Law UK. D140. http://www.scl.org.uk/design-risk-allocation-under-traditional-build-contract October 2012

Standards Australia (1997) AS4000 – 1997 General conditions of contract. Standards Australia

Standards New Zealand (2012) Release of draft standard for public comment - DZ3910 conditions of contract for building and civil engineering construction. http://www.shop.standards.co.nz/docserv/drafts/3910_pc_final_draft_V2.pdf October 2012

Appendix 1: Alternative contents structure for DZ3910 using a project management method

Part A: Agreement, Definitions, and Provisions

1. Contract Agreement (Second Schedule)
2. Specific conditions of contract (First Schedule)
3. Special conditions of contract and other conditions of contract (First Schedule)
4. Conditions of Tendering
5. Other Schedules:

5.1.1. Form of Contractor's performance bond (Third Schedule)
5.1.2. Form of Principal's bond (Fourth Schedule)
5.1.3. Form of Contractor's bond in lieu of retentions (Fifth schedule)
5.1.4. Form of producer statement – construction (Sixth schedule)
5.1.5. Information as to contract works insurance (Seventh schedule)
5.1.6. Information as to public liability insurance (Eighth schedule)
5.1.7. Information as to the Contractor's construction machinery insurance (Ninth schedule)
5.1.8. Information as to the Contractor's motor vehicle insurance (Tenth schedule)
5.1.9. Information as to Contractor's professional indemnity insurance (Eleventh schedule)

6. Law, currency and language
7. Application
8. Definitions
9. General

Part B Standard Terms of Contract

1. General obligations, contract administration, and variations

1.1. The Contract
1.1.1. Type of contract
1.1.2. Lump sum contract
1.1.3. Measure and value contract
1.1.4. Cost reimbursement contract

1.1.5. Road or term maintenance contracts
1.1.6. Local authority contracts
1.1.7. Evidence of contract
1.1.8. Documents prepared by the Engineer of Principal
1.1.9. Documents prepared by the Contractor

1.2. Contractor's duties
1.2.1.1. General responsibilities
1.2.1.2. Compliance with laws
1.3. Engineer's powers and responsibilities
1.3.1.1. Duties of the Engineer
1.3.1.2. Appointment of the engineer
1.3.1.3. Role of the engineer
1.3.1.4. Engineer's representative
1.3.1.5. Inspection, recording, measuring and testing
1.3.1.6. Removal and making good
1.3.1.7. Certification
1.3.1.8. Suspension of the work
1.3.1.9. Emergency work
1.3.2. Service of notices

2. Time
2.1. Commencement
2.2. Late supply by the Principal
2.3. Due date for completion
2.4. Contractor's obligations for the programme
2.5. Extension of time (change to adjustment of time)
2.6. Certification of Practical Completion
2.7. Damages for late completion
2.8. Bonus for early completion
2.9. Occupancy by the Principal

3. Financial
3.1. Indemnity and insurance
3.1.1. Indemnity
3.1.2. Contractor to insure Contract Works and Materials
3.1.3. Contractor to insure Contractor's construction machinery
3.1.4. Contractor to insure against public liability risks
3.1.5. Professional indemnity insurance for Contractor's design
3.1.6. Contractor's insurance general
3.1.7. Principal's option to insure Contract Works and Materials
3.1.8. Principal's option to insure public liability
3.1.9. Principal's insurances – General
3.2. Payments
3.2.1. Contractor's payment claims
3.2.2. Progress Payment Schedules
3.2.3. Retention monies
3.2.4. Final payment claim
3.2.5. Final Payment Schedule
3.2.6. Effect of Final Payment Schedule
3.2.7. Interest
3.2.8. Cost Fluctuations
3.2.9. Provisional Sums
3.2.10. Prime Cost Sums
3.2.11. Contingency sums

3.2.12. Goods and services tax
3.3. Variations
3.3.1. Variations permitted
3.3.2. Variation orders
3.3.3. Valuation of Variations
3.3.4. Daywork
3.3.5. Unforeseen physical conditions

4. Quality and other Obligations
4.1. Contractor's obligations for quality management
4.2. Period of defects liability
4.3. Defects liability
4.4. Defects Liability Certificate
4.5. Effect of Defects Liability Certificate
4.6. Guarantees
4.7. Contractor's representative
4.8. Control of employees
4.9. Care of the site and works
4.10. Possession of the site
4.11. Protection of persons and property
4.12. Setting out
4.13. Materials, labour and plant
4.14. Patents and royalties
4.15. Underground and above ground utilities
4.16. Treasure
4.17. Clearance of site
4.18. As-built drawings, operation and maintenance manuals

5. Subcontracting and other Contractors
5.1. General
5.2. Nominated subcontractors (change to Other contractors)
5.3. Contractors obligations for separate contractors

6. Termination
6.1. Default by the Contractor
6.2. Default by the Principal
6.3. Frustration

7. Resolution of Disagreement
7.1. General
7.2. Engineer's review
7.3. Mediation
7.4. Arbitration
7.5. Suspension during dispute
7.6. Award of interest

8. Appendix
A. Cost fluctuation by indexation
B. Construction contracts in public roads
1.B.1. Interpretation
1.B.2. Engineer's powers and responsibilities
1.B.3. Time for completion
C. Term maintenance contracts (including road maintenance contracts)
1.C.1. Interpretation
1.C.2. General obligations

1.C.3. Underground and above ground utilities
1.C.4. Engineer's powers and responsibilities
1.C.5. Time for completion
1.C.6. Defects liability
1.C.7. Payment
D. Obtaining of license by the Contractor

9. Guidelines (remove in preference of clarifying terms within body of contract)
10. Index

Appendix 2: AS4000-1997 Clause 34.1 – 34.5

34 Time and progress

34.1 Progress
The Contractor shall ensure that WUC reaches practical completion by the date for practical completion.

34.2 Notice of delay
A party becoming aware of anything which will probably cause delay to WUC shall promptly give the Superintendent and the other party written notice of that cause and the estimated delay.

34.3 Claim
The Contractor shall be entitled to such extension of time for carrying out WUC (including reaching practical completion) as the Superintendent assesses ('EOT'), if:
(a) The Contractor is or will be delayed in reaching practical completion by a qualifying cause of delay; and
(b) The Contractor gives the Superintendent, within 28 days of when the Contractor should reasonably have become aware of that causation occurring, a written claim for an EOT evidencing the facts of causation and of the delay to WUC (including extent).
If further delay results from a qualifying cause of delay evidenced in a claim under paragraph (b) of this subclause, the Contractor shall claim an EOT for such delay by promptly giving the Superintendent a written claim evidencing the facts of that delay.
34.4 Assessment
when both non-qualifying and qualifying causes of delay overlap, the Superintendent shall apportion the resulting delay to WUC according to the respective causes' contribution. In assessing each the Superintendent shall disregard question of whether:
(a) WUC can nevertheless reach practical completion without an EOT;
or
(b) The Contractor can accelerate, but shall have regard to what prevention and mitigation of the delay has not been effected by the Contractor.

34.5 Extension of time
Within 28 days after receiving the Contractor's claim for an EOT, the Superintendent shall give to the Contractor and the Principal a written direction evidencing the EOT so assessed. If the Superintendent does not do so, there shall be a deemed assessment and direction for an EOT as claimed. Notwithstanding that the Contractor is not entitled to or has not claimed an EOT, the Superintendent may at any time and from time to time before issuing the final certificate direct and EOT.

Appendix 3: ASTCC Clause 2.5

2.5 Adjusting the date for completing the work

2.5.1 The Contractor must notify the Contract Administrator in writing immediately and show the impact on the programme and date for completing the work:

(i) if the Contractor cannot start the work on the date for starting the work; or

(ii) whenever there is any disruption to the Contractor's progress of the work.

2.5.2 The Contract Administrator must then assess and decide if the date for completing the work is affected.

2.5.3 Provided the delay is not due to any breach by the Contractor, the Contract Administrator may adjust the date for completing the work in an adjustment of time certificate due to the following reasons:

(i) the act or failure to act by the Client, the Contract Administrator, or others engaged by the Client or anyone within any of their control;

(ii) exceptionally bad natural environmental event (including exceptionally bad weather) beyond the parties' control occurring at the project site;

(iii) a government authority action;

(iv) the Contractor suspends the work following an instruction from the Contract Administrator to suspend the work; or

(v) the Contractor suspends the work following non-payment as provided under the termination provision in this contract.

Appendix 4: DZ3910 Clause 10.3

10.3 Extension of time

10.3.1 'G'
The Engineer shall grant an extension of the time for completion of the Contract Works or for any Separable Portion if the Contractor is fairly entitled to an extension by reason of:
(a) The net effect of any Variation; or

(b) Weather sufficiently inclement to interfere with the progress of the works; or

(c) Any strike, lockout, or other industrial action; or

(d) Loss or damage to the Contract Works or Materials; or

(e) Flood, volcanic, or seismic events; or

(f) Any circumstances not reasonably foreseeable by an experienced contractor at the time of tendering and not due to the fault of the Contractor.

10.3.2 'G'

The Engineer shall not be bound to grant an extension unless:
(a) The Contractor notifies the Engineer that it claims an extension and states the grounds for the extension;

(b) The notice is given within 20 Working Days after the circumstances arise which are relied on as the grounds for extension, or as soon as practicable thereafter;

(c) The notice either gives details of the period of extension sought or is followed within a reasonable time by a further notice giving such details.

10.3.3
Where the effect of any ground for an extension is of a continuing nature, the Contractor may give more than one notice, each claiming a specific period of extension on that ground.

10.3.4
Upon receipt of notice of a claim for extension of time the Engineer shall investigate the claim. The Engineer shall within 20 Working Days or as soon as practicable thereafter determine whether or not the Contractor is fairly entitled to an extension and shall notify the Contractor of his or her decision.

10.3.5
Upon receipt of details of the period of extension sought by the Contractor the Engineer shall, if he or she has determined that the Contractor is fairly entitled to an extension, then determine the period of the extension and notify the Contractor of his or her decision as soon as practicable.

10.3.6
In any circumstances in which the Engineer determines that the Contractor, is entitled to an extension of time the Engineer may, subject to the agreement of the Principal and the Contractor, approve the taking by the Contractor of steps to accelerate the work. In such case the Engineer shall grant either no extension of time or such lesser extension as may be agreed. The Contractor shall be entitled to be paid such sum as may be agreed or failing agreement such sum as is reasonable to compensate the Contractor for the additional Cost of the agreed steps taken to accelerate the work and of the effect of those steps on other aspects of the work together with an allowance for profit. The sum payable shall be calculated in accordance with 9.3.6, 9.3.8, 9.3.9, and 9.3.14.

10.3.7
The Contractor shall not be entitled to compensation for time-related Costs where an extension of time is granted on grounds other than the net effect of a Variation.

Forecasting the Manpower Demand for Quantity Surveyors in Hong Kong

5

Paul H K Ho (City University of Hong Kong, Hong Kong)

Abstract

Recently, there has been a massive infrastructure development and an increasing demand for public and private housing, resulting in a shortage of qualified quantity surveyors. This study aims to forecast the demand for qualified quantity surveyors in Hong Kong from 2013 to 2015. Literature review indicates that the demand for quantity surveyors is a function of the gross values of building, civil engineering and maintenance works. The proposed forecasting method consists of two steps. The first step is to estimate the gross values of building, civil engineering and maintenance works by time series methods and the second step is to forecast the manpower demand for quantity surveyors by causal methods. The data for quantity surveyors and construction outputs are based on the 'manpower survey reports of the building and civil engineering industry' and the 'gross value of construction works performed by main contractors' respectively. The forecasted manpower demand for quantity surveyors in 2013, 2014 and 2015 are 2,480, 2,632 and 2,804 respectively. Due to the low passing rate of the assessment of professional competence (APC) and the increasing number of retired qualified members, there will be a serious shortage of qualified quantity surveyors in the coming three years.

Keywords: Hong Kong, Manpower forecast, Quantity surveyors

Introduction

Since the early 2000s, Hong Kong has suffered a great economic recession which affects nearly all economic sectors, including the construction industry. Construction related firms and professions have actively looked for business and job opportunities outside Hong Kong. The massive building and infrastructure developments in Macau, Mainland China and some overseas countries provide a good opportunity for quantity surveying firms to expend their business. Following a long hibernation period and the global financial crisis in 2008, the Hong Kong Special Administrative Region (HKSAR) Government has decided to revitalise the economy by promoting infrastructure development, sustaining employment and improving living standard. To fulfil this policy objective, the HKSAR Government has significantly increased its expenditure in capital works from $20.5 billion in 2007/08 to $62.3 billion in 2012-13 and will further increase over $70 billion in the next few years. In addition to the large-scale infrastructure projects, the government also provides public rental housing for low-income families, targeting to complete 75 000 units in the next five years from 2011-12. Besides, the HKSAR Government has also resumed the new Home Ownership Scheme, targeting to supply more than 17 000 flats over four years from 2016-17 (HKSAR Government, 2012). Furthermore, the demand for private residential properties has also increased significantly due to the modest economic growth, better living standard requirements and increasing number of households, thus attracting more developers to invest in private residential properties. As a result of the massive infrastructure development and the increasing demand for public and private housing, many consultancy firms, developers and contractors have experienced difficulties in recruiting quantity surveying staff.

In Hong Kong, the University of Hong Kong, the City University of Hong Kong and the Hong Kong Polytechnic University offer full-time surveying degree programmes which admit totally about 200 students per annum. All surveying programmes offered by these universities is a generic programme combining the building surveying, quantity surveying and general practice surveying disciplines. Some graduates may eventually opt for building surveying and general practice surveying disciplines. Thus, the number of graduates who finally ends up in the quantity surveying discipline is much smaller than the admission quota.

The Hong Kong Institute of Surveyors (HKIS) is the only professional organisation representing the surveying profession in Hong Kong. In order to become a qualified member of the HKIS, all surveying graduates are required to undergo at least two years professional training under the supervision of an experienced qualified surveyor before attending the assessment of professional competence (APC). If a probationer can pass the APC, he/she will become a professional member of the HKIS. Currently, there are about 1 180 probationers undertaking the professional training in the quantity surveying discipline. However, the average number of candidates passing the APC is only about 89 each year. The low APC passing rate creates a bottleneck for the supply of qualified members. In addition, some qualified members also retire after reaching the normal retirement age of 65. The net increase in the number of qualified members is nominal.

At present, there is a strong demand for the qualified quantity surveyors due to the significant increase in the construction workload. However, there is limited supply of qualified quantity surveyors due to the low APC passing rate and the retiring of qualified members. As a result, there has been an unhealthy competition of human resources in the quantity surveying field. It is necessary to understand the demand for, and supply, of quantity surveyors in the near future so that the relevant parties may take the appropriate action to resolve the human shortage problem. Therefore, the aim of this study is to forecast the manpower demand for professionally qualified quantity surveyors in Hong Kong in the coming 3 years. While this study primarily deals with a practical human shortage issue, it will also propose a theoretical forecasting model that can overcome the problem of insufficient input data, while providing a reasonably accurate forecast.

Literature Review on Forecasting Manpower Demand

Manpower forecasting models in construction can be divided into four levels: the national aggregate manpower level, occupational manpower level, regional manpower level and regional manpower by the occupation level (Briscoe and Wilson, 1993). There are four commonly used statistical approaches to forecast the aggregate manpower based on time series data: (1) single-equation regression model, (2) simultaneous-equation regression model, (3) autoregressive integrated moving average (ARIMA) model and (4) vector autoregression (VAR) model (Gujarati and Porter, 2009).

Over the past five decades, economists have developed a wide range of theoretical models to forecast the employment level (Wilson, 1980). Early versions of short-run employment functions were derived from neoclassical economics theories. The most seminal short-run employment model was proposed by Ball and St. Cyr (1966), who postulated that the employment level was determined by output via a production function, while the time trend reflected the productivity growth generated by the capital accumulation and technical progress. This autoregressive model included one lagged dependent variable among its explanatory variables as follows:

$$\log E_t = a_0 - a_1 t + a_2 \log Q_t + a_3 \log E_{t-1}, \quad (1)$$

where E = the employment level, Q = the product output and t = the time trend as a proxy for the capital input term. Ball and St. Cyr's model was considered as a short-term approximation since in the long-term the capital labour ratio would be influenced by factor price ratios.

Dhrymes (1969) argued that a profit-maximizing employer would have a desired level of labour input such that the marginal product of additional labour was related to the cost of that labour relative to the price of output. Dhrymes incorporated wages as one of the determinants of the desired employment level as follows:

$$\log E_t = a_0 + a_1 \frac{W}{P} + a_2 \log Q_t + a_3 \log E_{t-1}, \quad (2)$$

where E = the employment level, Q = the product output, W = the labour cost and P = the output price. In examining the forecasting ability of these two employment functions, Evans and Roberts (1975) concluded that both the Ball and St. Cyr's and Dhrymes' models seemed to have under-predicted the level of employment.

Early derived models are based on the assumption that employment is demand determined. However, traditional economics theory asserts that the labour market is determined by both the supply side and demand side which are eventually brought into balance through wage adjustments. On this basis, some economists derived the labour supply and demand functions by the simultaneous-equation regression approach. For instance, Beenstock and Warburton (1982) proposed a dynamic relationship between the labour supply and demand as follows:

$$L^d = \alpha_0 + \alpha_1 Q - \alpha_2 w \quad (3)$$

$$L^s = \beta_0 + \beta_1 P + \beta_2 w \quad (4)$$

where Q = the product output, w = the real wage rate and P = the population of working age. The equilibrium condition occurs when labour supply and demand is in balance (i.e. $L^s = L^d$). On substitution from equations (3) and (4), this implies the following equilibrium relationships for real wage rate and employment for the given level of economic activity and population of working age:

$$w^* = II_{10} + II_{11} P + II_{12} Q \quad (5)$$

$$L^* = II_{20} + II_{21} P + II_{22} Q \quad (6)$$

where $II_{11} = \frac{-\beta_1}{\alpha_2+\beta_2}$, $II_{12} = \frac{-\alpha_1}{\alpha_2+\beta_3}$, $II_{21} = \frac{\alpha_2\beta_1}{\alpha_2+\beta_3}$, $II_{22} = \frac{\alpha_1\beta_3}{\alpha_2+\beta_3}$.

By identifying the appropriate reduced-form equations, the relevant coefficients were estimated by the ordinary least squares. However, the parameters estimated from the simultaneous-equation regression model were unstable over time.

The autoregressive integrated moving average (ARIMA) approach focuses on analysing the probabilistic properties of time series data itself. The dependent variable is explained by its lagged values and stochastic error terms. It comprises five basic steps: (1) differencing the series so as to achieve stationarity, (2) identification of a tentative model, (3) estimation of the model, (4) diagnostic checking and (5) using the model for forecasting (Maddala, 2001). It is fairly accurate and reliable for short-term forecasts. However, there are two major weaknesses. First, ARIMA is not derived from any economic theory and cannot provide insight into underlying factors that cause changes in the time series. Second, because ARIMA is based on extrapolation of the past trend into the future, there will be a large error if a sudden change occurs in the future trend. Wong et al. (2005) applied the univariate ARIMA model to forecast five construction manpower time series in Hong Kong: labour productivity, employment level, wage level, unemployment rate and underemployment rate. Their study indicated that except for the employment level, the univariate ARIMA model could produce reasonably good forecasts for other variables. However, they also concluded that univariate projection is not appropriate for forecasting construction employment and that multivariate forecasting analysis would produce more accurate estimates.

The vector autoregression (VAR) approach resembles the simultaneous-equation regression model where several endogenous variables are considered together, but each is explained by its lagged values and the lagged values of all other endogenous variables in the model. Since all variables in a VAR model are usually derived from economic theory, the forecasts obtained can be better than those obtained from the simultaneous equation model (Gujarati and Porter, 2009). Ball and Wood (1995) argued that the UK construction employment was a function of the construction output, different types of output, cost of capital, construction labour wages and materials prices. By combining these variables, they proposed four general models relating the employment growth as follows:

$$\Delta L_t = \alpha_0 + \textstyle\sum_i \alpha_{1i}\,\Delta Y_{t-i} + \sum_i \alpha_{2i}\,\Delta W_{t-i} + \sum_i \alpha_{3i}\,\Delta CC_{t-i} + \sum_i \alpha_{4i}\,\Delta M_{t-i} + \varepsilon_t \qquad (7)$$

$$\Delta L_t = \beta_0 + \textstyle\sum_i \beta_{1i}\,\Delta YH_{t-i} + \sum_i \beta_{2i}\,\Delta YR_{t-i} + \sum_i \beta_{3i}\,\Delta YN_{t-i} + \sum_i \beta_{4i}\,\Delta W_{t-i} + \sum_i \beta_{5i}\,\Delta CC_{t-i} + \sum_i \beta_{6i}\,\Delta M_{t-i} + \varepsilon_t \qquad (8)$$

$$\Delta E_t = \delta_0 + \textstyle\sum_i \delta_{1i}\,\Delta Y_{t-i} + \sum_i \delta_{2i}\,\Delta W_{t-i} + \sum_i \delta_{3i}\,\Delta CC_{t-i} + \sum_i \delta_{4i}\,\Delta M_{t-i} + \varepsilon_t \qquad (9)$$

$$\Delta E_t = \gamma_0 + \textstyle\sum_i \gamma_{1i}\,\Delta YH_{t-i} + \sum_i \gamma_{2i}\,\Delta YR_{t-i} + \sum_i \gamma_{3i}\,\Delta YN_{t-i} + \sum_i \gamma_{4i}\,\Delta W_{t-i} + \sum_i \gamma_{5i}\,\Delta CC_{t-i} + \sum_i \gamma_{6i}\,\Delta M_{t-i} + \varepsilon_t \qquad (10)$$

where L = the total employment, E = number of employees, SE = the self-employment, YH = the output of house building, YN = the output of non-residential construction, YR = the output of repair and maintenance, Y = the total construction output, CC = the cost of capital, PW = the real product wage and W = the deflated construction wage. All variables are in natural logarithms. Contrary to the theoretical prediction, Ball and Wood's study indicated a weak link between the total output and employment because of poor data quality, arising particularly from the self-employed workers. They concluded that more accurate construction data would be necessary; if not available, site-based estimate of labour requirements would be required.

When compared the accuracy among the univariate ARIMA, multiple regression and VEC approaches in predicting the construction manpower demand, Wong et al. (2010) derived the following manpower demand functions:

Regression model: $$\log_e MD = 11.25 + 0.56\log_e Q - 0.59\log_e LP - 0.96v_t \qquad (11)$$

VEC model: $$d = -0.63 + 1.28q + 0.81rw - 0.41mp - 0.02br - 0.77lp \qquad (12)$$

where for the regression model: MD = the manpower demand, Q = the construction output, LP = the labour productivity and v_t is an autoregressive parameter; and for the VEC model: . d = the manpower demand, q = the construction output, rw = the real wage, mp = the material price, br = the interest rate, lp = the labour productivity and all variables are in natural logarithms. Their study found that the multiple regression model was the most accurate method (MAPE = 2.93%), followed by the VEC model (MAPE = 4.38%) and then ARIMA model (MAPE = 11.92%).

The statistical approaches reviewed above are appropriate for forecasting the aggregate manpower demand. However, for a specific occupational manpower, one commonly used forecasting method is the multiplier approach, which is based on the premise that each type of project will demand the same level of human resources per unit of construction output because there should be no significant change in their productivity in the short run

(Uwakweh and Maloney, 1991). The manpower requirement is formulated by first deriving its multiplier.

$$M_i = \frac{D_i}{E_i} \tag{13}$$

where M_i = the multiplier based on historic data, D_i = the total manpower employed in different types of projects, E_i = the total expenditure of different types of projects, and i = different types of projects such as building works, civil engineering works and maintenance works. Once the multiplier is derived from historic data, it can be ulitised to forecast the manpower requirement based on the projected expenditure for different types of projects. Ng et al. (2011) adopted this approach to forecast the manpower demands of construction related professionals in Hong Kong.

As the multiplier approach is based on the proposition that a direct relationship exists between the total expenditure of projects and the total manpower demand, it is very essential that the future expenditure of each category of projects can be accurately estimated in order to produce a reliable forecast. If there is any change in the construction method, construction price level, productivity, wage and working hour, the multiplier must be adjusted in order to produce an accurate forecast.

Proposed Forecasting Model

As revealed from the above literature, there are some sophisticated statistical methods for forecasting the manpower demand. However, all of these methods require correspondingly sophisticated, comprehensive and reliable data in order to produce an accurate forecast. Given these constraints, it is necessary to decide whether to formulate a stringent theoretical model first and make appropriate adjustments to cater for the non-availability of data, or to look closely at existing data and develop a theoretical model which can make the best use of the available data. While both of these approaches have its own merits, this study adopts the latter approach because of the non-availability of reliable data.

The above literature review indicates that one key variable commonly used by all researchers is the product/construction output. Other variables such as the labour wage, labour productivity, output price, material price and cost of capital have a certain effect on the overall labour demand. However, these variables may not be significant for a specific occupational manpower. For instance, the manpower demand for quantity surveyors does not much depend on the construction cost, material price and cost of capital. Indeed, based on the classical economics theory, these variables can be assumed to be fixed in the short term. Therefore, the short-term demand for quantity surveyors is a function of the value of various construction outputs as follows:

$$MD_t = f\,(OB_t, OC_t, OM_t) \tag{14}$$

where t = the time trend, MD_t = the manpower demand for quantity surveyors, OB_t = output value of building work, OC_t = output value of civil engineering work and OM_t = output value of maintenance work. This model specification is in line with Ball and St. Cyr's (1966) and Uwakweh and Maloney's (1991) models which have been used for forecasting aggregate and disaggregate labour demands, respectively.

The proposed forecasting model involves two steps. The first step is to estimate the future values of building, civil engineering and maintenance works. This can be done by univariate forecasting (or time series) methods because it only uses the past, internal patterns in data to forecast the future. Univariate methods include exponential smoothing models (such as the simple, Brown's linear trend, Holt's linear trend, simple seasonal, damped trend, and Winters' additive and multiplicative models) and univariate ARIMA models. Since different models may produce different forecasts, a search process is undertaken to find out the best-fitting model that produces the most accurate result.

The second step is to forecast the manpower demand for quantity surveyors based on the forecasted values of building, civil engineering and maintenance works. This can be done by multivariate forecasting (or causal) methods which make projections of the future by modelling the relationship between the dependent variable, MD_t, and the independent variables, OB_t, OC_t and OM_t. Multivariate methods include multiple regression, econometric, multivariate time series and few advanced techniques. However, most of these models require a large amount of statistical data in order to produce a reliable result. Due to this data limitation, it is found that the multiple regression model is the only feasible approach.

Data Sources

This study requires two types of time series data: the manpower demand for quantity surveyors and the gross value of various construction outputs. The data for quantity surveyors was based on the "Manpower Survey Reports of the Building and Civil Engineering Industry" published by the Vocational Training Council (VTC). The demand figures were compiled by summing the series of "number employed" and "vacancy". Since the VTC's manpower survey produced only biennial statistics, the manpower figure laid between two survey years was based on the simple average of two consecutive surveys. Figure 1 shows the manpower demand of professional quantity surveyors from 2000 to 2012.

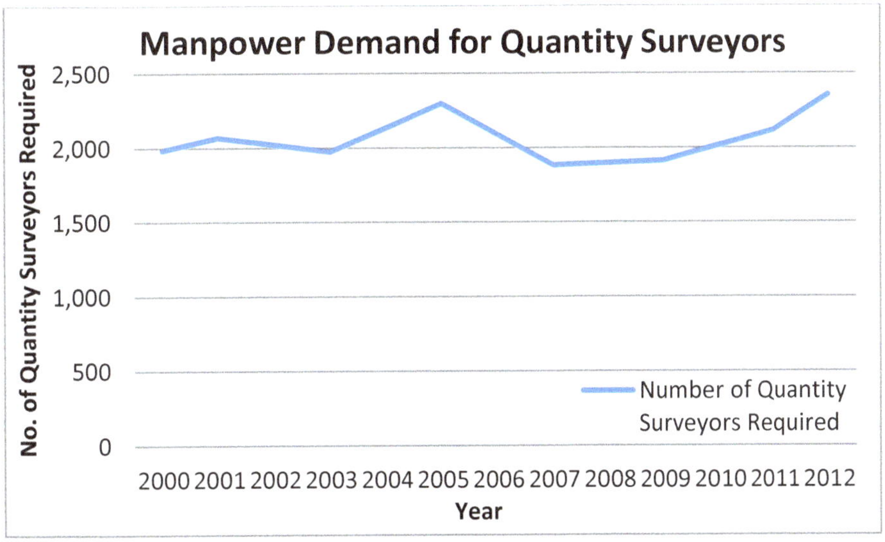

Figure 1 Manpower Demand for Quantity Surveyors

The construction output data was based on the "Gross Value of Construction Works at constant (2000) market prices performed by Main Contractors" published by the Census and Statistics Department of the HKSAR Government. The construction sector is divided into three main subsectors: building subsector, civil engineering subsector and maintenance subsector. It can be noted from Figure 2 that the construction output in both the building and civil engineering subsectors were declined substantially from its peak in 1997 to 2008 due to the Asian financial crisis in 1998 and the outbreak of SARS epidemic in 2003, but have significantly increased because the HKSAR Government has implemented a number of mega-infrastructural and housing projects to promote economic growth since 2009. On the other hand, the construction output in the maintenance subsector was gradually increased up to 2008, but has then declined in recent years.

It is revealed from previous studies such as Ball and St. Cyr (1966), Dhrymes (1969), Beenstock and Warburton (1982), Uwakweh and Maloney (1991), Ball and Wood (1995) and

Wong et al. (2010) that there is a direct relationship between the level of employment and the product output. As observed from Figures 1 and 2, both the manpower demand and total construction output curves are in downward and upward trends between 2000-2009 and 2010-2012, respectively, indicating a close relationship between the level of employment and construction output. Nevertheless, there are also few mismatches in some years between these two curves due to various reasons; for instance, the level of employment is only adjusted slowly over time rather than instantaneously. Based on this observation, it is found that these two sets of statistical data generally reflect the theoretical proposition, while there are also few abnormalities.

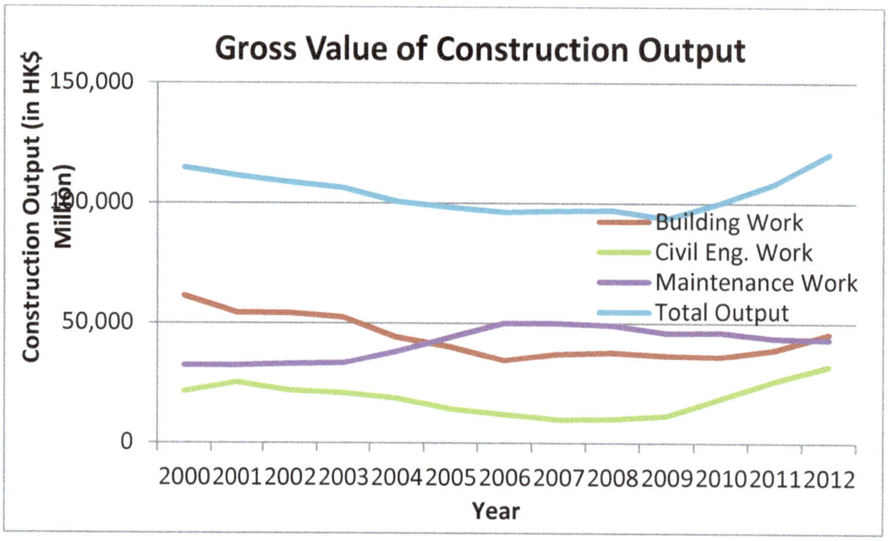

Figure 2 Annual Gross Value of Construction Output at Constant (2000) Market Prices

Findings and Discussions

The manpower forecast for quantity surveyors involves two steps. The first step is to estimate the gross values of various types of construction outputs by univariate time series methods, whereas the second step is to forecast the manpower demand for quantity surveyors by multivariate causal methods.

Forecast of Future Construction Output

First, the time series data in both normal and natural logarithm values are tested by univariate forecasting models. It is found that the data based on natural logarithm values produces a lower mean absolute percentage error (MAPE) and should be adopted. Second, since there are a number of univariate time series models, it is found that after testing all models, the Brown's double exponential smoothing is the best-fitting model for the building and civil engineering works time series, whereas the ARIMA (1,1,0) is the best-fitting model for the maintenance work time series. For both the building and civil engineering works time series, there is a linear trend but no seasonality. For the maintenance work time series, there are one order of autoregression, one order of differencing and zero order of moving average in the ARIMA model. Table 1 shows the model parameters of these three time series.

Model			Estimate	SE	t	Sig.
Log_e Building Work	No Transformation	Alpha (Level and Trend)	0.787	0.141	5.597	0.000
Log_e Civil Engineering Work	No Transformation	Alpha (Level and Trend)	0.951	0.123	7.724	0.000
Log_e Maintenance Work	No Transformation	AR Lag 1 Difference	0.603 1	0.231	2.614	0.024

Table 1 Model Parameters for Forecasted Construction Output

Table 2 shows the forecasted values of building, civil engineering and maintenance works (in natural logarithm) from 2013 to 2015. These values will be used for forecasting the manpower demand for quantity surveyors.

Model	2013	2014	2015
Log_e Building Work	24.648	24.763	24.878
Log_e Civil Engineering Work	24.413	24.639	24.865
Log_e Maintenance Work	24.479	24.474	24.471

Table 2 Forecasted Construction Outputs

Forecast of Future Demand for Quantity Surveyors

First, it is observed from Figure 2 that the time series data of the building, civil engineering and maintenance works are not linear. Thus, it is necessary to transform these data by natural logarithm. Second, the manpower demand function can take the following two forms:

$$\log_e MD_t = f(\log_e OB_t, \log_e OC_t, \log_e OM_t) \quad (15)$$

or

$$\log_e MD_t = f(\log_e MD_{t-1}, \log_e OB_t, \log_e OC_t, \log_e OM_t) \quad (16)$$

Equation (15) implies that the employers would hire and dismiss their employees instantaneously in response to the current construction workload. On the other hand, equation (16) implies that the employers would adjust the level of employment slowly over time in response to the current and future construction workload. After testing both manpower demand functions, it is found that equation (16) produces a R^2 higher than equation (15) and should be adopted.

With $\log_e MD_t$ as the dependent variable and $\log_e MD_{t-1}$, $\log_e OB_t$, $\log_e OC_t$ and $\log_e OM_t$ as independent variables, regression analysis is used to estimate the various coefficients. By using backward elimination method which evaluates all variables in the model and remove the one that results in the smallest change in R^2, and stops when there are no more variables that meet the criterion for removal, the model is summarised in Table 3. It is found that model 1 consists of $\log_e MD_{t-1}$, $\log_e OB_t$, $\log_e OC_t$ and $\log_e OM_t$ as independent variables, whereas model 2 consists of $\log_e MD_{t-1}$, $\log_e OC_t$ and $\log_e OM_t$ as independent variables. Both models 1 and 2 have the same R^2, but the adjusted R^2 of model 2 is higher than model 1. In model 1, the observed significance level is relatively small (p=0.092) so that the null hypothesis that the population values of all of the regression coefficients are 0 can be rejected. On the other hand, the null hypothesis in model 2 cannot be rejected because the observed significance level is high. Therefore, model 1 should be adopted.

Model	R	R Square	Adjusted R Square	Std. Error of the Estimate	Change Statistics				
					R Square Change	F Change	df1	df2	Sig. F Change
1	.799	.639	.433	.053	.639	3.097	4	7	.092
2	.799	.639	.504	.049	.000	.000	1	7	.985
3	.753	.567	.470	.051	-.072	1.605	1	8	.241

Table 3 Model Summary

	Model	Unstandardized Coefficients		t	Sig.
		B	Std. Error		
1	(Constant)	-3.672	15.712	-.234	.822
	Log_e Building Work	.006	.336	.019	.985
	Log_e Civil Engineering Work	.143	.053	2.680	.032
	Log_e Maintenance Work	.157	.337	.467	.655
	Log_e Manpower $Demand_{(t-1)}$	.516	.295	1.751	.123

Table 4 Coefficients

Based on the coefficients of model 1, the derived demand function for quantity surveyors is as follows:

$$\log_e MD_t = -3.6723 + 0.006\log_e OB_t + 0.143\log_e OC_t + 0.157\log_e OM_t + 0.516\log_e MD_{(t-1)} \quad (17)$$

By applying the above derived equation and based on the forecasted values of building, civil engineering and maintenance works, the forecasted manpower demand for quantity surveyors in the coming three years is shown in Table 5. The forecasted demand for qualified quantity surveyors in 2013, 2014 and 2015 are 2 480, 2 632 and 2 804 respectively, representing an average annual increase of 5.7%.

Year	Forecasted $\log_e MD_t$	Forecasted MD_t	Additional No. of QS	Percentage Increase
2013	7.815	2 480	128	5.2%
2014	7.875	2 632	152	5.8%
2015	7.938	2 804	172	6.1%

Table 5 Forecasted Manpower Demand for Quantity Surveyors

Supply of Quantity Surveyors

In theory, the supply of quantity surveyors is determined by (1) the number of students currently studying the surveying programme (i.e. potential members), (2) the number of probationers currently undertaking the professional training under the supervision of qualified quantity surveyors, (3) the number of existing qualified quantity surveyors (i.e. active members), and (4) the number of retired quantity surveyors (i.e. inactive members).

Currently, three local universities admit about 200 full-time surveying students each year. About 100-150 surveying graduates will join the quantity surveying discipline, while the remaining graduates will join the general practice or building surveying disciplines. In order to become a qualified member of the HKIS, a surveying graduate needs to apply to be a probationer and then undertake at least two years professional training. If a probationer passes the HKIS's assessment of professional competence (APC), he/she will become a qualified quantity surveyor. Therefore, there is a gradual increase in the number of qualified members each year.

According to the HKIS's membership statistics, the numbers of available qualified quantity surveyors, probationers and newly qualified members as at 1 January 2013 are shown in Figure 3. At present, there are 2 207 and 1 185 qualified quantity surveyors and probationers respectively. While there are a large number of probationers, only a small number of these probationary members can pass the APC each year. This reflects the relatively low passing rate of the HKIS's APC which has created a bottleneck for the supply of qualified members.

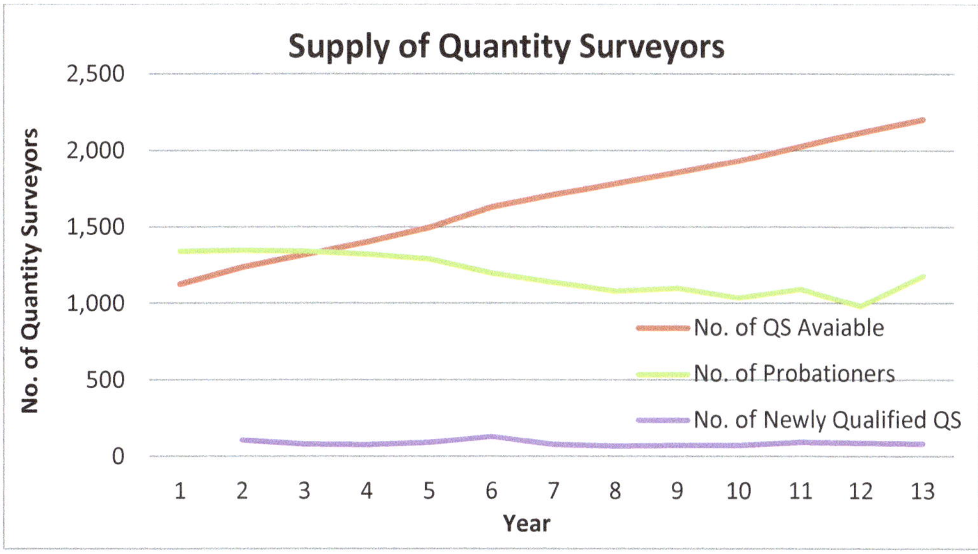

Figure 3 Supply of Quantity Surveyors

Currently, probationers are the main source of supply for qualified members. Based on the past three years' figures, the average number of probationers becoming qualified members (i.e. newly qualified members) is about 90. The past trend is fairly stable as shown in Figure 3. Assuming that the APC passing rate will remain unchanged, there will be a supply of 90 qualified members in coming three years. Based on the current number of qualified members and an increase of 90 members each year, the forecasted supply of qualified quantity surveyors in 2013, 2014 and 2015 are 2 297, 2 387 and 2 477 respectively.

Some of the qualified members would retire at the ages typically at 60-65, depending on individual's situation. Assuming the retirement age of 65 and based on the HKIS's membership profiles, the number of qualified members aged 66 and above will be 52, 59 and 71 in 2013, 2014 and 2015 respectively. After the adjustment of retired members, the number of qualified quantity surveyors will be 2 245, 2 328 and 2 406 in 2013, 2014 and 2015 respectively. The forecasted supply and demand of qualified quantity surveyors are summarised in Table 6.

Year	Forecasted Supply	Forecasted Demand	Shortage
2013	2 245	2 480	235
2014	2 328	2 632	304
2015	2 406	2 804	398

Table 6 Forecasted Supply and Demand of Quantity Surveyors

As shown in Table 6, there will be more and more serious shortage of qualified quantity surveyors in the coming three years because the increasing demand for quantity surveyors, the limited supply of newly qualified members and the retirement of more experienced members.

Conclusions

In view of the increasing workload available in the construction market, this study has investigated a practical issue of the shortage of qualified quantity surveyors in the short term. While there are many sophisticated models for forecasting the supply and demand of manpower, the availability of reliable data has created a great limitation on the choice of these models. As a result, this study has adopted a practical approach to develop a model by making the most efficient use of the available data. If some advanced modelling methods are to be used for improving the accuracy of forecasts, the data quality should be improved first; otherwise, the "garage-in and garage-out" situation would still exit.

As at 1st January 2013, there has been already a shortage of about 145 members. This study indicates that the annual demand for additional quantity surveyors would be 128, 152 and 172 members in 2013, 2014 and 2015 respectively, while the net annual supply would only be 38, 31 and 19 members in 2013, 2014 and 2015 respectively after deducting the retired members. It is clear that the gap between the supply and demand would become wider and wider in coming years as the construction workload will be kept on increasing.

In order to mitigate the manpower shortage, the universities should increase the admission quota for the surveying programme to ensure that there would be adequate supply of fresh graduates to cater for the industrial need. Since education and professional training would take totally at least six years, the increase in surveying students would only be useful in the medium term. In the short term, the HKIS should find ways to help their probationers pass the APC so that there would be more supply of qualified members. In this regard, the HKIS should organise some specific APC training courses (including the mock APC) in addition to the normal pre-qualification structured learning (PQSL) and continued professional development (CPD) seminars. In addition, the employers should also strengthen the on-job professional training for their young surveyors to ensure that they would undergo an adequate training to cover all required competency areas before attending the APC. It is noted that if half of the current probationers could pass the APC, it would be adequate qualified members for serving the construction industry. However, if the above measures still could not resolve the manpower shortage, it would be necessary to import some foreign quantity surveyors. Finally, the manpower supply and demand should be continually monitored so that an appropriate action would be taken to remedy the situation.

Due to the data limitation, this study is based on an assumption that local quantity surveyors would primarily serve the local construction industry. Nowadays, many local quantity surveyors are actually working in projects located in the Mainland China, Macau, Middle East and other countries. In the short term, this could be considered as a constant in the statistical estimate. However, as the globalization trend would increase, more and more quantity surveyors would work on overseas projects. This additional manpower demand should not be neglected and would be a good area for future study, subject to the availability of reliable data.

Traditionally, researchers spend a considerable effort to develop the best all-purpose forecasting model (i.e. one that can forecast all situations accurately). However, in order to overcome the data limitation problem, this study has adopted a different approach where various theoretical models are reviewed at the beginning, and the model that produces the most accurate result is chosen for carrying out the forecast. This study has demonstrated that the traditional goal for the development of all-purpose forecasting model may not be achievable and that some approaches are better than others in particular circumstances. A simple forecasting model may perform better than an advanced model. In addition, most statistical models are only appropriate for forecasting the aggregate manpower demand of individual industry. There are also few previous studies based on a specific occupational manpower. Therefore, this study has contributed to a better understanding of the theory and practice in forecasting specific occupational manpower in the construction industry.

Acknowledgement

I would like to thank the HKIS for providing the relevant membership statistics for this study.

References

Ball, J. and St Cyr, E. (1966) 'Short-term employment functions in British manufacturing', *Review of Economic Studies*, **33**, 179-207.

Ball M. and Wood, A. (1995) 'How many jobs does construction expenditure generate?' *Construction Management and Economics*, **13**, 307-318.

Beenstock, M. and Warburbon, P. (1982) 'An aggregative model of the UK labour market', *Oxford Economic Papers*, **34** (2), 253-275.

Briscoe, G. and Wilson, R. (1993) *Employment forecasting in the construction industry*, Avebury Publishing Co., Hants, England.

Dhrymes, P.J. (1969) 'A model of short-run labour adjustment', in Duesenberry et. al. (eds.), *The Brookings Model: Some Further Results*, North-Holland Publishing Co., Amsterdam, 110-150.

Evans, G.J. and Roberts, C.J. (1975) 'Short-run employment functions in simulation and forecasting', Journal of the Royal Statistical Society, **24** (4), 269-282.

Gujarati, D.N. and Porter, D.C. (2009) *Basic econometrics*, McGraw-Hill, New York.

Hong Kong Special Administrative Region Government (2012) *The 2012-13 Budget*, Financial Secretary.

Maddala, G.S. (2001) *Introduction to econometrics*, John Wiley & Sons, Chichester, England.

Ng, S.T. and others (2011) *Development strategy for professional resources of Hong Kong's construction and related engineering service sector*, A consultancy study for Development Bureau, the HKSAR Government submitted by the University of Hong Kong in association with the Hong Kong Polytechnic University and Ove Arup Hong Kong Limited, Hong Kong.

Uwakweh, B.O. and Maloney, W.F. (1991) 'Conceptual models for manpower planning for the construction industry in developing countries', *Construction Management and Economics*, **9** (5), 451-465.

Wilson, R.A. (1980) 'Comparative forecasting performance of disaggregated employment models', *Applied Economics*, **12**, 85-101.

Wong, J.M.W., Chan, A.P.C. and Chiang, Y.H. (2005) 'Times series forecasts of the construction labour market in Hong Kong: the Box-Jenkins approach', *Construction Management and Economics*, **23**, 979-991.

Wong, J.M.W., Chan, A.P.C. and Chiang, Y.H. (2010) 'Construction manpower demand forecasting: a comparative study of univariate time series, multiple regression and econometric modelling techniques', *Engineering, Construction and Architectural Management*, **18** (1), 7-29.

“Pack 'em, rack 'em and stack 'em”: The Appropriateness of the Use and Reuse of Shipping Containers for Prison Accommodation

6

Elizabeth Grant, (The University of Adelaide, Australia)

Abstract

Shipping containers are gaining increasing recognition for their apparent durability, adaptability, light weight, 'low cost' and ease of stacking, spurring a trend that has resulted in shipping container sculpture, homes, housing, hotels, and museums. The use of prefabricated, pre-manufactured and prototype building methods for prison construction has grown considerably as some jurisdictions attempt to deal with the construction of prisons with speed and economy. In the last three years, shipping containers have been used in the prison sector as a way of managing burgeoning prison populations. Recent prison developments in both Australia and New Zealand where shipping containers have been employed for prisoner housing are of considerable interest. In this article, the financial, functional, structural, technical, environmental and architectonic impacts of this approach are discussed.

Keywords: Prison construction, Shipping containers, Prison architecture, Material reuse, Building methods

Introduction

The advent of the shipping container revolutionised the transport industry (Levinson 2006) profoundly impacting the geography of production and distribution. Production became globalised as distribution systems were able to interact more efficiently (Theo and Jean-Paul; 2008; Notteboom and Rodrigue 2008) with the invention of the shipping container, which allowed goods to be moved between rail, road and sea transport with minimal handling. In 1970, container design was standardised worldwide through the introduction of the International Standards Organization (ISO) shipping container (Sawyers 2011). A typical ISO compliant container is constructed of corrugated weathering and corrosion resistant steel with double leaf doors at one end. The walls are built using a frame and filling design that allows for size and load bearing reinforcement. Both frames and fillings are designed to withstand dynamic and static loads.

The International Standards Organisation stipulates the standards for the dimensions, waterproofness, strength, handling and methods to secure shipping containers (Smith 2005). While there are some minor variations, containers generally measure 2.44 m wide by 2.44 m high, and either 6.1 m or 12.19 m long. Containers must be able to be stacked up to seven units high (Smith 2005). To conform to the standards, the eight corners of containers must have castings with openings for 'twistlock' fasteners. Containers fall into two categories depending on their use. When used solely for shipping, containers are generally referred to as an ISO container, a cargo container, or a 'Conex Box'. When used for building construction or storage, containers are referred to as an ISBU (Intermodal Steel Building Unit) Module or a 'GreenCube'.

When used for transport purposes and exposed to harsh sea conditions and handling, shipping containers have an average life span of between 10 and 15 years (Spurgeon et al. 2003). Some containers are lost at sea while others are written off in handling accidents (Muller 1999; Informa Group 2012). Around one million containers leave the transport sector annually with old, broken, obsolete, damaged or corroded boxes being scrapped and stockpiling as ubiquitous artefacts of industrial capitalism (Kalkin & McLean 2008). However, approximately 300,000 to 500,000 containers experience an afterlife (Spurgeon et al. 2003). These containers go to new industries that have arisen to sell or lease containers or modify them for other uses. One such industry is the modification of shipping containers for building purposes.

Initially modules were favoured for use in post disaster scenarios and in military and mining operations. Their value lay in the capacity of the units be easily transported and rapidly redeployed. In the last decade, there has been growth in the use of shipping containers in other contexts and they have been employed for a myriad of uses. Containers serve as sleep-outs, garden sheds, workshops, student and disaster housing, artists' studios, storage facilities, sales, site and ticket offices, lunch rooms, cafes, and ablutions blocks. A British country house used refrigerated containers to kill insects infesting antique carpets (CS Shipping Containers 2012). The Masakhane Container Mall was built from converted shipping containers in a black settlement in South Africa when there were inadequate funds to build a permanent mall (Keller 1993; Spurgeon et al. 2003). London's Container City (Chapa 2007; Urban Space Management Ltd 2012) and the University of Utrecht Student Housing (Brandt 2011) are interesting examples of large-scale cargo container complexes. Container City is bright and colourful and relies on component pieces rather than units to create adaptable liveable spaces (Chapa 2007) while the Utrecht student housing complex took a one container per unit approach (Brandt 2011). The growth of industries for container conversion has been phenomenal. There are many businesses solely focusing on adapting various grades of pre-used containers for use in the built environment. Other firms are also fabricating new containers for immediate use as accommodation.

A number of countries have also used shipping containers as 'quick fix' solutions for prison accommodation. In 1999, the Western Australian Government purchased 24 renovated containers to use as temporary prison cells to help alleviate overcrowding at the then Canning Vale (now Hakea) Prison. The twenty four renovated shipping containers were fitted out with a window, shelves, floor coverings and a television (Le Grand 1999) but had a relatively short life span. Reports cited that the containers "corroded [and] design problems created security risks" (Winter 2011). In 2002, the US Government adapted shipping containers for use as cells for holding foreign nationals at Camp Delta at the U.S. Naval Base, Guantánamo Bay. Camp Delta was a detention facility with the capacity to apparently hold 612 prisoners in separate units. Each unit was reportedly constructed of "...shipping containers cut halfway lengthwise with the pieces stuck together end to end. Cells were small, six foot eight inches by eight foot with steel beds fixed to the steel mesh walls" (Tucker et al. 2010 p. 502). The units are reported to measure 2.43m long, 2.07m wide and 2.43m tall and constructed with metal mesh material on the sides (Global Security 2011). Canover in 2003 described the layout of Camp Delta as consisting of "twenty-four ...cells, constructed out of Connex shipping containers placed end to end, situated opposite 24 others, and a roof with ventilators is constructed overhead; this assemblage of 48 cells constitutes a cellblock. So far, there are 19 of these cellblocks at Camp Delta, suggesting a capacity of approximately 1,000" (Conover 2003).

The conditions and practices at Guantánamo Bay have always been a point of contention and in 2004, it was reported that Pentagon planners were seeking funding to build a purpose built prison to replace the rows of rugged cells fashioned from shipping containers transported from South Asia (Kenny 2003; Rosenberg 2004). In his second day in office, President Obama issued a directive to cease detentions at Guantánamo Bay (Mendelson and CSIS Working Group on Guantánamo and Detention Policy 2008) and close the facility (Crook 2009). Due to the refusal of the US Senate to release funding for the phasing out of the Guantánamo Bay Facility, a prohibition of the use of federal funding to "release, transfer or incarcerate detainees detained at Guantanamo, Cuba to or within the United States"

(Corcoran 2011 p. 209) and the subsequent passing of Government bill banning the transfer of detainees to the United States (even for prosecution), operations at this point continue at the facility (Corcoran 2011).

More recently, shipping containers have been employed for prison construction in both New Zealand and Australia. With a paucity of research into the use and reuse of shipping containers for this specific use, it is a timely and important to consider the benefits and constraints of reusing shipping containers from a number of perspectives. This article will describe recent projects, discussing the cost effectiveness, building performance and the environmental and wider implications of taking this approach.

The Rimutaka Container Prison Project

The Rimutaka Container Prison Project was conceived to increase the capacity of an existing prison to cope with a burgeoning prison population. New Zealand's prison population peaked in 2010 at 8,816 people and forecasts suggest the prison population is expected to reach 10,314 by 2017 (Department of Corrections New Zealand 2010). With projections of an increasing prison population, New Zealand Corrections commenced investigating options to increase prison capacity. One initiative included the use of prefabricated or modular prison cells as a way of housing rising prisoner numbers.

The additional unit developed at Rimutaka Prison was designed to house 60 low to medium security prisoners using ex-shipping containers. The design of the unit followed the design principles used in many of New Zealand male prisons, where cells and ablution blocks are located around three sides of an open court yard. The entrance administration, staff facilities, control room and dining spaces are accommodated on the fourth side (Grant 2009; Department of Corrections 2012). This design of individual cells located around an external area has some disadvantages. The central external area is unusable during inclement winter and staff tend to have less interaction with prisoners. Most modern prisons in both Australia and New Zealand situate cells around a central internal living area (using the principles of third generation or podular design) to increase staff interaction with prisoners. The shipping containers are covered by a roof that encompasses the complex. In all, 23 high grade shipping containers were used to develop the unit and prisoners are accommodated in a mix of double and single bunked cells with each shipping container housing three cells (Department of Corrections New Zealand 2009).

The budget for the Rimutaka Prison Container Project was set at $5.4m (NZ) ($4.46m AUD) for infrastructure and services, plus a further $2m (NZ) ($1.65 AUD) for fencing and external security elements. While it is reported the project was completed under budget (Rawlinsons 2010), estimates of the actual cost of the refit of the containers and development vary slightly. One source suggests the capital cost of each bed lies somewhere between $53,000 NZ (Wheeland 2010) ($43798 AUD) and $70,000 NZ (Gower 2010) ($57846 AUD). The Rimutaka Prison Container Project immediately spurred interest by other correctional jurisdictions also examining cost-effective methods to house burgeoning prison populations.

The South Australian Developments

One such jurisdiction was the South Australian Department of Corrections. After the cancellation of the South Australian Prison Project in 2009 (where a major facility was to be built to replace ageing prison stock) the Department of Corrections needed to look at methods to increase beds at existing prisons. Given the apparent success of the Rimutaka Prison Container Project, the South Australian Government trialled shipping containers for the provision of additional low security prison accommodation at Cadell Training Centre (Wills 2011). The existing prisoner housing at Cadell Training Centre consists of two types. Some of the low security male prisoners live in standalone cottages in groups of four or five, each with their own bedroom and shared kitchen and bathroom facilities. Cadell Training

Centre also accommodates prisoners in one cell block (Grant 2008). The new shipping container accommodation followed a cellblock design and was constructed using new prefabricated standard 40 foot containers for the accommodation and association area, with a slightly wider container used as the hallway module. Each container houses six single cells installed with windows and doors, insulation, heating, plumbing, internal walls, roof and external decking and fitted with prison grade fittings (including high grade windows and doors) (Department of Correctional Services, SA 2011).

Shortly after the trial, the South Australian Government Minister announced that "using shipping containers in the construction of new prison cells [meant] the Government [could] deploy new prison cells to cope with any increase in demand in up to half the time with a cost saving of up to 40 per cent compared to traditional cellblock construction," and "construction time [was] only six months, as opposed to more traditional cell construction which could take more than 12 months" (Department of Correctional Services, SA 2011). The Department indicated that "due to the high security features of the modular, it is hard to make a precise cost comparison when compared to recent traditional builds, however the cost for one modular cell is about $75,000 (AUD) compared to the cost of a traditional build for a low security unit which ranges from $100,000 to $200,000" (Department of Correctional Services, SA 2011).

After the initial trial at Cadell Training Centre, the South Australian Government announced plans to use shipping containers to increase the capacity of an existing prison in Mount Gambier in the south east of the state by more than 100 prisoners in a $23m (AUD) development (Wills 2011). The new unit was planned to accommodate low to medium security prisoners and constructed as a single storey facility. The design consists of two wings of 54 beds in various configurations[1] with a central officer's station. Each wing is reported to be designed around a central external exercise yard and reports note that the unit is to be fully self-contained with ablutions, dining and recreation facilities as well as interview and programs rooms (Department of Correctional Services, SA 2012).

Given this low to medium security development is well outside the existing experience of the use of shipping containers in correctional settings, it appears prudent and timely to examine the benefits and constraints to using shipping containers as prison accommodation.

The Benefits of the Use and Reuse of Shipping Containers in Prison Settings

The use and reuse of shipping containers can be conceived as prefabrication, something that has been used in various formats for prison construction since the 1960s. In particular, precast concrete modules and prefabricated buildings have been increasingly used to construct prisons in Australia and internationally. Various authors have identified the benefits of using prefabricated construction methods for prison construction. In particular, time, cost, quality and productivity benefits through minimisation of on-site operations and duration, less congestion on site, improved health and safety along with greater and more predictable quality and potential greater adherence to budget may be identified as benefits. The shipping container construction model, however, does entail higher materials procurement costs and different on-site skill and equipment needs, but may reduce labour risks and associated costs. It may also improve quality consistency since modules are fabricated in a controlled environment.

The Constraints of the Reuse of Shipping Containers for Prison Construction

While there may be arguments for the reuse and refurbishment of shipping containers in areas where building materials are scarce, in most urban and rural areas, refurbishment

[1] The configurations of planned accommodation include 42 bunked cells, four dual cells (two beds on the floor per cell) and four quad cells (four beds on the floor per cell) with 16 beds for disabled prisoners (Department of Correctional Services South Australia 2011).

costs and logistics of transporting containers will not offset the environmental and economic benefits of using local materials and labour and developing a purpose designed building. This is particularly relevant in prison design where architecture has a profound effect on the experiences of people using the complex and the security and operation of the prison.

As previously mentioned, most Australian and many New Zealand medium and maximum security prisons are designed using the principles of unit management and third generation prison design, where cells are located around a shared interactive internal space. The salient architectural and operational innovations to accompany unit management and third generation design include normalising the environment with soft furnishings, dispersing the dining and recreation activities to the living units and fitting cells with showers and toilets to lessen the prisoner violence associated with shared amenities (Grant & Jewkes 2013). Prisons are thus designed to promote high levels of interaction between staff and prisoners and spaces are configured so prisoners live within semi-autonomous groups to maximise each individual prisoner's responsibility. Research has shown that prisons designed under the principles of unit management and third generation (or podular) design have decreased numbers of disturbances and assaults.

Shipping containers are most suitable for layouts of small rooms joined together. This layout conforms to the design model that New Zealand has developed where cells open directly onto a central secured courtyard. The modularity does not necessary fit into principles of unit management or third generation design principles. The constraints of using modular units in a design generally result in linear design which has been shown as a factor contributing to increased violence in prison settings. Linear design, in conjunction with the indirect or intermittent staff supervision that must accompany this planning model increases opportunities for both prisoner to prisoner and prisoner to staff violence (Wright & Goodstein 1991). From this perspective, the use of shipping containers in prison architecture limits the design and may be a barrier to innovative and safe design.

There are also questions about the usability of the central courtyard space in certain climates. While having a central court yard as an association and activity space for prisoners may be appropriate in Mediterranean climates, it is unlikely to be useable in other climates. The site planned for the new project in Mount Gambier experiences cool to cold winters with high average rainfalls with summer temperatures reaching over 40°C on certain days. It is likely there will be periods where the central courtyard will be unusable due to wet weather and at other times where due to the extreme heat where it is unsafe for prisoners to be in the courtyard (Grant et al. 2012). There is a capacity for prisoners to be effectively isolated and spend much of their time in individual cells rather than living and socialising as a group.

Building Performance

There are many factors which may indicate that shipping containers are not suitable from a building performance perspective. Shipping containers are prone to interstitial[2] and surface condensation[3] due to the resistive nature of the materials which tend to act as a vapour

---

[2] Interstitial condensation is a form of structural damping that occurs when warm, moist air (generally, from inside a building) penetrates into a wall, roof or floor and meets a cold surface. The air cools, lowering its capacity to carry moisture, resulting in surface condensation. Over time, this condensation may cause rotting of timber or corrosion of metal components. Resulting structural damage may occur without visible indications. This becomes a particular issue if the diffusion of the water vapour through the material is toward the cold side of the structure and if the insulation or thermal conductivity induces a significant temperature difference between the warm and cold sides, such as with heavy insulation or in air conditioned buildings.

[3] Surface condensation is the process by which a gas or vapour changes into a liquid and accumulates on the surface of a material. If the temperature of an object (e.g. metal, glass) falls below what is known as the 'dew point' temperature for a given relative humidity of the surrounding air, water vapour from the atmosphere condenses into water droplets on its surface. The 'dew point' varies according to the amount of water in the atmosphere (known as humidity). Even in cold conditions condensation occurs despite relatively low humidity.

barrier. Condensation will tend to have a detrimental effect on insulation and over time may cause break downs of installed insulation. Vapour barriers need to be used internally and adequate ventilation needs to be installed to alleviate interstitial condensation problems (Smith 2005). There is also a possibility of cold bridging if containers are situated adjacent to each other. Reverberation within the common parts is also a problem and may cause problems due to the sound reflective nature of the external materials and there is a need to consider the installation of a sound absorbent layer or coating to reduce the reverberation. These factors indicate the importance of the installation of insulation, condensation and sound barriers. Shipping containers are generally fitted with an internal frame and insulation. These installations in turn diminish the available floor space, raise the floor and reduce the ceiling heights.

It is also important to consider the structural performance of decommissioned ISO containers for building applications. Containers are decommissioned when they fail to conform to standards which may relate to superficial or structural damage. This damage may compromise the structural integrity or aesthetics of the building. Other areas of structural performance need also to be considered. Strength may also be compromised when the shell is modified, such as by penetrating the container to fit door or window openings. While containers are designed to carry heavy loads, and can support multiples of their own weight, they are not designed to carry loads such as large roofs. In general, containers resist the effects of harsh environments; they are vulnerable to degradation caused by natural elements and need to be rust proofed cyclically. Corrosion does, however, have the potential to have an effect on the structural integrity and the life span of the building.

The cost effectiveness of shipping container projects becomes debatable when refit requirements and additional developments are taken into account. Is it cheaper to design and make purpose built accommodation which fits the correctional operation regimes and needs of prisoners? One must also consider the need for governments to support local building industries. Given that in 2001 China produced 80 per cent of the world's shipping containers and Europe was the second largest producer of containers (Spurgeon 2003), the likelihood of sourcing locally manufactured containers may be difficult in Australia. While the prefabrication of special orders or retrofitting of containers may be able to be completed within the country in question, this takes work away from local building industries and potentially affects the local economy.

The Use and Reuse of Shipping Containers from an Environmental Perspective

The use and reuse of shipping containers for building projects is often linked to quasi-environmental arguments. Correctional agencies are generally obliged to report on the environmental impact and sustainability of potential projects and are required to reduce environmental impacts wherever possible. The use of shipping containers as building material is environmentally flawed from a number of perspectives. One of the environmental rationales often employed for the use of shipping containers is the apparent reduced levels of embodied energy when compared to other building materials. When new, the embodied energy of shipping containers is high in comparison to other building materials as they are constructed of materials which take large amounts of energy to produce and there are CO^2 emissions associated with transport. The level of embodied energy for shipping containers as a building component is thus generally calculated under the assumption that the unit has been pre-used for another purpose (Smith 2005 p. 4). In the most recent prison projects, new prefabricated containers were used (Smith 2005 p. 101). In some instances, arguments have been mounted that the transportation of the shipping container from the place of

---

Internal surface condensation can occur in areas of high humidity and/or during exceptionally cold weather while external surface condensation occurs in certain climatic conditions with high humidity levels and/or particularly cold weather.

manufacture to the site amounts to pre-use. This is fallacy - the calculated embodied energy levels of transported new prefabricated units should be calculated without discount.

The environmental performance of shipping container as a building component should be questioned. Extensive insulation is required to make the shells thermally comfortable and avoid corrosion and condensation. Most projects will require air-conditioning or other forms of heating and cooling. The constraints of the modular building system generally do not allow many of the design elements which would be incorporated into 'best practice' passive design to maintain thermal comfort. In turn, the energy consumption of prison accommodation constructed of shipping containers is likely to be high due to the reliance on mechanical heating and cooling.

The transformation of a steel box is an eco-romance where one takes the excesses of capitalism and through human ingenuity transforms them into a serviceable object. Using environmental rationales to justify reusing shipping containers is nonsensical. The environmental waste edict of reduce, reuse and recycle is a hierarchy where emphasis is placed on each action in order to promote ecological balance through conscious behaviour and choices. Unpacking this environmental argument, it becomes apparent that reusing shipping containers does not tackle the excessive global movement of goods that occurs to satiate our consumer society. The reuse of shipping containers simply reduces the evidence of our excessive consumerism and in turn our guilt. Employing human ingenuity to transform the steel boxes can also not be seen as a sustainable or environmental action. Peccei and King (1972) in the landmark publication, 'the Limits to Growth' gave warnings that technology could not function as a way of solving environmental problems. The world's consumption patterns needed to change.

Conclusion

The global fascination for constructing buildings out of recyclable materials is growing. At first glance, it appears that the shipping containers that litter industrial areas on the outskirts of every major city could be usefully and innovatively recycled to resolve overcrowding issues in prisons. On closer examination, shipping containers are simply not suitable for use as a building component in the prison context. To achieve the level of security and finish required, new containers must be used. In using new containers, environmental advantages of low embodied energy and the use of recycled materials are mitigated and the energy and operating costs are increased.

Prison overcrowding is a very serious issue. It has negative impacts on the prisoner, staff and the functioning of the prison, negating the health and well-being of offenders, undermining human dignity in prisons and rendering the safety and security of offenders and the community vulnerable. Placing more accommodation within existing prisons has been shown to impact prisoners in terms of the availability of programs, staffing levels, drug and alcohol treatment and health care provision, access to services and amenities (Harding 1987) and is only one method of tackling overcrowding. Other strategies include reducing the number of prisoners and building new prisons. It is interesting that South Australia have added beds and additional units in the wake of a major prison construction program being cancelled. This stop gap measure is unlikely to address issues within the system. Much of the literature on the topic concurs that overcrowding problems are essentially associated with problems of governance, weak economies and issues in criminal justice systems (Gadon et al. 2006).

One is left to wonder if the use of shipping containers is related to the potential time, cost, quality control and productivity benefits or to other agendas. This becomes particularly poignant in South Australia where the then Treasurer announced that the State Labor Government was happy to "pack 'em, rack 'em and stack 'em'" when cancelling a planned prison project. For many politicians, members of the public and, indeed, some prison staff, the provision of amenities in prisons should be "consistent with purposes of discipline and reformation" (Forsythe 2004 p.760). One is left to wonder whether the use of shipping containers for prison expansions is more about political agendas and displaying to the public a punitive attitude to prisons and prisoners alike.

Bibliography

Brandt, K. (2011) Plugging In: Reinterpreting the Traditional Housing Archetype within a Community using Shipping Containers, (Master of Science Dissertation), The University of North Carolina.

Chapa, J. (2007) 'Prefab Friday: Container City in London', *Inhabitat,* June

Conover, T. (2003) 'In the Land of Guantánamo', *New York Times Magazine,* June

Corcoran, E. (2011) 'Obama's Failed Attempt to Close Gitmo: Why Executive Orders Can't Bring About Systemic Change', *University of New Hampshire Law Review,* **9** (2), 207-235

Crook, J. (2009) 'Contemporary Practice of the United Stated Relating to International Law: General International and U.S. Foreign Relations Law: President Obama Orders Closure of Guantanamo Detention Facilities; Obstacles Remain', *American Journal of International Law,* **103** (2), 325-331

CS Shipping Containers (2012) *Shipping Containers and Container Conversions – 5 Unusual Requests,* www.csshippingcontainers.co.uk/shipping-container-conversions/shipping-containers-and-container-conversions-5-unusual-requests Date viewed 1st August, 2012

Department for Correctional Services, South Australia (2011) 'Modular Accommodation Unveiled At Cadell Training Centre', *Corrections Newsletter,* June

Department of Correctional Services, South Australia (2012) *Mount Gambier Prison Expansion, Department for Correctional Services,* www.corrections.sa.gov.au/dcs-projects/mgp-expansion Date viewed: 2nd November 2012

Department of Corrections, New Zealand (2009) *Media Release: New container unit to deliver much needed beds,* www.corrections.govt.nz/news-and-publications/media-releases/2010-media-releases/container_cells_open_for_business_at_rimutaka.html Date viewed: 13th July, 2012

Department of Corrections, New Zealand (2010) *Annual Report 2009-10* Wellington, Department of Corrections, Wellington

Department of Corrections, New Zealand (2012) *Rimutaka Prison* www.corrections.govt.nz/utility-navigation/find-a-facility-or-site/find_a_corrections_facility/rimutaka_prison.html Date viewed 13th July 2012.

Forsythe, B. (2004) 'Loneliness and Cellular Confinement in English Prisons 1878-1921', *British Journal of Criminology,* **44** (5), 759-770

Gadon, L. Johnstone, L. et al. (2006) 'Situational variables and institutional violence: a systematic review of the literature', *Clinical Psychology Review,* **26** (5), 515-534

Global Security (2011) *Military Guantanamo Bay - Camp Delta* www.globalsecurity.org/military/facility/guantanamo-bay_delta.htm Date viewed: 12th November 2012

Gower, P. (2010) 'A night in shipping container prison', *New Zealand News,* 2nd June 2010

Grant, E. & Jewkes, Y. (2013) 'Finally Fit for Purpose: The Evolution of Australian Prison Architecture', *The Prison Journal* (forthcoming)

Grant, E. Hansen, A. & Williamson, T. (2012) 'Design issues for prisoner health: Thermal conditions in Australian custodial environments', *World Health Design,* **5** (3), 80-85

Grant, E. (2008) Towards safer and more congruent prison environments for male aboriginal prisoners: a South Australian study (PhD Thesis) School of Architecture, Landscape Architecture and Urban Design, The University of Adelaide

Grant, E. (2009) *Report on 2008 Churchill Fellowship to investigate correctional facilities for Indigenous prisoners New Zealand, Canada & Denmark,* Churchill Fellowship Trust, Canberra

Harding, R. (1987) 'Prison overcrowding: Correctional policies and political constraints', *Australian & New Zealand Journal of Criminology,* **20** (1), 16-32

Informa Group, (2012) *Containerisation International Yearbook*, Informa Group, London

Joint Task Force Guantánamo and Department of Defense, US Government, (2003) *Camp Delta Standard Operating Procedures (SOP) Guantánamo Bay*, Department of Defense, US Government, http://www1.umn.edu/humanrts/OathBetrayed/SOP%201-238.pdf
Date viewed: 12th December 2012

Kalkin, A. & McLean, W. (2008) *Quik build: Adam Kalkin's ABC of Container Architecture,* London, Bibliotheque McLean

Keller, B. (1993) 'A South African Mall that Black know-how built', *The New York Times*, September, 3

Kenny, S. (2003) 'The Hicks Trial' in *Conference Proceedings from the National Forum On The War on Terrorism and The Rule Of Law*, NSW Parliament House, 10th November 2003

Le Grand, C. (1999) 'Jail Crush Spills to Sea Containers', *The Australian*, 11th February, 1999, 6

Levinson, M. (2006) *The Box: How the shipping container made the world smaller and the world economy bigger* Princeton University Press, Princeton

Meadows D & Club of Rome (1972) *The Limits to Growth; a report for the Club of Rome's project on the predicament of mankind* Universe Books, New York

Mendelson S & CSIS Working Group on Guantánamo and Detention Policy (2008) *Closing Guantánamo: From Bumper Sticker to Blueprint; A Report of the CSIS Human Rights and Security Initiative and the Working Group on Guantánamo and Detention Policy,* Center for Strategic and International Studies, Washington

Muller, G. (1999) *Intermodal Freight Transportation,* Eno Transportation Foundation Inc., Washington

Notteboom, T. & Rodrigue, J. (2008) 'Containerisation, Box Logistics and Global Supply Chains: The Integration of Ports and Liner Shipping Networks', *Maritime Economics and Logistics* **10** (1-2), 152-174

Rawlinsons (2010) *Rimutaka Container Prison,* www.rawlinsons.co.nz/our-projects/terrace-downs-high-country-resort-2/) Date viewed: 17th July, 2012

Rosenberg, C. (2004) 'Pentagon plans for long-term detention at Guantánamo', *Seattle Times,* 10th December 2004

Sawyers, P. (2011) *Intermodal Shipping Container Small Steel Buildings,* Paul Sawyers, Kentucky

Smith, J. (2005) *Shipping Containers as Building Components*, University of Brighton, Brighton

Spurgeon, K., Prozzi, J. et al. (2003) *The Secret Life of the Container: Evidence from Texas,* Transportation Research Board, Texas

Theo, N. & Jean-Paul, R. (2008) 'Containerisation, Box Logistics and Global Supply Chains: The Integration of Ports and Liner Shipping Networks', *Maritime Economics and Logistics,* **10**, 152-174

Tucker, S., Roberts, P. & Zinni A. (2010) *The Encyclopaedia of Middle East Wars: the United States in the Persian Gulf, Afghanistan, and Iraq conflicts*, ABC-CLIO, Santa Barbara

Urban Space Management Ltd, (2012) *Container City* www.containercity.com/home.html Date viewed: 1st August, 2012

Wheeland, M. (2010) 'A (Green) Prison Made of Shipping Containers', *GreenBiz*

Wills, D. (2011) Shipping containers for prisoners at Mt Gambier Prison, *AdelaideNow*, Adelaide

Winter, C. (2011) 'Union to put cell plan back in its box', *ABC News,* 4th August 2011

Wright K and L Goodstein (1991) 'Correctional Environments in Thompson', J. & Mays, L. (eds), *American Jails: Public Policy Issues*, Plenum Press, New York, 181-194

A Comparison of Subscription and Open Access Journals in Construction Management and Related Fields

7

Bo-Christer Bjork, (Hanken School of Economics, Finland)

Abstract

The Internet has profoundly changed the technical infrastructure for the publishing of scientific peer reviewed journals. The traditional business model of selling the content to subscribers is increasingly being challenged by Open Access (OA) journals, which are either run at low cost by voluntary academics or which sell dissemination services to authors. In addition authors in many fields are taking advantage of the legal possibilities of uploading free manuscript versions to institutional or subject-based repositories, in order to increase readership and impact. This study looks closer at the current publishing situation in construction management and related fields and compares empirical data about 17 OA journals and 16 traditional subscription journals. Of the articles published in 2011 in the subscription journals only 9 % could be found as OA copies. The overall OA availability (including articles in OA journals) was 14 % for Construction Management and Economics and 29 % for construction IT scholarship. The study also compared the scientific quality of OA journals to subscription journals using a number of citation indexes. Four of the OA journals are approaching the citation levels (using Scopus data) of the average subscription journals. OA journals are outperforming subscription journals by almost a year in speed of publication.

Keywords: Scholarly publishing, Internet, Open Access, Business model

Introduction

The emergence of the Internet has profoundly changed the way we do business in many industries as well as our everyday social life. Among the most affected industries are those in which the product can be digitalized and thus multiplied and disseminated at almost zero marginal cost. All media industries, including the publishing of peer reviewed scholarly journals, belong to this category. Scientific publishing differs from the other branches of media, such as news, films, music and literature, in that the interest of the stakeholders funding and producing the content is not in revenue from the sales of the content but in maximizing the dissemination and impact on other researchers, on innovations and in the end public welfare. Nevertheless the publishers of scientific journals have been slow in adapting their business models to take full advantage of the opportunities offered by the Internet. Around the turn of the millennium, all major publishers started to publish electronic versions of their journal articles in parallel with the printed issues, as well as market electronic licenses to large bundles of titles ("big deals"), but the fundamental business model of selling access to the content to subscribers remained the norm. The novel idea of Open Access and seeing journals as service providers to the authors was initially embraced by individual scholars or groups of scholars setting up journals run on a volunteer basis and later by companies funding the operations by charging the authors for publishing, not by the established publishers.

Open Access (OA) means that the full text of the articles is accessible with no charges and access restrictions to anybody with an Internet connection. In the ideal OA scenario any reference in a scientific publication would be just one mouse-click away. There are many arguments for Open Access, but the most fundamental is that OA maximizes the impact of research on science, technology and society, by providing more access and readership than the restricted subscription model. Considering the huge public expenditure on research globally, the subscription revenue of an estimated 8 billion USD to the publishers of around 25000 peer-reviewed journals (Ware and Mabe 2009), which would have to be recuperated by other means in fully Open Access scenario, is next to nothing.

Why has the industry been so slow in changing its business model if OA is as attractive as it seems. The basic reason is to be found in the high profit levels of the major publishers in a very oligopolistic market. Elsevier made a profit of 724 million GBP on revenues of 2 billion in 2010, a margin of 36% (Economist 2012), and many leading scientific societies have built up big publishing branches that are dependent on subscription revenue. Such publishers understandably perceive no need to change a very comfortable status quo. The situation is further "cemented" by the virtual monopoly of ISI's citation indexing service and the impact factors it publishes for a restricted class of mainly older established journals, which has contributed to very high barriers to entry in this market.

The pressure for OA has thus come from outside the core group of established publishers and increasingly also from research funders like the National Institute of Health in the USA, the single biggest funder of medical research in the world. Publishers have reacted by anti-OA lobbying and recently tried to introduce legislation in the US forbidding public research funders from requiring open access to the research results of their grantees. Due to widespread opposition from the academic world this initiative failed (Howard 2012).

Gold and Green OA

Open access can be achieved in a number of ways. Firstly the publisher can make the primary publishing site or parts of it openly accessible. Full Open Access where all articles are immediately free is often called "Gold" Open Access. This can be achieved using a variety of business models. The journal can be published only electronically and funded either with only voluntary labor, grants, using Open Source software etc, or by requiring the authors to pay for the publishing service in the form of Article Processing Charges (APCs). A recent empirical study shows that APCs fall in a wide range between 20 and 3000 USD, the average across some 100,000 articles published in 2010 being around 900 USD (Solomon and Björk 2012). The charges are typically highest in biomedicine journals, due to the better funding opportunities of the authors, and society journals and journals from the developing countries tend to have the lowest charges.

Another variant is where established journals, often society ones, have made the electronic version of the journals free while still getting subscription income from the printed version. Alternatively journals can open up the e-version with a delay so that readers requiring immediate access pay a subscription. A further variant is where subscription journals offer authors the possibility to open up their individual articles in otherwise subscription journals against a payment (hybrid OA). There are over 4000 journals allowing this, but the high price of usually around 3000 USD has resulted in a very low uptake of only around 1-2 % of eligible articles (Dallmeier-Tiessen 2010).

From a legal perspective there are two flavors to Open access publishing, Gratis OA and Libre OA. Gratis OA means just the technical barrier free accessibility, in Libre OA the usage and re-use rights have been precisely defined, often via the use of Creative Commons (CC) licences attached to the journals and articles. Libre OA is particularly important is such sciences where journal articles and related data sets can be harvested to build aggregated data collections or new services.

There are currently around 7600 journals registered in the Directory of Open Access Journals (DOAJ 2012), publishing an estimated 340,000 articles in 2011. In terms of the global number of articles, the OA share is currently around 10 %, and the volume is increasing by around 30 % per year (Laakso et al 2011).

The second route to disseminate the results openly is for the authors to make electronic copies of some version of the article manuscript available on a suitable web site ("green OA"). The three main options are on the author's own or his department's web pages, usually linked to his CV, in the institutional eprint repository of his university or in a subject-based repository for the research area in question. Only a minority of universities have repositories and subject-based repositories have become popular only in some research areas (i.e. high-energy physics and biomedicine). The copy which is posted can range from a copy of the manuscript as initially submitted to the exact published pdf-file, the most common one being the manuscript version as accepted for publishing, but without the final copyediting and layout of the publisher. This is something a majority of publishers allow in their copyright agreement with the authors (The Sherpa/Romeo website is a good source to find out about the policies of individual publishers or journals, Sherpa/Romeo 2012). In practice many academics post clearly illegal exact copies of the published articles.

The global share of articles for which green copies can be found is estimated to be around 12-15 % (Björk et al 2010, Gargouri et al 2010). The share is much higher in certain disciplines like High-Energy Physics and Medicine, and for authors from universities, which require their faculty to post green copies.

The OA situation in Civil Engineering Research

Open Access has not yet been widely embraced by civil engineering researchers. There were until the last couple of years only a few OA journals in construction, none of which yet indexed by Web of Science. The field lacks a subject-based repository of its own and only a small minority of researchers has access to institutional repositories. Unlike fields such as biomedicine, the funders of the research haven't yet started requiring OA. Recently the situation has started to change due to the fact that a number of commercial OA publishers have launched journals also in civil engineering and keep spamming academics with requests to join the editorial boards of and to submit papers to these author charge funded journals. This has created some confusion among academics due to the apparent dubious quality of many of these journals and also triggered a recent debate on the CNBR mailing list (CNBR 2012).

The aim of this study has consequently been to shed some light on the current situation of scholarship in civil engineering, in particular in construction management and economics as well as in IT in construction. The focus has been on OA issues, but the general quality in terms of citation rates of the journals has also been included, since the thrush of the critique against OA publishing has been that such journals are of lower quality and put the whole peer review system in peril. With the exception of some parameters, such as the acceptance rates of journals, a lot of the information could be extracted from the web sites of the journals as well from some databases like Journal Citation Reports (JCR), Google Scholar and the Directory of Open Access Journals (DOAJ). For some aspects synthesizing the information was relatively easy, for others time-consuming.

Journals included in the Study

All in all 33 journals were studied, 17 OA journals, 16 subscription journals. Originally there were 16 of both categories but the author became aware of the recently founded International Journal of Construction Supply Chain Management at a late stage of manuscript preparation and added it. All relevant journals specialized in either construction management and economics or the use of IT in construction were included. In addition some general civil engineering titles with broad scope were included in the list of OA journals, in order to demonstrate better the spectrum of open access journals and publishers. Sources that helped determine titles of interest included the list of CIB[1] encouraged and endorsed

[1] International Council for Research and Innovation in Building and Construction

journals, JCR rankings of journals in "Engineering, Civil", the Directory of Open Access Journals. The journals and some basic data about them are listed in table 1.

Name	Founded	OA from	Type of publish Commercial	Society	University	Scientists	Publisher	Country
OA journals:								
Journal of the South African Institution of Civil Engineering	1993	2010		√			S. A. Institution of Civil Eng..	South Africa
Journal of Construction in Developing Countries	1995	2006			√		Uni. Sains Malaysia Press	Malaysia
Journal of Information Technology in Construction	1996	1996				√	CIB	Netherlands
Electronic Journal of Structural Engineering	2001	2001				√	EJSE	Australia
Australasian Journal of Construction Economics and Building	2001	2010			√		Uni. of Technology Sydney	Australia
International Journal of Civil Engineering	2003	2006			√		Iran Uni. of Scie. and Techno.	Iran
Lean Construction Journal	2004	2004				√	Lean Construction Insititute	USA
The Open Civil Engineering Journal	2007	2007	√				Bentham	U.A.E.
the Open Construction and Building Technology Journal	2007	2007	√				Bentham	U.A.E.
Advances in Civil Engineering	2008	2008	√				Hindawi	Egypt
International Journal of Advanced Structural Engineering	2009	2012	√				Springer	Germany
International Journal of Civil and Structural Engineering	2010	2010		√			Integrated Publishing Ass.	India
Journal of Civil Engineering and Construction Technology	2010	2010	√				Academic Journals	Nigeria
ISRN Civil Engineering	2011	2011	√				Hindawi	Egypt
Journal of Civil and Environmental Engineering	2011	2011	√				OMICS	Egypt
Int. Journal of Construction Supply Chain Management	2011	2011				√	IJCSCM	New Zealand
Buildings	2011	2011	√				MDPI	Switzerland
Subscription journals:								
Journal of Construction Engineering & Management	1875			√			ASCE	USA
Building Research and Information	1973		√				Taylor & Francis	UK
Construction Management and Economics	1983		√				Taylor & Francis	UK
Facilities	1983		√				Emerald	UK
Journal of Management in Engineering	1985			√			ASCE	USA
Journal of Computing in Civil Engineering	1986			√			ASCE	USA
Computer-aided civil and infrastructure engineering	1986		√				Wiley & Blackwell	USA
Advanced Engineering Informatics	1987		√				Elsevier	Netherlands
Automation in Construction	1992		√				Elsevier	Netherlands
Engineering, Construction and Architectural Management	1994		√				Emerald	UK
Journal of Civil Engineering and Management	1995		√				Taylor & Francis	UK
Journal of Architectural Engineering	1995			√			ASCE	USA
Construction Innovation	2001		√				Emerald	UK
International Journal of Construction Management	2001				√		Chinese Res. I. of Con. Man.	China
Journal of Engineering Design and Technology	2003		√				Emerald	UK
Architectural Engineering and Design Management	2005		√				Taylor & Francis	UK

Table 1 The studied OA and subscription journals

In the table the publishers of the journals have been classified into four categories, commercial publishers, scientific societies, universities/research institutes and independent journals run be groups of academics. In some OA cases the classifications are a matter of choice. Journal of Information Technology in Construction for instance is formally published by CIB but in practice by a group of collaborating academics. The journals are listed in age order, the oldest ones first. Three publishers; Taylor & Francis, Emerald and the American Society for Civil Engineers, ASCE, dominate the subscription market. Commercial publishers have entered the OA market rather late. In the case of two of the commercial OA publishers (Bentham and Academic Journals) it was difficult to find out about the publisher's home country from the web site, but there are indications that the United Arab Emirates is the site of Bentham and Nigeria or Kenya of Academic journals (Beall 2011).

Table 2 shows the pricing policies of the journals. All prices are in US Dollars. For the Open Access journals the price indicated is the article processing charge (APC) that authors have to pay. Journal of Civil and Environmental Engineering has a three-tier price based on the country of origin of the corresponding author, with the lowest price level for developing country authors. Buildings has an introductory offer of no APC for submissions in 2012, but can be expected to levy an APC later on, based on pricing info for some other OA journal from the same publisher.

Pricing and information			
OA journals:	Article processing charge		
Journal of the South African Institution of Civil Engineering	Free		
Journal of Construction in Developing Countries	Free		
Journal of Information Technology in Construction	Free		
Electronic Journal of Structural Engineering	Free		
Australasian Journal of Construction Economics and Building	Free		
International Journal of Civil Engineering	Free		
Lean Construction Journal	Free		
The Open Civil Engineering Journal	800		
the Open Construction and Building Technology Journal	800		
Advances in Civil Engineering	600		
International Journal of Advanced Structural Engineering	Free		
International Journal of Civil and Structural Engineering	Free		
Journal of Civil Engineering and Construction Technology	550		
ISRN Civil Engineering	500		
Journal of Civil and Environmental Engineering	900-1800		
Int. Journal of Construction Supply Chain Management	Free		
Buildings	0		
Subscription Journals:	Article OA Option (hybrid)	Institutional Subscription	Pay per view
Journal of Construction Engineering & Management		908	30
Building Research and Information		1944	27
Construction Management and Economics	3250	3521	27
Facilities		10589	27
Journal of Management in Engineering		305	30
Journal of Computing in Civil Engineering		490	30
Computer-aided civil and infrastructure engineering		1316	?
Advanced Engineering Informatics	3000	1424	39
Automation in Construction	3000	964	39
Engineering, Construction and Architectural Management		2229	25
Journal of Civil Engineering and Management	3250	712	32
Journal of Architectural Engineering		317	30
Construction Innovation		599	25
International Journal of Construction Management		330	?
Journal of Engineering Design and Technology		419	25
Architectural Engineering and Design Management	3250	542	32

Table 2 Pricing information for the studied journals

For the subscription journals the hybrid OA charge has been shown for those journals that propose this option for authors wishing "to open" up their individual article. The Institutional subscription price, usually for electronic only subscriptions are shown. In the case of Emerald it was almost impossible to find pricing information, but a price list for 2009 with a wide ranges of prices was found on the web (Emerald 2008). The prices for two of the journals seem believable but the prices of Engineering, Construction and Architectural Management and in particular Facilities do not look credible. It seems that Emerald prefers to have potential customers contact them by email and then negotiate prices on case-by-case basis.

OA journals:	Articles published 2011
Journal of the South African Institution of Civil Engineering	7
Journal of Construction in Developing Countries	14
Journal of Information Technology in Construction	44
Electronic Journal of Structural Engineering	8
Australasian Journal of Construction Economics and Building	21
International Journal of Civil Engineering	35
Lean Construction Journal	9
The Open Civil Engineering Journal	25
the Open Construction and Building Technology Journal	23
Advances in Civil Engineering	29
International Journal of Advanced Structural Engineering	0
International Journal of Civil and Structural Engineering	86
Journal of Civil Engineering and Construction Technology	18
ISRN Civil Engineering	17
Journal of Civil and Environmental Engineering	5
Int. Journal of Construction Supply Chain Management	4
Buildings	1
	325

Subscription journals:		OA posting allowed (delay months)	OA copies 2011 articles	OA %	OA % 10 most cited papers	OA % More than 100 GS
Journal of Construction Engineering & Management	125	0	7	6	50	26
Building Research and Information	43	12	2	5	30	7
Construction Management and Economics	86	12	2	2	10	7
Facilities	38	0	3	8	10	0
Journal of Management in Engineering	28	0	3	11	0	4
Journal of Computing in Civil Engineering	49	0	5	10	30	12
Computer-aided civil and infrastructure engineering	43	12	6	14	40	2
Advanced Engineering Informatics	67	0	9	13	80	8
Automation in Construction	112	0	26	23	70	5
Engineering, Construction and Architectural Management	36	0	5	14	10	0
Journal of Civil Engineering and Management	57	12	0	0	80	1
Journal of Architectural Engineering	16	0	3	19	20	0
Construction Innovation	27	0	2	7	20	0
International Journal of Construction Management	20	?	0	0	10	0
Journal of Engineering Design and Technology	21	0	0	0	10	0
Architectural Engineering and Design Management	19	12	0	0	30	0
	787		73	9		72

Table 3 Article volumes and share of subscription journal article openly available

The institutional subscription prices should only be seen as indicative since most subscriptions are sold as part of bigger package deals and since there may be discounts for some subscribers. A key lesson seems to be that there seems to be little correlation between journal quality and volume with the price. Earlier studies have shown that society journals tend to be much cheaper than equivalent commercial journals (European Commission 2006). The pay-per-view prices for access to individual articles are in much more narrow range, 25-39 USD. Interestingly Emerald also offers articles to be rented for 24 hours in non-printable versions for USD 3.99.

The first column in Table 3 shows information about the number of articles published in 2011. Understandably some of the OA journals, especially the recently founded ones have low publication volumes. Many of the subscription journals are quarterly journals with a volume of around 20-30 articles per year, but there are a couple of journals publishing more than one hundred articles per annum.

The second column ("OA posting allowed") indicates the copyright restrictions of the journals. None of the journals allow the posting of the PDFs as published. In the column the restrictions concerning the so-called personal versions, the manuscripts as accepted for publishing, are indicated. A zero means that authors can post this version immediately, 12 that this can be done one year after publishing. For the International Journal of Construction Management it wasn't possible to find such information.

Columns 3 and 4 indicate the number and percentage of green OA copies found using Google Scholar (GS) for the 2011 articles. The advanced search facility in GS makes it possible to search for articles published in a named journal in a given year. Openly available full text copies are clearly shown in a separate column. All indentified copies were downloaded, studied and classified according to the location of the copy and the type of version. Of the copies found it turned out that around a third were clearly illegal copies (downloaded published versions or proof versions the authors had received from the publishers). Very few copies fulfilled all the stipulations in the copyright agreements with the publisher in the sense of indicating the status of the copy as well as the definitive published version. An example of how this should be done is shown in figure 1. In practice the likelihood of publishers following up on the compliance to these rules not to mention trying to enforce them by sanctions is extremely low.

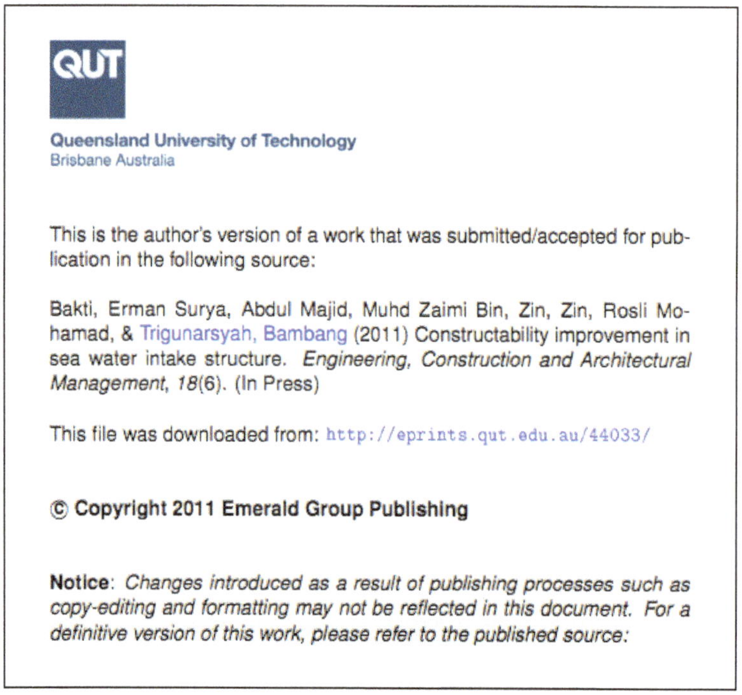
QUT

Queensland University of Technology
Brisbane Australia

This is the author's version of a work that was submitted/accepted for publication in the following source:

Bakti, Erman Surya, Abdul Majid, Muhd Zaimi Bin, Zin, Zin, Rosli Mohamad, & Trigunarsyah, Bambang (2011) Constructability improvement in sea water intake structure. *Engineering, Construction and Architectural Management, 18*(6). (In Press)

This file was downloaded from: http://eprints.qut.edu.au/44033/

© Copyright 2011 Emerald Group Publishing

Notice: *Changes introduced as a result of publishing processes such as copy-editing and formatting may not be reflected in this document. For a definitive version of this work, please refer to the published source:*

Figure 1 An example of how a green copy should ideally be posted, in this case in an institutional repository

An interesting observation is that authors in the construction IT journals seem to be more active in posting OA copies of their articles. The OA availability of highly cited articles from any year was additionally checked in two ways, for the ten most cited articles in each journal (column 5) and for all articles that had more than 100 citations in GS. For most journals highly cited articles were more commonly available than 2011 articles in general and the average availability of articles with >100 citations was 40 %.

Quality Labels and Citation Metrics

Table 4 contains data related to the quality and the citation rates of the journals. CIB encouraged journals are marked with an √, including all the subscription journals except the one's published by the ASCE and Advances in Engineering Informatics, as well as three of the OA journals. Most of the OA journals are listed in the Directory of Open Access journals, which applies some quality checks for inclusion. The criteria for the publisher being accepted as a member in the Open Access Scholarly Publishers' Association OASPA are more stringent, and include criteria related to the trustworthiness and transparency of the peer review process.

OA journals:	CIB endorsed	DOAJ	OASPA member	GS h5-index	articles 5 years	in SCOPUS Scimago 2-year	in ISI JCR 2-year	ISI self-citation %
Journal of the South African Institution of Civil Engineering		√		-	-	0.58		
Journal of Construction in Developing Countries	√	√		3	51	√		
Journal of Information Technology in Construction	√	√	√	13	181	0.96		
Electronic Journal of Structural Engineering		√		5	46	0.44		
Australasian Journal of Construction Economics and Building	√			-	-	√		
International Journal of Civil Engineering				3	158	0.64	0.54	
Lean Construction Journal		√		3	20			
The Open Civil Engineering Journal				4	66	0.11		
the Open Construction and Building Technology Journal		√		2	80	0.15		
Advances in Civil Engineering			√	3	94			
International Journal of Advanced Structural Engineering		√		-	-			
International Journal of Civil and Structural Engineering		√		-	-			
Journal of Civil Engineering and Construction Technology				-	-			
ISRN Civil Engineering		√		-	-			
Journal of Civil and Environmental Engineering				-	-			
Int. Journal of Construction Supply Chain Management				-	-			
Buildings			√	-	-			
Subscription journals:								
Journal of Construction Engineering & Management				19	682	1	0.67	29
Building Research and Information	√			21	222	1.06	1.25	33
Construction Management and Economics	√			16	531	1.05		
Facilities	√			11	177	0.44		
Journal of Management in Engineering				14	349	0.96	0.75	10
Journal of Computing in Civil Engineering				15	254	1.1	0.9	14
Computer-aided civil and infrastructure engineering	√			21	239	3.38	3.17	49
Advanced Engineering Informatics				20	238	2.02	1.4	56
Automation in Construction	√			28	540	2.03	1.31	25
Engineering, Construction and Architectural Management	√			12	181	√		
Journal of Civil Engineering and Management	√			18	235	3.92	3.71	30
Journal of Architectural Engineering				6	121	0.29		
Construction Innovation	√			7	114			
International Journal of Construction Management	√			-		0		
Journal of Engineering Design and Technology	√			-				
Architectural Engineering and Design Management	√			5	93	0.17		

Table 4 Quality assurance and performance

The only practical way to empirically estimate the scientific quality or impact of a journal is by counting citations per published article. Three sources for citations have been used; Web of Science, Scopus and Google Scholar. The best-known statistics for this purpose are provided by the ISI for those journals indexed in its Web of Science in the Journal Citation

Reports. Since ISI is quite conservative in including new journals into this index only some of the journals in this study have so-called impact factors, of which the two-year impact factor is the best known.

The clearly higher JCR impact factors of some of the journals in this study should be regarded with some skepticism. Ideally the relative shares of different citing journals should reflect where the scholarship in the field is to be found. But in some cases the journal itself can provide a disproportionate share of the citations, thus improving the rating. Especially for journals with initially low average citation numbers the impact factors can dramatically rise from one year to the other as a result of a lot of citations in the journal itself or in other closely connected ISI-indexed publications. To highlight this the share of journal internal citations that contributed to the two-year impact factor in the 2010 Journal Citation reports was calculated and is shown in the last column of table 4. For most journals the shares varied between 10-33 % but both Advanced Engineering Informatics with 56 % and Computer-Aided Civil and Infrastructure Engineering with 49 % had much higher rates.

An alternative to Journal Citation reports is offered by the similar citation metrics available from the openly accessible SCImago web site (SCImago 2012), which uses data from the Scopus citation index as a base. Scopus includes around 50 % more journals and 20 % more articles than ISI, and since there is a high correlation between its figures and the ISI impact factors it provides a better tool for comparisons. In our case it was possible to get Scopus two-year citation averages for 19 of the journals, compared to 9 impact factors from JCR.

A newcomer to the family of journal level metrics is the h5-index, which Google Scholar has started to publish in April 2012. This index counts the number of citations in Google Scholar to articles published in the most recent five year in a particular journal, and calculates the highest number where there are at least x articles with at least x citations. It is the same metric (Hirsch index) that has been proposed for authors, but here applied to journals. The h5-index can be regarded as a better measure of overall journal performance compared to the traditional impact factor, in particular for smaller volume journals were one or two highly cited articles can raise the average impact factor a lot. The only drawback is that for journals of equal average level of citations this index tends to grow higher with the number of published articles. Not all the journals in this study could directly be found in the Google Scholar reports, but for such journals that were older than five years it was easy to hand-count the citations in the general GS search engine and construct the figure manually.

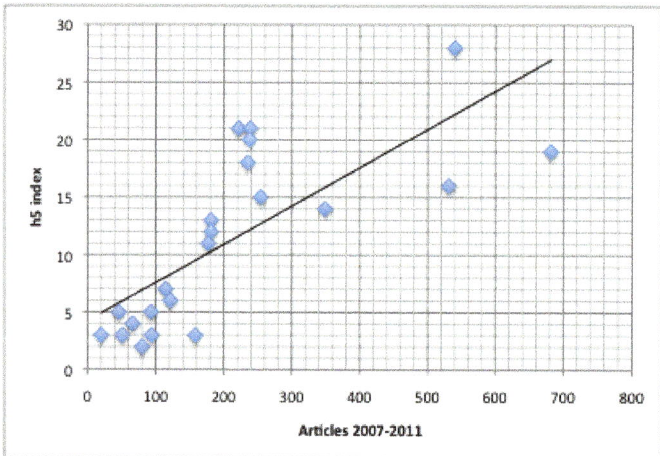

Figure 2 H5 index for citations in Google Scholar. The journals can be identified by comparing the x- and y-values with table 4. Journals above the line perform above average

Of these three sources of citations Google Scholar is the most useful for this study in reflecting research impact, since comparable data can be obtained for almost all the journals in this study (except the newly started ones). Google scholar also catches citations in for instance conference proceedings, which in our field are an important part of the scholarship

The citation data for the journals are shown in the table 4 above. If a figure is not given in the table for some journal the symbol "–" denotes that the figure isn't relevant (the journal is too young to have data)

and "√" that the journal has been accepted for indexing in the data base but that impact factors cannot yet be calculated. An interesting way to illustrate the h5-index is in the form of a graph with the index on the vertical axis and the number of published articles on the horizontal axis (Figure 2). The journals situated above the line can be regarded as having better than average quality. It is easy to identify the different journals by comparing with the values in Table 4.

Acceptance Rates and Time from Submission to Publication

In addition to the scientific quality or prestige of a journal authors are usually, prior to submitting their manuscripts to it, also very interested in finding out the acceptance rate and the expected time from submission to publication. The only way to find information on the acceptance rate is if the publisher in question has put this information on the web. The OA publisher Hindawi does this regularly for its journals. In other cases it was possible via a Google search ("name of journal" + "acceptance rate") to find figures quoted in for instance editorials. Table below 5 includes some data found this way.

Journal	Acceptance Rate	Source
Advances in Civil Engineering	33	Hindawi 2012
Australasian Journal of Con. Economics and Building	30	Data from publisher
Computer-aided Civil end Infrastructure Engineering	35	Easa & Hoel 2008
Construction Management and Economics	48	CME 2010
Journal of Computing in Civil Engineering	46	Abudayyeh et al 2006
Journal of Information Technology in Construction	36	Data from publisher
All 30 ASCE journals	56	Sinha 2004

Table 5 Acceptance rates of some of the journals in the study

The differences are not that big, there is no journal that is characterized by a very selective review process. The acceptance rates might be higher for some of the recently launched APC funded OA journals which advertise very rapid lead times from submission to publication on their web sites.

Data about the lead times from submission to publication can be obtained in two ways. Firstly from statistics provided by the publisher or reports in earlier studies and secondly for those journals that publish submission dates of the individual articles by counting the lead times article by article (a rather painstaking process). Table 6 contains information on the lead times for some journals.

Journal	Publication lead time	Source
Advances in Civil Engineering	6.7	Hindawi 2010
Automation in Construction	18.7	Björk et al 2005
Construction Innovation	18	Björk at al 2005
Journal of Information Technology in Construction	6.7	Björk et al 2005
Open Civil Engineering Journal	9.5	Counted from journal web site

Table 6 Average lead times from submission to publication for some of the journals in the study

Of these the OA journals have much shorter times than the subscription ones. In addition to these the Australasian Journal of Construction Economics and Building has stated in announcements that it strives to publish submissions within 6 months. International Journal of Civil and Structural Engineering, based in India, says on its web pages: "Rapid publication, your article will be online within 4 weeks of time after submission date". It was not, however, possible to check if this was true, but if it were, it would not make the journal a credible peer reviewed one. For journals of equal scientific integrity (meaning that the review process takes approximately the same time) the difference in speed between OA and traditional journals is largely due to the fact that subscription journals have a very fixed issue schedule and volume, and that it is in the editors' interest to keep a fairly sizeable buffer of publishing-ready papers waiting in line, in order to avoid possible shortages. In implicit recognition of this problem most traditional journals have nowadays started posting the electronic versions of the articles earlier than the actual publication in the paper version ("in press").

Accessibility of the Subscription Journals to Scholars and Practitioners

Since I'm working in a business school the portfolio of electronic subscription journals that our library provides is limited in its coverage of fields outside the core business management and economics disciplines. Of the 16 subscription journals included in this study I have full access to seven, access to articles published more than 12 months ago in the case of four journals and no access for the remaining five journals. All the journals with full or partial access are included in large e-licenses provided by the major publishers or aggregators (i.e. Elsevier SD Freedom Collection, EBSCOhost Business Source Complete).

This kind of situation if probably not uncommon for many researchers in our field. Those who are fortunate to work in the civil engineering departments of major universities in the leading industrialized countries might even have full immediate access to all the journals, but many researchers, in particular in the developing countries, not to mention practitioners all over the world, have much more limited access. It is not only the price of the subscriptions that is a barrier, it's also the transaction cost of setting up and paying subscriptions, in particular for single journals. An email sent to a number of former colleagues in Scandinavia confirmed this pattern; Two PhDs working in organizations funding construction research had no access at all. A researcher at a big research centre with several hundreds of researchers in civil engineering had full access to 10 and delayed access to 4 journals, another one at national research centre access to 7 out of 16. Two researchers at major technical universities had immediate access to 14 out of the 16 journals.

The ease of access to the core journal literature in our field could easily be studied more systematically via a web survey sent out to members of professional societies in our field or email lists like CNBR, asking respondents about their paid access to the journals in question and whether they us pay-per-view services in case they don't have access. It would in particular be interesting to check the availability of the journals that CIB endorses to the researchers in its member organizations.

Repositories for Green OA copies

The only subject based repository in our field is the ITC Digital Library for construction IT research (ITC 2012) which was started in 2003 and is hosted at the University of Ljubljana. It partly uses the same software as the Journal of Information Technology in Construction. The bulk of the content consists of the papers presented at the CIB Working Commission W78 conferences since 1984, but otherwise the service has not become popular as a place where researchers post manuscript copies of their journal papers.

Some of the universities of researchers in our field can use institutional repositories, for instance provided by Loughborough University and Queensland University of Technology (for a list of such repositories see either OpenDOAR 2012 or ROAR 2012). In some cases this technical infrastructure is also supported by university policies requiring faculty to deposit copies of their manuscripts for journal articles, whenever legally possible (Bath,

Salford, Reading, QUT). A list of such mandates is available at (ROARMAP 2012) or (Sherpa/Juliet 2012).

Conclusions

The journal scenery in construction management as well as construction IT can be characterized as rather "flat" in the sense of several journals of relatively equal scientific quality competing for manuscripts. This is in contrast with some other scientific fields where there is a single top journal with very rigorous peer review and low acceptance rate and below that a clear hierarchy of journals, in terms of quality and prestige. Most journals in our field have an acceptance rate in the range 30 -50 % and given that a share of papers of very low quality or with the wrong topic are desk rejected by the editors all reasonable papers are likely to be accepted, if not in the first journal submitted to, then at least in the second. The journals are pretty much run by the editors-in-chief, who in many cases have been in charge of the journals for over a decade, and there is no rotation at regular intervals, a practice common for instance in top management journals. Editorial boards are rarely consulted for advice and have often remained almost the same for decades, since the journals were founded. There is strong overlap in the composition of these editorial boards and many academics in our field sit on the editorial boards of several of these journals.

Since only one of the OA journals is indexed in ISI it is difficult to compare the quality with subscription ones using that data. Scopus and Google Scholar offer better possibilities for this. The five OA journals for which Scopus data is available have two year citation counts in the range of 0.41 – 0.96. Although there are four subscription journals with clearly higher averages, the remaining twelve fit in the range 0.17 – 1.10. It can be noted that a recent study has shown that Scopus indexed OA journals overall receive around two thirds as many citations as subscription ones (Björk and Solomon 2012). The global average is lowered by the many old journals that have made the e-version free. When comparing only journals founded after 2001 the difference more or less disappears. Unfortunately the very low grade journals from a handful of publishers frequently spamming academics with emails is giving OA a bad image, but their publication volumes are so low that their overall share of published OA articles is only around 2-3 %.

In the old printed-only environment journals that were not successful in attracting subscribers went out of business after a few years. In the new situation with electronic delivery as the main channel journals published by major publishers are rarely cancelled. This is partly because university libraries concentrate their subscription budgets to huge e-licenses with a few publishers, thus obscuring the marginal revenue of individual titles and partly because the marginal cost of keeping journals running is relatively low due to the economies of scale in using the workflow and production systems of a publisher. The trend thus seems to be that only subscription journals published by major publishers have the "staying power" to survive using this business model.

The situation can best be illustrated by the development of a number of journals from small publishers. Construction Innovation originated as the International Journal of Construction Information Technology, a journal published by the University of Salford between 1992 and 2000. The journal changed name to Construction Innovation as well as focus when Arnold Publishers took over in 2001. In 2006 Emerald acquired the journal. A competing journal was Computer Integrated Design and Construction published by Millpress from 1999 to 2001 and then by the Institution of Structural Engineers from 2003 to 2004 under the title International Journal of IT in Architecture, Engineering and Construction (IT-AEC). Due to an insufficient number of subscribers the journal was merged with the Open Access Journal of Information Technology in Construction (ITcon) in 2005.

ISI started to index several journals published by the Vilnius Gediminas Technical University (Lithuania) around 2008, as part of a campaign to index more regional journals from countries outside the leading Anglo-Saxon ones. In 2010 Tailor & Francis signed a co-publication agreement with the university including several of the journals that were due to receive impact factors in 2011 (ATG 2010]. The Journal of Civil Engineering and Management which had formerly been known as Statyba – Civil Engineering, was one of these journals which achieved very high impact factors near the top ranking in their respective disciplines in the first year of receiving an IF.

All the OA journals started before 2007 are published either by scientific associations or by independent groups of academics. Some have been OA from the start (for instance Journal of Information Technology in Construction and Electronic Journal of Structural Engineering) while others have started out as print journals but later converted to OA. The costs of going OA have been kept very low by using third part free OA portals like Scielo (Journal of the South African Institution of Civil Engineering) or by using Open Source software for running the journal (Australasian Journal of Construction Management and Building).

After 2007 a number of commercial publishers funding their journals with article procession charges have entered the market. A rather curious journal is the International Journal of Advanced Structural Engineering. It is part of a group of nine open access journal published by Springer in close co-operation with the Islamic Azad University, South Tehran branch. The university provides funds to cover all costs of publication in these journals, including the Article Processing Charges (APC's) for all authors, which Springer charges for its other OA journals.

OA journals rarely go out of business, since the cost of running them are very low. They can more easily than subscription journals survive bad years with few published articles. The only OA journal in civil engineering or a related field to have disappeared from the scene so far is the International Journal of Design Computing, published by the Faculty of Architecture of the University of Sydney between 1997 and 2004. The journal encouraged interactive media features in articles but only a handful of articles were published per year. Fortunately the National Library of Australia has ensured the long-term preservation of the articles.

Only five of the OA journals are specialized in construction management and economics or construction IT. If the article volumes of these in 2011 are compared with the number of articles in the thirteen corresponding subscription journals, the overall gold OA percentage would be 10.0 %. The green OA share of the articles in the 13 journals was 10.7 %. These figures are well in line with global averages for articles published in 2008 reported in Björk et al (2010).

Of particular interest is that the overall OA availability of the articles was 14 % for construction Management end Economics and 29 % for construction IT. These figures can be compared to 21 % for Information Systems research (Björk, Laakso and Welling 2011), 26 % for Library Information Systems (Way 2010) and 36 % for Business and Management (Lyons and Booth 2011).

Discussion

The open access of construction management as well as construction IT journal literature could be enhanced with only a modest level of effort. The green route is the most direct. In almost all cases the authors could legally upload the manuscript copies as soon as they have been accepted for publishing and in the case of one publisher with a delay of one year. This should not take more than a couple of hours per article and is in the best interest of the authors as well as their institutions and funders. In many cases the employing universities already have suitable repositories and if not authors can put the copies on their home pages.

Unfortunately construction management lacks a suitable subject-based repository and the existing repository for construction IT is not much used. Setting up such a repository would not be difficult to do, there are couple of widely used open source platforms available exactly for the purpose (D-space, E-prints). Such a platform could also become a very useful source for conference papers, in particular if the organizers would use it as the primary publishing platform.

Journal of Information Technology in Construction is well established as the primary Open Access alternative in construction IT field, and has proven that given the support of the research community it is possible to sustainably run a journal that does not need to charge the authors. There are three existing Open Access journals within the field of construction management, of which both *Lean Construction Journal* and *International Journal of Construction Supply Chain Management* have a rather narrow focus and the *Australasian journal of Construction Management and Building* a regional basis. Establishing a global CME journal with a broad scope using the Open Access business model would be a very valuable service to the research community. This could be done either based on the Open Source – Volunteer model or in partnership with a reputable OA publisher using article processing charges in the 500-900 USD range. Although major commercial publishers would probably be very interested in adding such a journal to their rapidly growing OA offering, the higher APCs that they tend to charge would be likely to deter submissions, at least in our field. The absolutely key issue in launching and sustaining such a journal would be the organization of the collective effort in such a way that a strong international team of editors and associate editors supported by an active editorial board would shape the peer review and other policies of the journal, and in engaging the research community in the field to submit good manuscripts and to review them.

Acknowledgements

The inspiration for the study presented in this article was a recent debate on the CNBR mailing list started by Peter Love's message (23.3. 2012) with the subject header "open Access: caution", in which colleagues were warned not to publish in open access journals. The motivation for his advice was the poor quality of some journals from commercial OA publishers that currently spam academics with emails asking them to submit papers and to join their editorial boards. I took part in the debate presenting some empirical facts about the prevalence and quality of OA in general, but it also arouse my interest in studying the current situation in our field more closely and to share the findings with my colleagues in an efficient openly accessible way.

References

Abbudayeh, O., Dibert-De Young, A., Rasdorf, W., Melhem, H. (2006) 'Research Publication Trends and Topics in Computing in Civil Engineering', *Journal of Computing in Civil Engineering*, **20** (1), 2–12
OA copy available from http://www.lib.ncsu.edu/resolver/1840.2/554

ATG (2010) Taylor & Francis and VGTU Journals forge co-publishing partnership, 26.7.2010, *Against the Grain*, e-newsletter,

http://www.against-the-grain.com/2010/07/taylor-francis-and-vgtu-journals-forge-co-publishing-partnership/

Beall, J. (2010) '"Predatory" Open Access Publishers', *The Charleston Advisor*, 15.2.2010, eprints.rclis.org/bitstream/10760/14576/1/predatory.pdf

Björk, B-C., Laakso, M. and Welling, P. (2011) 'Open Accessibility to Information Systems Research Articles', European Conference on Information Systems, ECIS 2011, Helsinki, Finland, OA copy available from

http://is2.lse.ac.uk/asp/aspecis/20110153.pdf

Björk, B-C. and Solomon, D. (2012) 'Open Access versus subscription journals – A comparison of scientific impact', Submitted to BMC Medicine, April 2012

Björk, B-C., Turk, Ž, Holmström, J. (2005) 'The Scholarly Journal Re-engineered: A Case Study of an Open Access Journal in Construction IT',

Electronic Journal of Information Technology in Construction, **10**, 349-371, http://www.itcon.org/

Björk, B-C., Welling P., Laakso, M., Majlender P., Hedlund T. and Gudnasson, G. (2010) 'Open access to the Scientific Journal Literature: Situation 2009', *PLoS ONE,* **5** (6). e11273. doi:10.1371/journal.pone.0011273

CME (2012) 'Construction Management and Economics: 1992-2008', Publisher's webpages, www.tandf.co.uk/Journals/pdf/announcements/rcme_stats.pdf

CNBR (2012) Co-operative Network for Building Researchers, email discussion list, tech.groups.yahoo.com/group/cnbr-l/

Dallmeier-Tiessen, S., Goerner, B., Darby, R., Hyppoelae, J., Igo-Kemenes, P., Kahn, D., et al. (2010) 'Open Access Publishing - Models and Attributes, SOAP project report', Max Planck Society digital library, http://edoc.mpg.de/478647

DOAJ (2011) 'Directory of Open Access Journals', http://www.doaj.org/

Economist (2012) 'The price of information - Academics are starting to boycott a big publisher of journals', Feb 4th 2012, *The Economist,* http://www.economist.com/node/21545974

Easa, S., Hoel, L. (2008) Introduction (guest editorial), *Computer-Aided Civil and Infrastructure Engineering*, **23** (7), 481-482

Emerald (2008) 'Emerald Journals 2009 pricelist', Publisher's webpages,

http://www.emeraldinsight.com/products/subs/pricelist_2009.xls?PHPSESSID=flogh5cbaco82i9h1b6p56uke7

European Commission (2006) 'Study on the Economic and Technical Evolution of the Scientific Publication Markets in Europe', European Commission, Directorate General for Research, Brussels, http://ec.europa.eu/research/science-society/pdf/scientific-publication-study_en.pdf.

Gargouri, Y., Hajjem, C., Larivière, V., Gingras, Y., Carr, L., Brody, T., et al. (2010) 'Self-Selected or Mandated, Open Access Increases Citation Impact for Higher Quality Research', *PLoS ONE* **5**(10): e13636. doi:10.1371/journal.pone.0013636.

Hindawi (2012) Statistics about the publishers journals, http://www.hindawi.com/statistics/

Howard, J. (2012) 'Legislation to Bar Public-Access Requirement on Federal Research Is Dead'. *The Chronicle of Higher Education*, 27 Febr 2012. http://chronicle.com/article/Legislation-to-Bar/130949/

ITC (2012) 'ITC Digital Library for construction IT research', http://itc.scix.net/

Laakso M, Welling P, Bukvova H, Nyman L., Björk B-C, Hedlund T. (2011) 'The Development of Open Access Journal Publishing from 1993 to 2009'. *PLoS ONE* **6**(6): e20961. doi: 10.1371/journal.pone.0020961.

Less, T., Adeli, H. (2010) 'Computational Earthquake Engineering of Bridges', Scientia Iranica Transaction A – Civil Engineering, **17**(5), 325-338

Lyons, C., Booth, A. (2011) 'An Overview of Access in the Fields of Business and Management' *Journal of Business and Finance Librarianship*, **16** (2), 108-124

OpenDOAR (2012) 'The Directory of Open Access Repositories', http://www.opendoar.org/

ROAR (2012) Registry of Open Access Repositories. http://roar.eprints.org/

ROARMAP (2012) Registry of Open Access Repository Material Archiving Policies, http://roarmap.eprints.org/

SCImago (2012), 'SCImago Journal & Country Rank', http://www.scimagojr.com/

Sherpa/Juliet (2012) 'Research funders' open access policies'.http://www.sherpa.ac.uk/juliet/

Solomon, D., and Björk, B-C. (2012) 'A study of Open Access Journals using article processing charges', Accepted for publishing, Journal of the American Society for Information Science and Technology, OA copy available from www.openaccesspublishing.org/apc2/

Way, D. (2010) 'The Open Access Availability of Library and Information Science Literature', *College & Research Libraries*, **71** (4), 302-309

http://crl.acrl.org/content/71/4/302.full.pdf+html

Sinha, K. (2004) 'Recent Changes in Journal Editorial Board', *Journal of Transportation Engineering*, **130** (3), 271

Ware, M., Mabe, M. (2009) 'The stm report - An overview of scientific and scholarly journals publishing', International Association of Scientific, Technical and Medical Publishers, http://www.stm-assoc.org/2009_10_13_MWC_STM_Report.pdf

Time for a Real Shift to Relations: Appraisal of Social Network Analysis Applications in the UK Construction Industry

8

Ximing Ruan, (University of the West of England, UK)
Edward Ochieng, (Liverpool John Moores University, UK)
Andrew Price, (Loughborough University, UK)
Charles Egbu, (University of Salford, UK)

Abstract

The *Social Network Analysis* (SNA) has been adopted in the UK construction management research and generated meaningful insights in analysing project management organisations from network perspectives. As an effective tool, social network analysis has been used to analyse information and knowledge flows between construction project teams which are considered as the foundation for collaborative working and subsequently improving overall performance. Social network analysis is based on an assumption of the importance of relationships among interacting units. The social network perspective encompasses theories, models and applications that are expressed in terms of relational concepts or processes. Many believe, moreover, that the success or failure of organisations often depends on the patterning of their internal structure. This paper reviewes existing literatures on SNA applications in construction industry from three leading construction management journals. From the review, the research proposed some advance in the application of SNA in the construction industry.

Keywords: Social network analysis, Project organisations, Integration, Knowledge integration, Construction industry

Introduction

Social network analysis has been adopted as an analytical tool in the research into construction industry to provide indications of knowledge integration collaborative working and effective communication (Loosemore 1998; Chinowsky *et al.*, 2008; El-Sheikha and Pryke 2010; Larsen 2011). Though it is generally agreed that this tool is effective in analysing project performance, there are some challenges in applying it in consideration of the special characters of construction industry. Social network analysis provides a method to understand informal networks within and between organisations and manage the informal networks systematically (Cross and Prusak, 2002). Social capital makes an organisation, or any collaborative group, more than a collection of individuals' intent on achieving their own private purposes. Social capital bridges the space between people, its characteristic elements and indicators include high levels of trust, robust personal networks and vibrant communities, shared understandings, and a sense of equitable participation in joint enterprise - all things that draw individuals together into a group. This kind of connection supports collaboration, commitment, ready access to knowledge and talent, and coherent organisational behaviour (Cohen *et al.* 2001).

The key difference that distinguishes social network analysis from other analysis and management methods is that social network analysis draws attention to informal networks in working places. Basically, social network analysis is used to measure and visualise relationships and flows between people, groups, organisations, computers or other information/knowledge processing entities, which complies with Drucker's assertion that '*it is this informal organization, rather than management, which actually determines rates of output, standards, job classification and job content.*'(p174). McCarty *et al.* (2001) explain that *Social Network Analysis* (SNA) is both a theoretical perspective and a set of methods. In terms of theory, SNA extends and complements traditional social science by focusing on the causes and consequences of relations between people and among sets of people rather than on the features of individuals. In terms of method, SNA focuses on the measurement of relationships between people. In use of relational concepts, the following are important:

- Actors and their actions are viewed as interdependent (rather than independent) autonomous units. Relational ties (linkages) between actors are channels for transfer or "flow" of resources (either material or nonmaterial);
- Network models focusing on individuals views of the network structural environment as providing opportunities for or constraints on individual action;
- Network models conceptualise structure (social, economic, political, and so forth) as lasting patterns of relations among actors (Wasserman and Faust, 1994).

This paper evaluates the current application of social network analysis in construction context. Some suggestions and recommendations are made at the end in attempt to advance the application of social network analysis in construction industry. The next section introduces some key concepts in adopting network perspective in research in construction industry. The third section evaluates the key contributions from existing literature in social network analysis in the UK construction industry. Discussions and final recommendations are presented at the end.

Network Perspective

Traditionally the research into the UK construction industry focused on how to improve the competitive advantages of individual organisations. Not surprisingly these researches paid close attention to the attributes of individual organisations. The individual organisations are analysed from a range of facets including overall strategy, human resource management in turbulent project processes, leadership and team working etc. Following two government reports in 1994 and 1998 calling for changes, the UK construction industry has been attempting to improve its overall performance by introducing some managerial concepts from manufacturing industry (Latham 1994; Egan 1998). These management frameworks include Supply Chain management in construction and later Lean Construction, Partnering, Quasi-firm, and Knowledge Management. In order to achieve the benefits from the different management approaches, all the participants in the construction project process are expected to display cooperative and collaborative working patterns based on trust and mutual support. Following the introduction of the above management approaches, the research in construction industry started shifting from individual organisational attributes towards network perspective, which means the research and analysis into construction organisations shift from pure economic dimension towards research taking into account of social dimension. The organisational behaviours in construction project process are analysed in their social context rather than economic motivations.

Though the management approaches listed above take into account the social dimension, their main emphases are still on business transactions in order to maximum the economic benefits. Tichy *et al.* (1979) advocates the network perspective in the study of organisational behavior and emphasises the social dimension rather than business benefits. According to Seufert *et al.* (1999), the term 'network' designates a social relationship between actors. Actors in a social network can be persons, groups, but also collectives of organisations, communities or even societies. The relationships evolving between actors can be categorised according to content (e.g., products or services, information, emotions), form (e.g., duration and closeness of the relationship) and intensity (e.g., communication-frequency). Since the boundaries of networks are difficult to determine, blurred boundaries are constructed socially by the network members. Grandori and Soda (1995) considered networks as nexuses of integration mechanisms encompassing all the range of organisational inter-firms coordination and cooperation. Network perspectives build on the general notion that economic actions are influenced by the social context in which they are embedded and that actions can be influenced by the position of actors in social networks (Gulati, 1998).

Similarly, Powell (1990) asserted that a network is a form of organisation in business practice. Network relations are perceived as the most central feature in moving the industry forward in the perspective of the "network form" (Powell, 1990). The functional explanation for the emergence of the "network form" claims that the opportunities for innovation exist within firms as well as in their external networks, and the ability of an organisation to learn results from its capacity to utilise both its own internal capabilities and the opportunities within its network through various forms of collaboration. In the context of networking, "network forms" of organisation have been described as a source of value for the firm (Kogut, 2000). Network based industries can be expected to generate new forms of collaboration, in which network partnerships will be based on maximising resource utilisation subject to the equitable distribution of returns, rather than on individual firms maximising their profits (Miles, *et al.*, 1998). Some authors (Thorelli, 1986; Siebert, 1991; Sydow 1992) viewed networks as a hybrid form of organisation since networks have features of market and hierarchy. This form of organisation demonstrates stronger incentives and adaptive capabilities than hierarchies, but can offer more administrative control than markets (Williamson, 1994).

Due to the division of labour and accompanying fragmentation, specialisation and distribution of knowledge, it becomes a requisite to integrate and thus share the diversity of complementary knowledge in order to produce complex products and services, this feature is enhanced in the construction industry. An organisational setting is implemented or emerges because none of the actors involved can produce the collective outcome individually (Boer *et al*, 2002). Therefore, Reich (1991: p.81) regards a firm as *a facade, behind which an array of decentralised groups and subgroups exist continuously contracting with similar diffuse working units all over the world.* This perspective reinforces the significance of the relationships between organisations in order to manage business successfully, and calls for a network perspective to improve business management performance.

Similarly, Nohria and Eccles (1992) advocate a network perspective on organisational management for five important reasons:

- All organisations are social networks and therefore need to be addressed and analysed in terms of a set of nodes linked by social relationships;
- The environment in which an organisation operates might be viewed as a network of other organisations;
- Organisations are suspended in multiple, complex, overlapping webs of relationships and we are unlikely to see the overall pattern from the point of view of one organisation;
- Actions (attitudes and behaviour) of actors in organisations can best be explained in terms of their position within networks of relationships;
- The comparative analysis of organisations must take into account their network characteristics.

By taking the network perspective, managerial concerns shift from the consideration and protection of the boundaries of a firm to the management of and care for relationships (Seufert *et al.* 1999). In the empirical study of strategic alliances, Gulati (2000) points out that network facilitate firms in the network in gathering superior information, thus reducing the informational asymmetries which otherwise may cause higher transaction cost. Furthermore, networks can mitigate transaction costs by making opportunism more costly because of reputational effects. Businesses take time to build reputations in society, but this reputation can be damaged overnight. Consequently, it is essential that firms exercise caution in protecting their reputation, and the network can help create a strong disincentive against opportunistic behaviour (Wang *et al.* 2004). Wang *et al.* (2004) viewed inter-firm networks as recourses which enable business to sustain above-average performance because they meet three criteria established by Barney (1991): resources being valuable, resources being rare and resources being imperfectly imitable. While the ties between the participating organisations become less important in traditional analysis approaches, as illustrated in Figure 1, the ties become imperative and a dominating issue in evaluating business performance in a network approach, which is illustrated in Figure 2.

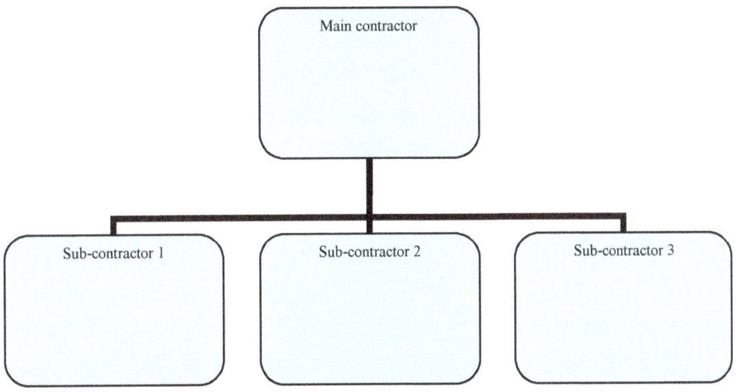

Figure 1 Traditional analysis of business management

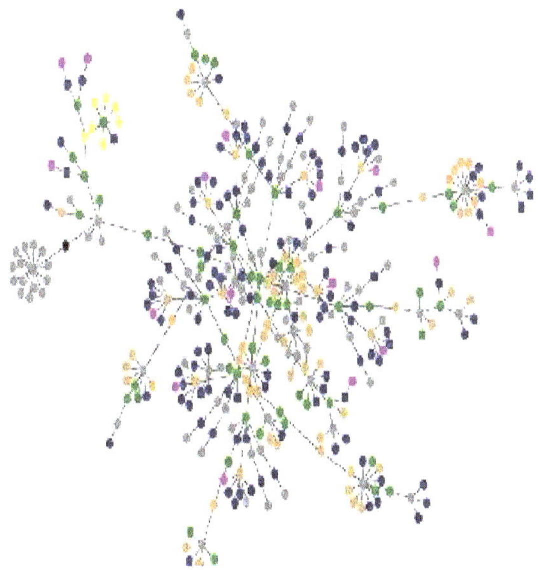

Figure 2 Boundaries between organisations

In Figure 2, all the boundaries between the organisations become less important and the network is considered as an overarching organisation within which the individual organisations function and pursue diversified goals. Network organisations are characterised by five key organisational principles:

- **Unifying purpose:** common views, values, and goals hold a network together. A shared focus on desired results sustains synchronised operations and network directionality;

- **Independent members:** each member of the network, whether an individual, company, or country, can continue to sustain its independent existence while benefiting from being part of the whole;
- **Voluntary links:** the distinguishing feature of networks is their links, as partners join forces voluntary;
- **Multiple leaders:** fewer bosses, more leaders. Each person or group in a network has something unique to contribute at some point of the process. With more than one leader, the network as a whole has greater resilience;
- **Integrated levels:** networks are multilevel, not flat. Networks operate on different levels, i.e. co-operation between organisations, departments and people (Lipnack and Stamps, 1994).

Though the five principles are useful for identifying networked organisations, they fail to explain how to create a successful networked organisation, as most of the features concluded above are developed without formal organisational management intervention. At organisational strategic level, inter-firm relationships have increasingly become a core component. They constitute valuable capital because they provide access to capabilities and resources that may otherwise be unavailable (Koka and Prescott, 2002). In regard to organisations in project management practices, Jones (1999) noted that the project organisation is best thought of as a project network, because projects are normally produced through a dynamic network of transactions involving specialised firms, subcontractors and freelancers. But this perspective has not been widely accepted by practitioners in the construction industry who constantly display adversarial relationships and work more competitively than collaboratively. The social network analysis perspective, along with its analytical tools, has been introduced into construction project management discipline which shadows a different light to understand the interactions between organisations in project processes.

Method

It is becoming difficult to ignore the social interactions and their impacts on analysing project teams. Firms have to take into account the reactions of relevant project participants and stakeholders in formulating their project plans and decisions. Not surprisingly, Social Network Analysis, which provides an effective tool to analyse the social interactions, has attracted considerable attention in analysing project team practices, revealing hidden working ties which are not visible from formal organisational chart. So far most of the attention on project teams has been drown on the formal organisational institutions. This section starts with the data collection process which presents the research results from three leading construction management journals in order to examine the current application of social network analysis in construction industry. The results show that the application of Social Network Analysis in construction industry is still limited, and the application of social network analysis and theories from network perspective in construction industry needs to be advanced.

Basic Concepts

The social network model was created based on criticisms of neo-classical economics models proposed by Williamson (1994), in which transaction activities are directed by bounded rationality and opportunism (Gordon and McCann 2000). The 'social-network model' proposes that there is more order to inter-firm interactions and less order to intra-firm interactions than the economic models would imply (Granovetter, 1985). Social networks of certain strong interpersonal relationships can be stronger than firm boundaries, with the result that many inter-firm social interactions may be stronger than their intra-firm counterparts (Gordon and McCann 2000). People use their social network as important resources for information and knowledge to accomplish their job. Freeman (2005) suggests that social network analysis focus on two structural patterns of the ties that link social actors. The patterns can reveal subsets of actors that are organised into cohesive social groups and divulge subsets of actors that occupy equivalent social positions or roles. Therefore, the actor-by-actor ties provide basic data for social network analysis.

The broad majority of social network studies use either "whole-network" or "egocentric" designs. Whole-network studies examine sets of interrelated objects or actors that are regarded for analytical purposes as bounded social collectives, although in practice network boundaries are often permeable and/or ambiguous. For example, supportive connections in a group of researchers, helping activities in a classroom of people, by measuring relationships between all units, the analysis reveals the properties of relationships in the whole network. The boundary of "whole-network" analysis is pre-set by researchers. The second one is the "ego-centric" approach. Starting from a specific unit for research, this method collection relational connections with other "alters" (chosen according to the research aims), which later determine the final boundary of the whole network. It is a method to study how relational influences produce efforts on the individual.

Data Collection

Research papers were searched from three leading construction journals, namely Construction Management and Economics, Journal of Construction Engineering and Management, and Engineering, Construction and Architectural Management. The first two are ranked as two star journals and the last one as a one star journal. The set duration for this search is from 10/2002-10/2012, and 'Social Network' was used in searching titles, abstracts and key words from the three journals.

Data Analysis

Two-hundred and seventy-seven papers were found from Construction Management and Economics, however, further reading shows that most papers merely used 'social' and 'network' separately without any discussion or application of Social Network Analysis. For instance, some papers used social housing or business networking were included in the research results. Eventually only fourteen papers from Construction Management and Economics are selected for this study. Journal of Construction Engineering and Management, and Engineering, Construction and Architectural Management have four papers respectively which met the criteria of this study, which result in only twenty-two papers from these three leading construction management journals for further study.

- ***Engineering, Construction and Architectural Management (ECAM)***

Strictly speaking, none of the four papers from our refined search in ECAM used Social Network Analysis. However, they all approached project management problems by a network perspective, which pays close attention to relationships between participants in light of the project performance and related problems. Only the first two papers were taken into account in our review as the third paper is a literature review on contractor-supplier relationships, and the fourth paper focused how business networks contribute to organisational business development rather than relating network theory to project process. Table 1 summarises the two papers used network approach to investigate project process. Although the approach used in the paper by Sandhu and Helo (2006) is similar to social network analysis, no SNA literature was referenced and the ties are dependencies between actors, resources and activities.

Paper	Research focus	Whether SNA was applied	What measurements were used	How the network was established/defined
Sandhu, M and Helo,**P. (2006)** A network approach to project business analysis	To describe and analyse project businesses from a network approach, present a network-based framework on understanding key factors contributing to project success	No	N/A	Working relationships and inter-dependencies between participants in project process
Davis, **P.R. and Walker, D.H.T. (2009)** Building capability in construction projects: a relationship-based approach	How relationship-based approach and social capital can contribute to improvement of project performance	No	N/A	Working relationship in project process

- ***Journal of Construction Engineering and Management***

All the four papers from our refined search adopted Social Network Analysis in their studies which are summarised below in Table 2:

Paper	Research focus	Whether SNA was applied	What measurements were used	How the network was established/defined
Paul Chinowsky; James Diekmann; and Victor Galotti (2008) Social network model of construction	The focus on project outcome should shift focus of construction project management from efficiency of projects to high performance project teams. The social network model adopts a network perspective in managing project team performance with special focus on knowledge and information sharing	Yes	Network Density and actor distance	Communication between Project team members
Paul S. Chinowsky1; James Diekmann; and John O'Brien (2010) Project Organisations as social networks	How to improve team effectiveness by building up inter-organisational knowledge exchange network	Yes	Network Density	Knowledge exchange activities in project process
Heedae Park; Seung H. Han, M.ASCE; Eddy M. Rojas; JeongWook Son; and Wooyong Jung (2011). Social network analysis of collaborative ventures for overseas construction projects	Inter-firm collaboration networks and their effect on organizational performance in the construction industry,	Yes	Density, directional ties, betweeness centrality, closeness centrality	Collaborative working relationships between firms
Brad W. Wambeke, M.ASCE; Min Liu, A.M.ASCE; and Simon M. Hsiang (2012) Using pajek and centrality analysis to identify a social network of construction trades	To identify key subcontractors in a project process by social network analysis	Yes	Degree centrality and Eigenvector Centrality	Task dependency between project main contractor and subcontractors

Although the four papers used social network analysis, it is difficult to ignore the fact that all the four papers were published from 2008-2012.

- ***Construction Management and Economics(CME)***

There are fourteen papers from *Construction Management Economics* applied social network analysis in their studies into construction project management, which are summarised in Table 3.

Paper	***Research focus***	***Whether SNA was applied***	***What measurements were used***	***How the network was established/defined***
Stephen D. Pryke (2004) Analysing construction project coalitions: exploring the application of social network analysis	How network approach can assist in understanding coalitions in construction projects?	Yes	Centrality	Information exchange
Alexander Styhre, Per Erik Josephson and Ingeborg Knauseder (2004) Learning capabilities in organisational networks: case studies of six construction projects	How organisational learning capacities are developed during construction project management process from a network perspective	No		Knowledge sharing between project participants at organisational level
Stephen D. Pryke (2005) Towards a social network theory of project governance	To understand project governance from a social network analysis approach by measuring network density and actor centrality	Yes	network density and actor point centrality	Contractual, performance incentive and information exchange
Alexander Styhre (2008) The role of social capital in knowledge sharing: the case of a specialist rock construction company	To examine the knowledge acquisition patterns within a connected network	No	N/A	Knowledge sharing activities
Catherine Comet (2009) Social capital and profits of small firms in the French construction industry	How can social capital and social networks improve companies' performance in construction industry	No	N/A	Collaboration between project participants
Alexander Styhre and Pernilla Gluch (2010) Managing knowledge in platforms: boundary objects and stocks and flows of knowledge	How knowledge is stored and shared between project participants	No		Knowledge integration and sharing between project participants
Ahmed El Sheikh and Stephen D. Pryke (2010) Network gaps and project success	to examine project inter and intra networks for effective project governance with combined use of gap analysis and SNA	Yes	Degree Centrality and Network density	
Abu Naser Chowdhury, Po-Han Chen and Robert L.K. Tiong (2011) Analysing the structure of public–private partnership projects using network theory	How to identify the important players in PPP projects with network theory	Yes	Betweenness Centrality, Closeness centrality	Contractual relationships
Graeme D. Larsen (2011) Understanding the early stages of the innovation diffusion process: awareness, influence and communication networks	To investigate how actors in an innovation diffusion network become aware of an innovation and then how their opinion is influenced	Yes	Density of the whole network	Diffusion channel of innovation in construction projects

Cont'

Paper	*Research focus*	*Whether SNA was applied*	*What measurements were used*	*How the network was established/defined*
Ximing Ruan , Edward G. Ochieng , Andrew D.F. Price and Charles O. Egbu (2012) Knowledge integration process in construction projects: a social network analysis approach to compare competitive and collaborative working	How knowledge is integrated between project participants according to their social network patterns	Yes	Betweenness Centrality, Closeness centrality, Network Density	Knowledge contribution towards variation order requirements
Rayyan Alsamadani , Matthew Hallowell and Amy Nicole Javernick-Will (2012) Measuring and modelling safety communication in small work crews in the US using social network analysis	To investigate the frequency and structure of effective safety communication within small project teams	Yes	Degree centrality, Network density, and betweenness centrality	Communication between participants

Discussions on SNA Applications in the UK Construction Industry

From the tables which summarised the research papers adopted social network analysis, it is clear that the application of social network analysis is relatively new in the UK construction industry (El-Sheikha and Pryke, 2010), and the research from a social network analysis approach has favored whole-network method by which the relationships are measured by cohesion, density and distances. From the twenty-two papers selected for this study, it is evident that most of the ties and connections in these social networks were established by information/knowledge exchange or by task/contractual relationships in workplace during project processes. Although the literature sections of these papers placed great emphasis on social settings and project context in the study of issues relating to project performance, the '*social*' interactions outside of the project processes and working place were not investigated. Social context should take into account social interactions which take place in social events in which social capital accumulate. Also the measurements adopted by these papers are limited to degrees, centralities and densities. The density is used as a measurement to reflect overall network collaboration or cohesion and centralities are used as a measurement to reflect relational patterns among those papers. However, the real meaning of such social constructs can be argued in practice (Larsen, 2011).

During the past ten years, only twenty papers were published in the three leading construction management journals used social network analysis as a tool to analyse different aspects of project performance. All of these papers used social network analysis established the ties between actors by a kind of relationships, namely contractual/intensive connections, communication and information channels, and knowledge contribution activities. Among these papers, several authors made significant contributions towards the application of SNA in construction project management. Pryke has studied a range of issues using SNA, including project governance (Pryke, 2005), project success factors (El-Sheikh and Pryke, 2010), and project coalitions (Pryke, 2004). Other papers used similar approaches to map the project management process from a social network analysis perspective and analysed different aspects by the established networks. The following discussion focuses on Pryke's papers but the arguments can be extended to other papers.

Pryke (2004) considers degree centrality as an effective means to measure power and influence of an organisation. The degree centrality can be understood as in a given network, an organisation has higher degree centrality if it has more ties than other organisations. Pryke (2004) suggested that a multi-layer of interdependent networks within project coalitions should be explored as the following:

- Networks of contractual relationships;
- Networks of performance incentives;
- Networks of information exchange, sub classified into:
    - Client requirements;
    - Design activities;
    - Progress management; and
    - Financial management (p.795, Pryke, 2004).

However, Pryke ignored the fact that the effective application of social network relies on *social* interactions based on informal relationships. For example, social network can analyse how some people can become powerful with their social ties, even if they do not have any formal authority that can be found from an organisational chart. These kinds of informal relationships, which social network analysis investigates, do not follow the formal relationship channels such as organisational charts and business contracts. Networks contractual relationship and performance incentives should be applied to determine the whole network boundaries rather than to establish the ties between firms as those relationships can be defined by traditional analytical methods. The critique of Pryke's paper can also be applied to other papers which did not take into account the informal social interactions outside of working places, as some papers adopted SNA measurements without strong justifications. Similarly, in Chinowsky's study (2010), it is not clear why the density is used as a single measurement to benchmark the team performance, and why the effectiveness of the ties between dyads are not considered.

In a recent research into project success in the UK construction industry, El-Sheikha and Pryke (2010) adopted social network analysis to map the communication pattern which has significant impact on client satisfaction and subsequently project success. El-Sheikha and Pryke (2010) consider that the application of social network analysis can be used to analyse the systems associated with delivering project through:

- *Addressing problems at early stages and thus minimising any client surprises;*
- *Identifying and improving information exchange networks (Pryke, 2004) and reducing any possible discontinuity in financial incentives; and*
- *Achieving a balance between social and scientific aspects in management of projects.*

However, social network is analytical in nature and can indicate the social network pattern and attributes under investigation, but it can be asked how this analytical tool can be used to 'improve information exchange networks' and 'reducing discontinuity in financial incentives' as social network analysis by itself does not provide any control methods. El-Sheikha and Pryke (2010) constructed information exchange networks between a range of project actors comprising of architect, construction project manager, employer's agent, quantity surveyor, contractor, service engineer and structural engineer. Not only are the networks differentiated by their foci on project time, cost and scope, but also the informal and formal discussions are distinguished to construct two networks. Nevertheless, it is not clearly explained in their paper what kind of discussion should be considered 'informal' if the discussions are about project delivery which is directly associated with contractual relationships and responsibilities. Meanwhile, the density of communication in this review is considered as a major means to evaluate the relationships between different actors, without acknowledging that communication frequency is not always directly linked with support and close relationships. Some actors in a network can communicate less than others due to closer relationships and shared understanding.

It is also worth mentioning that among all these papers, Chowdhury *et al's* study (2011) is the only one which adopted two-mode Social Network Analysis approach to establish the network under investigation. This two-mode approach establishes the ties between members by mutual participations on particular events, rather than direct ties such as communication between members. Basically the two-mode social network analysis has two sets of actors and the ties between actors in one set are created by investigating their relationships to the other set. For example, a group of people can be considered as one set of actors and a range of social events can be considered as the other one. The ties between people are created by whether they are present at those events. The reason behind this approach is that '*individuals, by their agency, create social structures while, at the same time, social structures develop an institutionalized reality that constrains and shapes the behavior of the individuals embedded in them*' (Hanneman and Riddle, 2005). Therefore, data like these involve two levels of analysis (or two "modes"). Normally, the term "affiliation" is used to describe which actors are affiliated (present, or members of) which macro structures. This approach can look into the social dimension and how individuals make meanings of the events or groups in a social context.

The Way Forward

The paper has reviewed relevant concepts in social network analysis and its application in the UK construction context. The UK construction industry has adopted a number of management methods and analytical tools from other disciplines and industries, such as Total Quality Management, Supply Chain Management. Not surprisingly, it takes some efforts to apply those methods in the construction context due to the special character of construction management. It is generally agreed that social network analysis provides a quantitative tool, which is positivism in nature, to analyse interactions between actors in project management process in construction industry; the actors can be either individual persons or organisations. The connections of the social network between all the participants are established by the participant themselves in the context of project process and there is no exploration of 'social' dimension in larger social context, where participants establish or reinforce their social ties in social interactions. It can be said that the analysis of social networks in construction industry merely measures relations at the micro level where the ties of individuals are examined, the macro structure, such as subgroups of special interests, cannot be revealed.

This reflects that the current applications of social network analysis are limited to one-mode social networks, where a set of actors are linked by predefined ties. From the studies reviewed, the ties can be information, discussion, supportive activities etc. The application of two-mode social network analysis has not been fully explored in construction industrial context with only one of the reviewed papers having adopted the two-mode approach.

Furthermore, social network analysis should be considered as a complementary tool to be used in combination with other research methods in order to reveal the real meaning of the data presented in social diagrams (Larsen, 2011). For instance, the density of a network is positively related to collaborative working, trust, and frequency of communications in these papers. However, this can be negatively related to these factors as tacit understanding between members can result in a network with less density if this network is measured by communication frequency. Although SNA is embedded in the positivist perspective (Larsen, 2011), the data collection process and interpretations are more interpretative in nature as the ties established by support, trust, collaborative working and effective communications are subjectively perceived and socially constructed. The reconciliation of the both positivism and interpretivism in social network analysis should be justified in future studies.

Conclusion and Recommendations

Though the social network analysis is considered as a relatively new approach in the construction industry, it meets the trend that the research foci in construction management is shifting from economic activities to include more social dimensions. However, this paper reveals that the application of the social network analysis in construction industry needs further development in terms of research design and data interpretations to serve as a strong analytical tool. Moreover, the current application of social network analysis in the construction industry is limited by focusing only on ties established *within* workplace, the ties created or reinforced in social contexts remaining unexplored, and as a consequence the strengths of the social network analysis are not fully explored. Therefore, further research is needed to advance the application of social network analysis in construction industry by developing stronger justification for the selection of measurements, the adoption of two-mode analysis and taking into account more of the social context to realise the full potential of social network analysis in the construction industry.

In summary, there is a need for:

- Strong justification of the adoption of social network analysis as an effective tool in order to interpret the real meaning of these connections in real life.
- The use of two-mode social network analysis which takes into account social events should be advanced further.
- Taking into account social interactions outside the work place in data collection in order to justify the use of SNA which places emphasis on social context.
- Use social network analysis as a complementary tool in combination with other tools in order to generate meaningful insight.

References

Alsamadani, R., Hallowell, M. and Javernick-Will, A.N. (2012) 'Measuring and modelling safety communication in small work crews in the US using social network analysis', *Construction Management and Economics,* 1-12

Barney, J. B. (1991) 'Firm resources and sustained competitive advantage', *Journal of Management*, **17** (1), 99–120

Boer, N. I., van Baalen, P. J., and Kumar, K. (2002) 'The implications of different models of social relations for understanding knowledge sharing', Paper presented at the 3rd *European Conference on Organisational Knowledge, Learning, and Capabilities*, ALBA, Athens, April, 3-4

Chinowsky, P. S. Diekmann, J., and Galotti, V. (2008) 'The Social Network Model of Construction', *Journal of Construction Engineering and Management*, **134** (10), 804-810

Chinowsky, P., Diekmann, J., and O'Brien, J. (2010) 'Project Organisations as social networks', *Journal of Construction Engineering and Management,* **136**, 452–458

Chowdhury, A.N., Chen, P.H., Tiong R.L.K. (2011) 'Analysing the structure of public–private partnership projects using network theory', *Construction Management and Economics*, **29** (3), 247–260

Cohen, W.M., Malerba, F., Mowery, D. and Winter, S. (2001) 'Introduction to the special issue in honor of Richard Nelson', *Journal of Industrial and Corporate Change*, **10**, 549-559

Comet, C. (2009) 'Social capital and profits of small firms in the French construction industry', *Construction Management and Economics*, **27** (4), 411-418

Cross, R., Parker, A., Prusak, L and Borgatti, S. (2001) 'Knowing what we know: Supporting knowledge creation and sharing in social networks', *Organisational Dynamics* **3** (2), 100-120, re-printed in Darden Learning Primer and Creating Value with Knowledge: Insights from the IBM Institute for Knowledge-Based Organisations

Cross, R. and Prusak, L. (2002) 'The people that make organisations stop – or go', *Harvard Business Review*, **80** (6), 104-112

Drucker, P. (1950) *The New Society: The Anatomy of industrial order*, New York: Harper and Brothers

Egan, J. (1998) *Rethinking Construction.* Department of the Environment, HMSO, UK

El-Sheikha, A and Pryke, S. (2010) 'Network gaps and project success', *Construction Management and Economics,* **28** (12), 1205-1217

Freeman, L. C. (2005) 'Graphic techniques for exploring social network data' in *Models and Methods in Social Network Analysis,* eds, Carrington, P. J., Scott, J. and Wasserman, S., Cambridge University Press, NY

Grandori, A. and Soda, G. (1995) 'Inter-firm networks: Antecedents, mechanisms and forms', *Organization studies,* **16** (2), 183-216

Granovetter, M. (1985) 'Economic action and social structure: The problem of embeddedness', *American Journal of Sociology,* **91** (3), 481-510

Gordon, I. R. and Mccann, P. (2000) 'Industrial clusters: Complexes, agglomeration and/or social networks?', *Urban Studies,* **37** (3) 513-532

Gulati, R. (1998) 'Alliances and networks', *Strategic Management Journal,* **19** (4), 293-317

Gulati, R., Nohria, N. and Zaheer, A. (2000) 'Strategic Networks', *Strategic Management Journal,* **21** (3), 203-216

Hanneman, R. A. and Riddle, M. (2005) *Introduction to social network methods,* University of California, Riverside, CA

Jones, B. (1999) 'Bounded rationality', *Annual Review of Political Sciences,* **22**, 297-321

Kogut, B. (2000) 'The network as knowledge: Generative rules and the emergence of structure', *Strategic Management Journal,* **21**, 405–425

Koka, B. R. and Prescott, J. E. (2002) 'Strategic alliances as social capital: A multidimensional view', *Strategic Management Journal,* **23**, 795-816.

Larsen, G. (2011) 'Understanding the early stages of the innovation diffusion process: awareness, influence and communication networks', *Construction Management and Economics,* **29** (10) 987-1002

Latham, M. (1994) *Constructing the Team* Department of the Environment, HMSO, UK

Lipnack, J. and Stamps, J. (1997) *Virtual Teams: Reaching across Space, Time, and Organizations with Technology,* J. Wiley, New York, Chichester

Loosemore, M. (1998) 'The influence of communication structure upon management efficiency', *Construction Management and Economics,* **16** (6), 661-671

McCarty, C., Killworth, P.D., Bernard, H. R., Johnsen, E. and Gene A. S. (2001) 'Comparing two methods for estimating network size', *Human Organisation,* **60**, 28-39

Miles, G., Miles, R. E., Perrone, V. and Edvinsson, L. (1998) 'Some Conceptual and research barriers to the utilization of knowledge', *California Management Review,* **40** (3), 281-288

Nohria, N., Eccles, R. G. (1992) *Networks and organisations: Structure, form, and action,* Harvard Business School Press, Boston, Mass

Park, H., Han, S., Rojas, E., Son, J., and Jung, W. (2011) 'Social network analysis of collaborative ventures for overseas construction projects', *Journal of Construction Engineering Management,* **137** (5), 344–355

Pryke, S. D. (2004) 'Analysing construction project coalitions: Exploring the application of social network analysis', *Construction Management and Economics,* **22** (8), 787-797

Pryke, S.D. (2005) 'Towards a social network theory of project governance', *Construction Management Economics,* **23** (9) 927–39

Powell, W. W. (1990) 'Neither market nor hierarchy: Networks forms of organisation', *Research in Organisational Behavior,* **12**, 295-336

Reich, R. B. (1991) *The Work of Nations: Preparing Ourselves for 21st-Century Capitalism,* A.A. Knopf 1991, New York

Seufert, A., Von Krogh, G. and Bach, A. (1999) 'Towards knowledge networking', *Journal of Knowledge Management,* **3** (3), 180-190

Siebert, H. (1991) 'Konomische analyse von unternehmungsnetzwerken' in *Managementforschung*, **1**, eds, Staehle, W. H. and Sydow, J., DeGruyter, Berlin, New York, 291-311

Styhre, A., Josephson, P.E and Knauseder, I (2004) 'Learning capabilities in organizational networks: case studies of six construction projects', *Construction Management and Economics,* **22** (9), 957–966

Styhre, A. (2008) 'The role of social capital in knowledge sharing: the case of a specialist rock construction company', *Construction Management and Economics,* **26** (9), 941-951

Styhre, A and Gluch, P. (2010) 'Managing knowledge in platforms: boundary objects and stocks and flows of knowledge', *Construction Management and Economics,* **28** (6), 589-599

Sydow, J. (1992) *Strategische netzwerke, Evolution und organisation,* Gabler, Wiesbaden

Thorelli, H. B. (1986) 'Networks: Between markets and hierarchies', *Sloan Management Review,* **7**37-51

Tichy, N., Tushman, M. and Fombrun, C. (1979) 'Social networks analysis for organizations', *Academy of Management Review,* **4,** 507-519

Wambeke, B., Liu, M., and Hsiang, S. (2012) 'Using pajek and centrality analysis to identify a social network of construction trades', *Journal of Construction Engineering Management*, **138** (10) 1192–1201

Wang, Y., Ahmed, P. K. and Worrall, L. (2004) 'Understanding inter-firm network: A theoretical review', *International Journal of Management Concepts and Philosophy,* **1** (1), 173-198

Wasserman, S. and Faust, K. (Eds.) (1994) *Social Network Analysis: Methods and Applications,* Cambridge University Press, Cambridge

Williamson, O. E. (1994) Comparative economic organisation: The analysis of discrete structural alternatives, *Occasional Papers- International Center for Economic Growth,* ALL

Ruan, X., Ochieng, E.G., Price, A.D.F and Egbu, C.O. (2012) 'Knowledge integration process in construction projects: a social network analysis approach to compare competitive and collaborative working', *Construction Management and Economics*, **30** (1), 5-19

The Relevance of the ISO26000 Social Responsibility Issues to the Hong Kong Construction Industry

9

Lisa Robyn Barnes and Norman Croker, (University of Newcastle, Australia)

Abstract

Corporate Social Responsibility ("CSR"), the responsibility of a corporation for the impacts of its decisions on society and the environment, originated as a term in the 1950's (Carroll 1999). However, ISO 26000 "Guidance on Social Responsibility" published in November 2010, redefines "Social Responsibility" ("SR") broadly, by reference to SR principles, core subjects and issues. The Hong Kong Construction Industry ("HKCI") sector is critical to Hong Kong's economy, accounting for 5.6% of the GDP and 9.2% of employment in 1999 (Tang 2001). In 2011, the gross value of construction work amounted to HK$215.0 billion and contributed HK$65.4 billion (HKSARCSD 2013a) to the total GDP of HK$1,823.2 billion (HKTDC 2013; HKGov 2013). Provisional quarterly data for 2012 indicates an increase in the gross value of construction works for 2012 of 25% to be now in excess of HK$250 billion (HKSARCSD 2013b). The aim of this research is to investigate which ISO 26000 SR principles, core subjects and issues are relevant to HKCI firms. The principle findings are that: the reported level of HKCI CSR activity relative to the seven core subject of ISO 26000 is a function of company size; many SME's consider that many of the ISO 26000 SR issues are irrelevant to them; but for the large construction contractors, with employee numbers greater than 200, the SR issues in ISO 26000, particularly those relating to the health and safety in the workplace and to the environment, are not only relevant but in many cases the corporate SR activities exceed legislated requirements.

Keywords: Construction Industry, Corporate Social Responsibility

Background

Corporate Social Responsibility ("CSR"), as a term and concept, is quite modern although societal and academic interest in the business and society relationship has a long history. As a term, CSR's use has expanded greatly since its modern inception in the 1950's in America such that it is now well-established, not only in America but also world-wide, whereby:

> *"companies integrate social and environmental concerns in their business operations and in their interaction with their stakeholders on a voluntary basis" (Knopf et al. 2011, p. 9).*

However, it remains an inherently complex and multifaceted concept (Cochran 2007), with a wide variety of meanings and numerous definitions, such that the term has been criticised as being a vague and unclear concept (Votaw 1972; Coelho, McClure & Spry 2003). In America, CSR developed as an explicit activity, although in Europe "*there was a tradition of CSR-relevant institutions (referred to as 'implicit CSR') long before the concept of CSR was discussed explicitly*" (Knopf et al. 2011, p. 9). With the growth of globalisation, the American model expanded internationally for numerous reasons (Hatcher 2003; Chaudhri 2006; Jones 2009; Ahmad 2011). The expansion of CSR as a concept and management imperative globally has resulted in firms increasingly showing interest in, and reporting on, CSR as the concept matures and stakeholders increasingly demand such behaviour by firms, particularly over the last decade when various institutions and governments have produced CSR-related standards, guides and policies with over 400 CSR Standards now in place world-wide (Diao 2010).

Corporate Social Responsibility: ISO 26000

The Social Responsibility ("SR") of an organisation is defined by the International Organisation for Standardization ("ISO") in ISO 26000 "Guidance on Social Responsibility", published in November 2010 (pp. 3-4) as the:

"responsibility of an organization for the impacts of its decisions and activities on society and the environment, through transparent and ethical behaviour that

- *contributes to sustainable development, including health and the welfare of society;*
- *takes into account the expectations of individuals or groups that have an interest in any decision or activity of the organisation (stakeholders);*
- *is in compliance with the applicable law and consistent with international norms of behaviour; and*
- *is integrated throughout the organisation and practiced in its relationships".*

ISO 26000 (2010) was qualified as being "not a management system standard [nor] intended or appropriate for certification purposes" (p. 1) and further defines SR broadly by reference to seven principles, seven inter-related core subjects and their 36 related SR issues (pp. 10-68), as shown in figure 1.

Figure 1 The inter-relationship of Seven Core Subjects (ISO 26000, 2010, p 20)

The Hong Kong Construction Industry ("HKCI")

The construction industry ("CI") comprises:

> *"all those organisations and persons concerned with the process by which building and civil engineering works … are procured, produced, altered, repaired, maintained and demolished. This includes companies, firms and individuals working as consultants, main and sub-contractors, material producers, equipment suppliers and builder's merchants. The industry has close relationships with clients and financiers"* (CIB TG 29, 1998, Agreed International Definition in Hillebrandt 2000, p. 4).

Although the CI is economically important, its process are geographically and functionally fragmented (O'Brien & Al Soufi 1993); it is "motivated by turnover, profit and shareholder returns"; and slow to adopt CSR (Murray & Dainty 2007 p. 3). Indeed, it is the CI's product characteristics, being large heavy and expensive; the separation of design from construction; and the method of price determination, generally through tendering and negotiation; which determine the structure of the industry and its geographical and functional fragmentation (Hillebrandt 2000, p. 5).

The CI is important world-wide because (a) its product is large and therefore significant to the world's economies; (b) it produces investment goods which can contribute to world economic well-being; and (c) infrastructure construction is vital for development, particularly in developing countries (Hillebrandt 2000). The HKCI is critical to Hong Kong's economy, accounting for 5.6% of the GDP and 9.2% of employment in 1999 (Tang 2001), which fell to 3.2% of GDP in 2004, following the SARS pandemic but has subsequently recovered (Rowlinson 2007). HKCI activities are labour-intensive, dangerous and polluting; it is fragmented; adversarial and employs non-value adding multi-layered subcontracting (Tang 2001). The HKCI is characterized by a small number of large local and overseas contractors which depend on sub-contracting (Rowlinson 2007), a practice that can dissipate espoused CSR policies (Green 2007).

Gross value of construction works performed HK$ M	No. of establish-ments (% of Total)	Number of persons directly engaged (Ave. No)	Payments to employees and labour-only sub-contractors HK$ M	Consumption of materials, supplies, utilities, & fuels, etc, HK$ M	Payment to fee sub-contractors HK$ M (Ave % of gross)	Miscell-aneous operating expenses HK$ M	Gross value of construction works performed HK$ M
0-9,999	20,849 (89.0%)	71,045 (3.4)	3,435	5,220	5,044 (27.2%)	963	15,452
10,000-19,999	1,789 (7.6%)	22,684 (12.7)	4,835	5,587	7,915 (35.5%)	1,930	22,294
20,000-49,999	294 (1.3%)	13,412 (45.6)	4,767	2,696	1,485 (12.7%)	823,	11,671
50,000-99,999	281 (1.2%)	19,395 (69.0)	5,270	3,757	7,754 (40.5%)	1,144	19,144
100,000+	203 (0.9%)	43,809 (215.8)	21,874	21,084	61,394 (53.0%)	9,167	115,870
Total	23,417	170,345	50,625	44,400	91,082 (42.4%)	17,540	214,976

Table 1 Principal statistics for the construction sector by gross value of works performed, 2011 (All figures in HK$ millions) (Source: HKSARCSD 2013a, p. 13)

Most sub-contracting firms in the HKCI are part of the small to medium ("SME") sector, which accounts for 98% of Hong Kong's business establishments, defined as "economic units under a single owner or control in one kind of economic activity" (HKSARCSD 2013a, p. 42). As shown in Table 1, in the construction industry, only 2% of all firms have a gross value of construction works exceeding HK$ 50 million (£ 4.2 million) and average employee numbers greater than 50. The remainder, on average would be "small to medium firms". Less than 1% of firms have a gross value of construction works exceeding HK$ 100 million, but they account for 54% of the gross value of construction works performed; they sub-contract 54% of their works to fee sub-contractors and, on average, would be "large firms" with employee numbers exceeding 200. Similarly, firms having a gross value of construction works between HK$ 50 to 100 million account for a further 9% of the gross value of construction works performed, sub-contract 40% of their works to fee sub-contractors and, on average, would be "medium" with employee numbers between 50 and 100 (HKSARCSD 2013a). If it is assumed that the gross value of sub-contracted works (HK$ 91 million) is sub-contracted to firms in the HK$ 0 to 50 million per annum category, then the actual value of work directly performed by this sector amounts to 2/3rds of the total gross value of construction work. Thus, at least 98% of construction industry firms fall in the category of SME firms and account for at least 65% of the gross value of construction works performed (albeit with the majority of that value being as sub-contractors to medium or large firms). However, most existing efforts to engage Hong Kong's SME's in environmental and social performance issues have had little impact on SME's to date (Studer et al. 2008).

SME's and CSR

During the past decade, the evolving CSR literature, and the imperatives of increased requirements for accountability and sustainability moved the main discussion from a "whether or not to" to a "how to" implement CSR debate, but this debate has primarily centred on large firms with very little focus on SME's, on the implicit assumption that small businesses are essentially the same as big businesses but on a smaller scale. Welsh and White (Welsh & White 1981, p. 18) argued some time ago that "a small business is not a little big business" due to small business's "*resource poverty*", arising primarily out of: a tendency to be in highly fragmented, very competitive industries; inability to afford professional staff; higher relative cost to comply with legislated regulations; and, severe restraints to raise finances, all of which issues can be attributed to SME's in the CI. Although few articles have explored this difference between SME's and large businesses during the interim 30 years have been identified, it was recently argued that stakeholder theory might generally be appropriate for CSR research in large corporations but it is inapplicable for SME's for which resource theory /social capital principles are more appropriate (Perrini 2006; Perrini, Russo & Tencati 2007; Russo & Perrini 2010)

The Relevance of ISO 26000 (2010) SR issues to the CI: a Research Problem

Although there are, as yet, but a few journal articles that address the recent publication of ISO 26000 (n = 21 to 2012), it has been found that "the most important barrier and difficulty for CSR implementation is the general lack of knowledge or awareness of CSR [in Iran]" (Valmohammadi 2011, p. 260). Moreover, there is a paucity of articles which address CSR in the HKCI, although it is a very important industry for many national economies. No articles identifying which issues are important for individual industries by reference to a large sample population have been identified. This research will investigate the level of knowledge and awareness of CSR in the HKCI by reference to ISO 26000 (2010), and by reference to all available HKCI companies in publicly available lists and for which postal addresses can be readily identified.

Therefore, the focus of this research is to fill the literature gap relating to CI practitioners' perceptions of which CSR issues identified in ISO 26000 (2010), if any, are applicable to the HKCI. This leads to the Research Problem of:

How does, and what is the extent to which, CSR as now defined by ISO 26000 apply to CI firms, particularly to HKCI firms?

The relevant factors identified in the literature generate the following research question and hypothesis:

RQ1 To what extent do HKCI firms and their management consider that HKCI firms are meeting or exceeding their legislated and voluntary CSR obligations as described in ISO 26000 (2010)?

RH1 Large (or main contractor) HKCI firms and their management consider that HKCI contractors currently meet but do not exceed legal obligations except in the specific areas of health and safety.

Methodology

Mauch and Birch (1998, pp. 116-122), identify 19 'methodologies' including the five 'research strategies' used in the social sciences as described by Yin (2003) of: experiments, surveys, archival analysis, histories and case studies. Fellows and Liu (2003) assert that the majority of research in the social sciences and management spheres involves two of these strategies, being questionnaire surveys and case studies. From these conditions, the survey approach is applicable to "who, what, where, how many and how much" questions and the case study approach is applicable to "why and how" questions (Yin 2003, p. 5). RQ1 falls in the category of "how many and how much" and requires original empirical data to be sourced; and, a nomothetic style questionnaire methodology is appropriate. Therefore, the adopted methodology consisted of a questionnaire to a broad cross section of predominantly contractors and sub-contractors from lists of HKCI firms identified on publicly available websites.

Establish the Survey Frame and the Sample Population

Hong Kong is a Special Administrative Region of the People's Republic of China and a very high density city-state with a well-advanced and organised CI, subject to a degree of government regulation. To large extent, then, Hong Kong forms the area frame for the study. There are a number of organisations that maintain publically available lists of CI firms on their websites. The target population for the questionnaire survey was therefore decided to be the senior management of those HKCI listed on such organisations' websites. Among these, the Hong Kong Construction Association Ltd. ("HKCA"), with a membership of approximately 320 contractors is the peak body for the industry. A separate body, the Hong Kong Federation of Electrical and Mechanical Contractor Ltd. ("HKFEMC") with a membership of approximately 150 represents the specialist (generally smaller) contractors in those specialist disciplines. Other organisations include the Hong Kong General Building Contractors Association ("HKGBCA"), the Hong Kong Bar-bending Contractors Association, the Hong Kong Electrical Contractors Association, the Hong Kong Construction Sub-Contractors Association ("HKCS-CA") and so on. Some of these associations maintain websites and some do not. The Hong Kong Construction Industry Council ("HKCIC") is a government legislated peak body for the CI in Hong Kong responsible, in part, for industry training and administration and maintains a list of "Voluntary Registered Sub-Contractors" ("VRS", subsequently renamed "list of Registered Sub-Contractors)".

In addition, the Hong Kong Government's Works Bureau ("Works Bureau") maintains a register of approximately 250 contractors which are pre-qualified to tender for government projects. This register encompasses contractors which are registered to tender for projects with a value less than HK$20 million (Class A), between HK$20 million and HK$50 million (Class B) and greater than HK$ 50 million (Class C). A separate register of approximately 500 contractors, which are pre-qualified to tender for specialist projects and as subcontractors for the specialist elements of larger projects as sub-contractors to main contractors, is maintained by the Works Bureau. Some contractors are listed on both registers and some are also members of the HKCA and the HKFEMC, or both associations, and some are also HKCIC registered sub-contractors. Of the other associations, most of the members listed on the websites which were available (such as the HKGBCA) were included in the other lists and /or for the remainder, identifying their postal addresses proved to be prohibitively time-consuming and they were not included as a source of potential invitees. Others, such as the HKCS-CA either had no website or no member lists.

A combined dataset of the members of the HKCA, the HKFEMC, the contractors listed on both registers of the Works Bureau, those listed in the HK Yellow Pages ("HKYP") and those identified in the HKCIC VRS register formed the list frame for the research and those who responded formed the sample population. Thus, there is an element of convenience sampling but the sampling also includes an element of volunteer sampling in that not all the potential invitees "volunteered to respond".

Background to the Analysis

The questions for the questionnaire were designed from first principles to gauge the relevance of each of the core issues in ISO 26000 (2010). In addition to the 44 questions relating to the seven principles and seven core subjects of ISO 26000 (2010), seven demographics questions were developed to potentially permit analysis of different segments, company types and position-related views. One question related to company size to potentially differentiate between large business views and actions on CSR and those of SME's. Different countries classify small and medium sized businesses by different "head count" numbers. Therefore this question aimed to permit comparison between sizes consistent with Australia (Small Enterprise < 20; Medium Enterprise < 200 (ABS 1999)), the European Union and the UK (Small Enterprise < 50; Medium Enterprise < 250 (EC 2003)), and Hong Kong (Small and Medium Enterprise (Manufacturing) < 100; Small and Medium Enterprise (Non-manufacturing) < 50 (HKSARCSD 2013a)).

Therefore, the initial dissemination was through the HKCA to its members and subsequent disseminations of questionnaires to the companies on the other lists were performed by the researchers. The purpose was primarily exploratory as there is little (or no) existing literature which investigates the relative importance of factors for and against adopting CSR practices in the CI (or the HKCI) and none that explores the relevance of ISO 26000 (2010) to the CI.

Quantitative Study – HKCI Questionnaire

An initial review of the number of publicly available lists of HKCI companies on the websites of the HKCA (HKCA 2012), the HKFEMC (HKFEMCL 2012) the Works Bureau's Lists (HKGDevB 2012) and the HKYP (HKYP 2012) was performed to determine how many HKCI contracting firms would be reasonably accessible, in that their particulars (names and phone numbers as a minimum) were available. Some of the large international CI firms are not listed on the Hong Kong Stock Exchange and it is, therefore, not a reliable source of identification of large CI firms. The lists contained by these organisations contain many duplicates, both internally due to companies being listed in several sub-categories and externally between the lists. In order to estimate the number of CI firms, each list was carefully examined and all duplicates removed with the resultant expected sample population estimated to be approximately 1,900. It should be noted that, because the firms in the first 4 lists in Table 2, below, account for nearly all the larger HKCI firms, those in the HKYP and HKCIC VRS (HKCIC 2012) lists are almost exclusively SME's.

Name of List	Number in List	Duplicated / no postal address	Net number in list
HKCA	320	-	320
HKFEMCL	148	-	148
HKGDevB: Contractors	259	146 / 0	113
HKGDevB: Specialists	553	190 / 0	363
HKYP: Construction Contractors	1,555	435 / 300	820
HKCIC VRS	~ 7,000	~ 5,911	1,089
Totals	~ 9,835	~ 6,982	2,853

Table 2 List of Contractors in public domain issued with invitations to participate

Assuming 85% of the listed companies could be contacted, and a response rate of approximately 10% could be achieved, then this would achieve the targeted total of 150. However, even if the result was at 5% (n = 75) it was considered that this would be sufficient based on an effective 1-5 rating scale (range of 4 units) divided by 4 on the assumption of a normal distribution divided by (say) a precision of 1/4 unit using the formula minimum sample size = [(degree of confidence {95%} x variability {range/4} / desired precision {1/4 unit}]^2 (Hair et al. 2003, p. 219): $[(2*1)/(0.25)]^2 = 64$.

Source of List	English	Chinese	Discard	Total
Hong Kong Construction Association (n = 320 – 4.70%)				
Hard Copy Response	6	3		9
Web-based Electronic Response	(8-5 = 3)	(6-3 = 3)	8	14
Net Total	9	6	8	23
Federation of Electrical and Mechanical Contractors (n = 148 – 10.81%)				
Hard Copy Response (including email)	(7-1 = 6)	7	1	14
Web-based Electronic Response	(3-1 = 2)	(3-2 = 1)	3	6
Net Total	8	8	4	20
HK Gov't List of Approved Contractors (n = 113 – 7.08%)				
Hard Copy Response	(3-2 = 1)	(5-1 = 4)	3	8
Web-based Electronic Response	2	(4-3 = 1)	3	6
Net Total	3	5	6	14
HK Gov't List of Approved Specialist Contractors & Suppliers (n = 363 – 4.96%)				
Hard Copy Response	5	(14-1=13)	1	19
Web-based Electronic Response	0	0	0	0
Net Total	5	13	1	19
HK YP & CIC-VRS (Combined n = 1,909 – 4.03%)				
Hard Copy Response	24	56 (-9)	9	80
Web-based Electronic Response	(5-3 = 2)	(6-2 = 4)	5	11
Net Total	26	51	14	91
Hard copy (n = 116 / 2,853, – 4.07%)	**42**	**74**	**14**	**130**
Web base (n = 18 / 855 – 2.11%)	**9**	**9**	**19**	**37**
Total (n =134 / 2,853 – 4.70%)	**51**	**83**	**33**	**167**

Table 3 Schedule of responses from Survey

In all, 2,853 invitations to participate were posted, as shown in Table 2. As can be seen from Table 2, the number for which no postal address could be found for the HKYP was far greater than the assumed 15%. In fact, most of the addresses for the HKYP entries were found through the HKCIC VRS list. Therefore, for both the HKYP and HKCIC VRS mailing lists, which were prepared together, the governing parameter became whether the entries identified that the entry represented a "sole trader" or an "unincorporated business name". All such entries were discarded on the basis that they did not represent "corporations" as were entries for which only residential as opposed to commercial addresses were found. 960 follow-up emails were issued and 105 automatic responses were received from "Postmaster" that the emails were not deliverable to the nominated email addresses, leaving a net 855 apparently delivered emails. 129 hardcopy responses, 1 email response and 37 web-surveys were received. Three response envelops were received with contents apparently intended for someone else. Most of the hardcopy responses were very well and fully completed. However, 11 were either incomplete or a single column was checked for all answers and were discarded. Similarly, although the surveytool was configured to require all answers to be completed prior to moving to the next "sheet" just over half the web-surveys were "drop-outs" meaning that the respondent had partially completed it and exited and these were also discarded. Therefore, the final net response number was 134 (1 email, 115 hardcopy and 18 web-surveys) for a total of 4.70%, as tabulated below.

Although the smaller number of the web-based survey responses showed greater variability, the trend line was, by inspection, exactly the same and so the two files were exported to Microsoft Excel and combined.

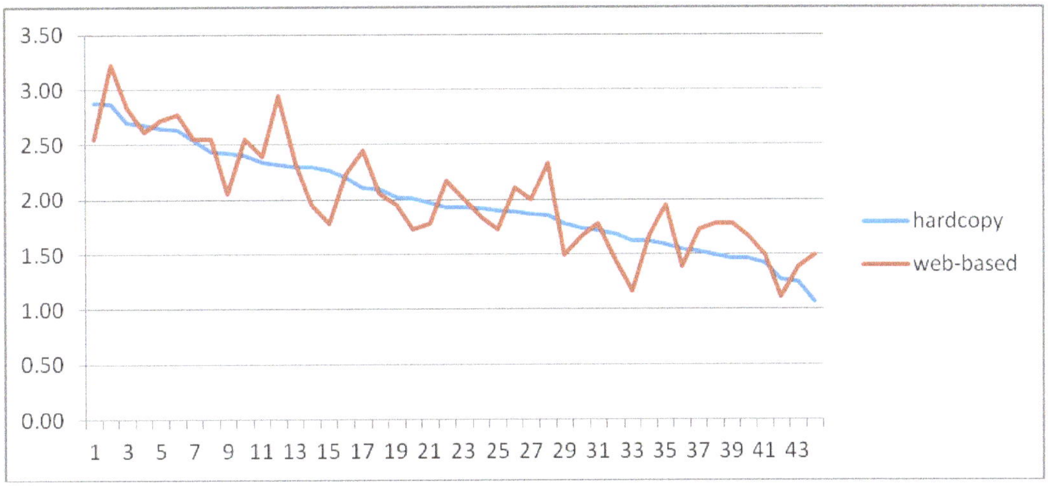

Figure 2 Comparison of hardcopy (n=116) and web-based (n=18) responses

The CSR principles together with core subject #1, Corporate Governance, and the six remaining CSR core subject incorporated into ISO 26000 (2010) and the questionnaire, can be used to form seven "constructs" or concepts from the elements comprising the questions in those sections to test which sections are, in the views of the respondents, more or less relevant to the CI. These were identified and used as the variable names, and the labels and values were then assigned. Decimal definition was set at 0; there was no incorrect data; there were no incorrectly coded items and no negatively worded items.

The 8 variables representing the responses in relation to Principles; 8 variables representing Human rights; 5 variables representing Labour practices; 4 variables representing Environmental practices; 5 variables representing Fair operating practices; 7 variables representing Consumer issues; and the 7 variables representing Community issues; being each of the seven potential constructs were examined. Each set was tested using Analyse ➔ Scale ➔ Reliability Analysis and can be summarised as follows:

	Principles (Including Governance)	Human Rights	Labour Practices	Environment	Fair Operating Practices	Consumer Issues	Community Involvement
N of Cases (Excluded)	134	134	134	134	134	134	134
N of Items	8	8	5	4	5	7	7
Alpha	0.895	0.911	0.857	0.820	0.807	0.903	0.894

Table 4 Summary of Reliability Scales for 7 Constructs (Alpha)

A review of the statistics in Table 4 and the Item-Total statistics shows that all of the Cronbach's Alpha figures for the 7 constructs indicate that there is "*very good*" correlation for combining the variables in each construct to form a single construct variable (0.8 <0.9) or "*excellent*" (> 0.9) (Hair et al. 2003, p. 172). Review of the Item-total statistics for each of the constructs revealed that, eliminating only impacts of biodiversity for the Environment (0.820 to 0.821) and health impacts for Community issues (0.894 to 0.898) improved the Cronbach's Alpha. The decision was made not to delete these items at this stage of the analysis. Review of the Inter-Item Correlation Matrix revealed that the vast majority of all inter-item relationships were above 0.3.

Testing of the Seven Constructs

Each of the 7 constructs were examined.

Firstly assumptions were tested.

i. Sample size

The sample size of 5 per variable and over 100 subjects (to be acceptable) criteria are both met (Coakes & Steed 2007, p. 123).

ii. Normality, Linearity and Outliers

The data was tested individually for these assumptions. A number of the variables were found to be skewed, some quite heavily skewed (Principle5, Labour4, Labour5 and Practices2) However, Factor analysis is robust for normality and therefore these variable were not adjusted (Coakes & Steed 2007). Six variables contained between 4 and 8 outliers. The total number of outlier cases amounted to 28 with 6 cases appearing twice, 2 cases appearing 3 times. However, rather than removing all of these cases from the data set they were retained. Linearity is assumed as per Coakes and Steed (2007, p. 123).

Job description	77	78	83	86	92	96	130	131
Project work	94	96	111	129				
Internal discipline	94	97	111	129				
Infringement of rights	97	112	121	131				
Social dialogues	82	84	85	86				
Political involvement	94	120	121	131				

Table 5 List of outliers in the data

Testing the Relationship between the Constructs and overall CSR

To test the relationship and whether all the constructs are significantly related to each other, liner regression was used sequentially to determine the relationship of each of the seven constructs to the other six constructs using Analyse → Regression → Linear and then again in a stepwise sequence to test the contributing relationship for the seven constructs to overall CSR performance; then the seven constructs were analysed using Analyse → Correlate. It was found that Human Rights and Labour Practices are the strongest indicators for CSR Principles; Operating Practices and CSR Principles are the strongest indicators for Human Rights; the Environment and CSR Principles are the strongest indicators for Labour Practices; Labour Practices and Operating Practices are the strongest indicators for the Environment; Human Rights and the Environment are the strongest indicators for Operating Practices; Operating Practices and the Community are the strongest indicators for Consumers; and Consumers and the Environment are the are the strongest indicators for the Community. All constructs are correlated to each other. CSR Principles and Human Rights issues are the strongest indicators /contributors to overall CSR performance.

Is overall CSR dependent on Size or Company Type?

The combined construct was then tested as the dependent variable against both size and company type using SPSS 14 Analyse ➔ General Linear ➔ Model Univariate. The results indicated that the reported degree of corporate social responsibility was dependent on size of the respondent's company and not whether the respondent was working for a main contractor or a subcontractor F = 5.212.

Source	Type III Sum of Squares	df	Mean Square	F	Sig.
Corrected Model	57021.911(b)	21	2715.329	2.325	.002
Intercept	380101.151	1	380101.151	325.485	.000
Size	24345.045	4	6086.261	5.212	.001
Company	1296.784	4	324.196	.278	.892
Size * Company	9635.044	13	741.157	.635	.820
Error	130793.522	112	1167.799		
Total	1261726.000	134			
Corrected Total	187815.433	133			

a Computed using alpha = .05

b R Squared = .304 (Adjusted R Squared = .173)

Table 6 Tests of Between-Subjects Effects Dependent Variable: All_responses

Demographics

Respondents were also asked to provide their name, organisation type and size, job title or position, respondent's position, age, education and profession, and nearly all obliged. The 134 respondents were classified accordingly. Age was classified as 20-40, 40-50 and over 50 and using the median for each grouping as the average, the average age for the main study population was 49. Sub-contractors (59) and Main contractors (42) accounted for 75% of the employer types. 47% (64) were employed by organisations with less than 50 employees and 25% (34) were employed by organisations with more than 200 employees. Engineers accounted for nearly 40% of the respondents but most (74) reported "other". Nearly all (129) were qualified to Higher Certificate level, 118 reported that they were involved in project work.

Of interest, is that the data collection into descriptive statistics evidences that:

1 Labour practices issue – Occupational Health and Safety scores highest and is the only issue when the averages score of those who considered it a relevant issue (n=130) is higher than the legal compliance of "Incorporates". The other two issues of Legally Recognised Employees and Skills Training for Employees are the next two highest issues.

2 Labour Practices overall with a score (mean) of 2.86 and an average of 128 respondents considering the construct overall to be relevant was the highest scoring construct. Consistent with the UK and Australian studies (Jones, Comfort & Hillier 2006; Petrovic-Lazarevic 2008), the Environment's first two issues of reducing pollution and reducing waste are rated very highly but the greenhouse gasses and biodiversity items have the effect of lowering the overall rating of the Environment to make Consumers, surprisingly, the next most relevant construct, apparently on the basis that the HKCI respondents have taken the next entity up the "construction food chain", i.e. their client, as the consumer and rated fair trading, dispute resolution and safety highly.

3 The construct that the respondents rated as least relevant was community involvement with 31% responding (on average, n = 42) that the issues forming this construct are irrelevant, although Fair Operating Practices (particularly relating to political involvement and Property rights) had the overall lowest average.

ISO 26000 Classifications	Less than 20	20 to 50	50 to 100	100 to 200	Over 200	All responses
SR Principles	2.15	2.55	2.67	2.84	3.11	2.65
Human Rights	1.83	1.86	2.26	1.99	2.41	2.09
Labour Practices	2.43	2.82	2.93	2.85	3.28	2.86
Environment	2.03	2.42	2.70	2.44	3.09	2.57
Fair Operating Practices	1.69	2.12	2.23	2.27	2.87	2.25
Consumer Practices	2.26	2.54	2.72	2.47	3.16	2.66
Community Involvement & Develop't	1.93	2.16	2.23	2.44	2.70	2.30
Survey totals	**2.06**	**2.35**	**2.53**	**2.46**	**2.93**	**2.48**

Table 7 CSR Grading (mean) by classification summary of responses

Comparison between SME to Large Companies

Table 8 shows the mean scores for each of the constructs and demonstrates and supports the major findings of this research, being that the reported level of CSR activity is a function of size of company and not necessarily a function of whether the company is a main contractor or a subcontractor, the major findings of this project,. Surprisingly the "100 to 200" employees group scored lower than the "50 to 100" employees group in Human Rights, Labour Practices, The Environment and Consumer Practices leading to an overall lower result. However, the results support the statistical analysis of the seven constructs that the independent variable is size alone.

For companies with an employee base of more than 200, the overall mean score is 2.93 ("Fully incorporates ISO 26000 issues, on average") and four of the constructs are between incorporates and exceeds – indicating going beyond the expected /legal requirements. On the other end of the scale, for companies of with less than 20 employees, the overall mean score is 2.06 (just "Accepts") and three of the constructs are below "Accepts".

In summary, this analysis has identified that the large contractors with employee numbers greater than 200, including the large sub-contractors, consider that they currently meet legal obligations and actually do exceed legal obligations in the specific area of health and safety but although there is a focus on environmental matters it is generally driven by compliance. In doing so it has answered Research Question RQ1 and found support for the primary component of Research Hypothesis RH1 in respect of the differentiating factor being size but not whether the company is a main contractor or a sub-contractor. The differentiating factor identified in this research is solely to do with corporate size.

Discussion, Contribution and Suggestions

Many of the issues identified in ISO 26000 are relevant to the CI in Hong Kong. However, as noted by ISO 26000 itself, not all of the issues and core subjects are perceived as being relevant by all of the industry segments. In particular, most of the human rights issues were found to be not relevant to many of the respondents. The survey respondents, unexpectedly, rated the CSR issues associated with consumer issues relatively highly. This would appear to be because the CI relationships are almost exclusively contractual up and down the supply chain. It would appear that the respondents to the survey interpret the consumer or customer as being the next client up the supply chain and have rated the issues to do with dispute resolution particularly and dealing with complaints from their "consumers" very highly.

However, although ISO 26000 only has one issue to do with safety in employment, this single issue scored most highly. Another focus of the respondents was environmental performance, but there was only minor perceived relevance to the issues of greenhouse gases and biodiversity. This research has clarified to an extent the findings of Jones et al. (2006) that the fiercely competitive nature of the CI generates a focus on profit maximisation, thus potentially inhibiting CSR especially for smaller non-listed, but nevertheless incorporated, firms and that the health and safety environment and environmental management systems components of contractors' policies are influenced by the legislated requirements in Hong Kong.

This research has added to knowledge in the following areas

1 It is the first industry wide, state-wide analysis of the implementation of CSR activities across a range of corporate sizes within a single industry and therefore it is the first research which can categorically confirm based on empirical research that small companies are not little big companies when it comes to implementing CSR.

2 It has identified in general terms the most relevant and least relevant core subjects and issues in ISO 26000 to the Hong Kong Construction Industry, which may be generalisable to other construction industries.

It is suggested that this research has developed a now-proven research instrument in the survey questionnaire form for assessing what core issues and core subject might be relevant to a company or industry. Future research would be to apply this questionnaire to another industry and in another location, across a broad range of corporate sizes to confirm the findings of this research that, in CSR activities, "size is the only thing that matters".

Acknowledgement
The Authors would like to thank the Hong Kong Construction Association for its support of this research.

References

ABS 1999, *Small and medium enterprises business growth and performance survey Australia*, viewed 13 November 2012 http://www.abs.gov.au/ausstats/abs@.nsf/DetailsPage/8141.01997-98?OpenDocument

Ahmad, NMB 2011, 'The Economic Globalisation And Its Threat To Human Rights', *International Journal of Business and Social Science,* **2** (19).

Carroll, AB 1999, 'Corporate social responsibility', *Business and Society,* **38** (3), 268-295.

Chaudhri, VA 2006, 'Organising Global CSR', *Journal of Corporate Citizenship,* **23**, 39-51.

Coakes, S & Steed, L 2007, *SPSS 14.0 Analysis without Anguish,* John Wiley and Sons, Australia.

Cochran, PL 2007, 'The evolution of corporate social responsibility', *Business Horizons,* **50** (6), 449-454.

Coelho, PRP,McClure, JE & Spry, JA 2003, 'The social responsibility of corporate management: A classical critique', *Mid - American Journal of Business,* **18** (1), 15-24.

Diao, Y 2010, *Research on International Standardization of Corporate Social Responsibility,*

EC 2003, *The New SME Definition; User Guide and Model Declaration*, European Commission, viewed 24 April 2009 http://ec.europa.eu/enterprise/enterprise_policy/sme_definition/index_en.htm

Fellows, R & Liu, A 2003, *Research Methods for Construction* 2edn, Blackwell Science, Oxford.

Gill, RWT & Leinbach, LJ 1983, 'Corporate Social Responsibility in Hong Kong', *California Management Review,* **25** (2) 107-123.

Green, SD 2007, 'The evolution of corporate social responsibility in construction', in M Murray & A Dainty (eds), *Corporate Social Responsibility in the Construction Industry*, Taylor and Francis Books Ltd., London.

Hair, JFJ, Babin, B, Money, AH & Samouel, P 2003, *Essentials of Business Research Methods,* John Wiley and Sons, Inc, Hoboken.

Hatcher, M 2003, 'New corporate agendas', *Journal of Public Affairs (14723891),* **3** (1), 32.

Hillebrandt, P 2000, *Economic Theory and the Construction Industry* 3edn, McMillan Press Limited, London.

HKCA 2012, *Hong Kong Construction Association*, viewed 23 November 2012 http://www.hkca.com.hk/

HKCIC 2012, *Construction Industry Council Voluntary Registration Scheme*, viewed 23 November 2012 http://www.hkcic.org/eng/main.aspx

HKFEMCL 2012, *The Hong Kong Federation of Electrical and Mechanical Contractors Limited*, viewed 23 November 2012 http://www.hkfemc.org/

HKGDevB 2012, *List of Approved Contractors*, viewed 23 November 2012 http://www.devb.gov.hk/en/construction_sector_matters/contractors/index.html

HKGov 2013, *Facts about Hong Kong*, viewed 13 January 2013 http://www.gov.hk/en/about/abouthk/facts.htm

HKSARCSD 2013a, *Key statistics on business performance and operating characteristics of the building, construction and real estate sectors in 2011 [29.11.2012]*, HKSAR, viewed 11 January 2013 http://www.censtatd.gov.hk/press_release/pressReleaseDetail.jsp?charsetID=1&pressRID=3057

HKSARCSD 2013b, *Report on the Quarterly Survey of Construction Output*, Census and Statistics Department HKSAR, viewed 28 May 2013

HKTDC 2013, *Economy Research*, viewed 13 January 2013 http://hong-kong-economy-research.hktdc.com

HKYP 2012, *Yellow Pages List of Construction Contractors*, viewed 23 November 2012 http://www.yp.com.hk/iypbusiness_e10/en/html/browsing_search/browsecat_result_douinfo.aspx?basket_id=19408&category_id=19430

ISO 2010, *ISO 26000 - Guidance on social responsibility,* International Organization for Standardiation; ISO copyright office, Geneva.

Jones, MT 2009, 'Disrobing the emperor: mainstream CSR research and corporate hegemony', *Management of Environmental Quality,* **20** (3), 335.

Jones, P,Comfort, D & Hillier, D 2006, 'Corporate social responsibility and the UK construction industry', *Journal of Corporate Real Estate,* **8** (3), 134-150.

Knopf, J, Kahlenborn, W, Hajduk, T, Weiss, D, Feil, M, Fiedler, R & Klein, J 2011, *Corporate Social Responsibility: National Public Policies in the European Union*, European Commission, European Union, viewed 28 December 2012 http://ec.europa.eu/social/main.jsp?langId=en&catId=331&newsId=1012&furtherNews=yes

Mauch, J, E. & Birch, JW 1998, *Guide to the Successful Thesis and Dissertation,* 4 edn, Marcel Dekker, Inc., New York.

Murray, M & Dainty, A 2007, 'Corporate social responsibility: Challenging the construction industry', in M Murray & A Dainty (eds), *Corporate Social Responsibility in the Construction Industry*, Taylor and Francis Books Ltd., London.

O'Brien, M & Al Soufi, A 1993, 'Electronic data interchange and the structure ot the UK construction industry', *Construction Management and Economics,* **11**, 443-453.

Perrini, F 2006, 'SMEs and CSR Theory: Evidence and Implications from an Italian Perspective', *Journal of Business Ethics,* **67** (3), 305-316.

Perrini, F,Russo, A & Tencati, A 2007, 'CSR Strategies of SMEs and Large Firms. Evidence from Italy', *Journal of Business Ethics,* **74** (3) 285-300.

Petrovic-Lazarevic, S 2008, 'The development of corporate social responsibility in the Australian construction industry', *Construction management and Economics,* **26** (2), 93-101.

Rowlinson, S 2007, 'Corporate social responsibility in Hong Kong and the Asia Pacific construction industry', in M Murray & A Dainty (eds), *Corporate Social Responsibility in the Construction Industry*, Taylor and Francis Books Ltd., London.

Russo, A & Perrini, F 2010, 'Investigating Stakeholder Theory and Social Capital: CSR in Large Firms and SMEs', *Journal of Business Ethics,* **91** (2), 207-221.

Studer, S, Tsang, S, Welford, R & Hills, P 2008, 'SMEs and Voluntary Environmental Initiatives: A Study of Stakeholders' Perspectives in Hong Kong', *Journal of Environmental Planning and Management,* **51** (2), 285-301.

Tang, HY-y 2001, *Construct for Excellence,* Construction Industry Review Committee of the HKSAR, Hong Kong.

Valmohammadi, C 2011, 'Investigating corporate social responsibility practices in Iranian organizations: an ISO 26000 perspective', *Business Strategy Series,* **12** (5), 257-263.

Votaw, D 1972, 'Genius Becomes Rare: A Comment on the Doctrine of Social Responsibility Pt. I', *California Management Review,* **15** (2), 25-31.

Welsh, JA & White, JF 1981, 'A small business is not a little big business', *Harvard Business Review,* **59** (4), 18-27.

Yin, RK 2003, *Case Study Research Design and Methods,* 3 edn, Sage, Thousand Oaks, California.

Relationship between Degree of Risk, Cost and Level of Compliance to Occupational Health and Safety Regulations in Construction

10

Abimbola Windapo (University of Cape Town, South Africa)

Abstract

This paper investigates the role of statutory health and safety (H&S) regulations in managing construction project risks. The study examines whether the decision made by contractors to comply with the regulations, the cost of compliance and savings of H&S regulatory requirements is influenced by the degree or level of risk, which the regulations are trying to prevent. The rationale for the examination stems from previous studies which establish that building designers and contractors perceive the cost of complying with regulations as additional burdens, which they have to conform to, and which are in some cases unnecessary, and also the fact that construction related injuries and fatalities are on the increase. Qualitative and quantitative data obtained from a descriptive survey and H&S site audit by the Master Builder Association of the Western Cape (MBAWC) were used as the measurements of risk, level of compliance to regulations, cost of compliance and savings. By correlating the quantitative and qualitative data, there is empirical evidence to support a negative relationship between the degree of risk, level and cost of compliance and cost savings. Based on the study's findings, this paper concludes that the decision made by contractors to comply with H&S regulatory requirements is influenced by the perceived cost saving on account of compliance and that cost savings are influenced by the probability of accident occurrence which is an element of the degree of risk which the regulation is trying to prevent or control.

Keywords: Accident, Cost, Health and Safety, Regulations and Risk

Introduction

Construction industries worldwide are notorious for unacceptably high accident and fatality rates, both of which were noted by Ulang et al. (2010), Sidumedi (2009) and Health and Safety Executive (HSE) (2004) in comparison to other industries. According to Odeyinka et al. (2005), construction workers are six times more likely to be killed at work than those in other industries. In South Africa, construction industry records show that work related deaths, occupational diseases and injury claims absorb a significant proportion of the Gross National Product (Benjamin and Greef 1997) even though there are concerns that the reports in South Africa fail to capture the full number of accidents (van Huyssteen et al. 2009; and Sidumedi 2009). Warwick (2011), Baxendale and Owain (2000), and Coble et al. (2000) attribute accidents and fatalities associated with and reported in the construction industry to non-compliance by contractors with H&S regulations on construction sites.

Regulations are intended to address and satisfy the public mandate for managing the risks and benefits of technology (Wolski, et al. 2000), designed to improve the performance of individual and organizational behaviour in ways that reduces social harms (Coglianese, et al. 2003), and stimulates the building designer and contractor to follow strict sets of rules with regard to the way in which structures are built (HSE 2004). However, in instances of poor quality housing a high number of accidents and fatalities (Construction Industry Development Board (cidb) 2009a), building failure (Ayininuola and Olalusi 2004), failed or unfinished construction projects, which are common in the construction industry, suggest that construction industry stakeholders are not complying with regulatory requirements

(Warwick 2011; Baxendale and Owain 2000; and Coble et al. 2000). The reasons for lack of compliance include regulatory authorities imposing regulations that are in practice unattainable or prohibitively expensive (Williams 1995); building designers and contractors viewing regulations as additional burdens with which they have to comply (Gann et al. 1998) and that while regulations only apply in some cases, the extra precautions are considered an unnecessary cost (Emrath 2011).

While Weil (2001) questions whether the standards complied with by contractors are linked with the underlying physical hazards and cost of accidents (CoA), or whether the H&S standards most cited by government regulation bodies diverge from the major sources of fatalities and injuries on construction projects. McConnaughey (1978) in a previous study maintains that the level of safety or performance outcomes which a regulation is intended to provide, and thus the resulting costs and benefits derived from its implementation, relates to the degree or level of risk which the regulation attempts to prevent or control.

These arguments by Weil (2001), Gann et al. (1998), Williams (1995), McConnaughey (1978) and the fact that it is not known whether one of the contributing factors to the escalating number of site accidents is that too much attention is placed by construction site management on lower risk activities and operations instead of on the activities that have the potential to cause the most injuries, form the basis for this study. The objective of this paper is therefore to fill the knowledge gap and to establish whether there is a relationship between the degree of risk which the regulation attempts to prevent/control, the level of compliance with the OHSA regulatory requirements, the cost of compliance/implementation and cost savings when the OHSA requirements are complied with.

The research framework for this study is shown in Figure 1. Before these correlates are determined and discussed, the paper will first of all review the concept of OHSA regulations, the measurement of construction risks, the level of compliance, and costs and benefits of compliance to OHSA regulations.

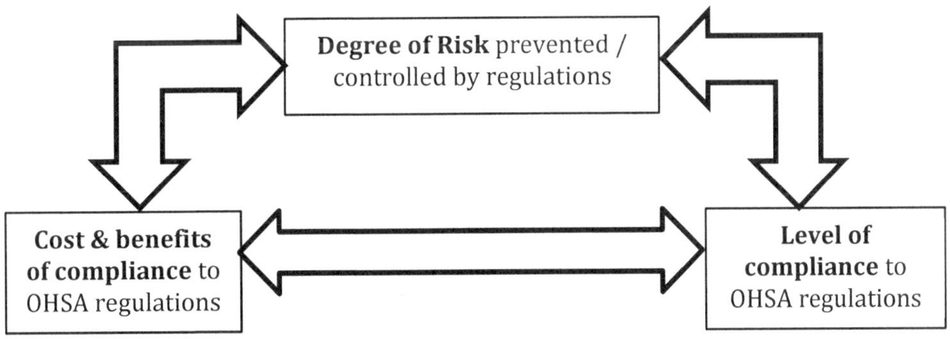

Figure 1 Research framework

Concept of Regulations

The state is a potential resource or threat to every industry in the society, the construction industry being no exception (Stigler 1971). The state establishes technical standards, permits construction activities and enforces state regulations (Winch and Campagnac 1995; and Ang et al. 2005). The government regulations are the outcome of legal efforts, strongly influenced by governmental and social forces, evaluated against the consequences of the process and are often enacted to reduce the probability of negative performance outcomes in terms of quality, health and safety issues on construction projects. Regulations come in many forms and can be imposed by government at different levels (Emrath 2011) for the reason that people habitually look to government to take the lead in creating a scheme of prevention, protection and control (Spence 2004).

Overview of OHSA Regulations in South Africa

According to cidb (2009a), the activities in the construction industry due to its poor H&S performance record are the subject of various legislative and institutional frameworks in South Africa of which the primary objective is the prevention of accidents and their consequences in terms of injury, disablement, fatality and ill health within the work environment. The cidb report notes that South Africa is not lacking in terms of H&S legislation and that South Africa's legislative framework addresses H&S at three levels firstly in terms of the national constitution, then in terms of Acts such as the Occupational Health and Safety Act No. 85 of 1993 (OH&S Act), and the complementary Compensation for Occupational Injuries and Diseases Act No 130 of 1993 (COID Act). These are followed by a range of regulations promulgated under OHSA, in particular the Construction Regulations promulgated in July 2003. Although it cannot be quantified, cidb (2009a) alleges that the Construction Regulations have had a positive impact on reducing H&S accidents.

The OHSA No. 85 of 1993 of South Africa stipulates the steps to be taken in order to ensure a safe and healthy work environment for all employees on a construction site (Hermanus, 2001) and contractors are obliged to comply with the requirements of the Act. However, Windapo (2011), Bettesworth (2011), MBAWC (2011) and cidb (2009a) noted that building contractors in South Africa do not comply fully with H & S regulations, and that even though H&S issues have seen some improvement over the years, the numbers of people that get injured or die on construction sites in South Africa is still high (Department of Labour (DoL) 2007; and cidb, 2009b).

Level of Compliance to OHSA Regulatory Requirements on Construction Sites

Weil (2001) notes that the H&S standards cited most frequently diverge from the major sources of fatalities and injuries on construction projects and that it is not known if the standards frequently complied with are linked with underlying physical hazards, and are in turn associated with injuries and illnesses. According to Weil (2007), the number and severity of H&S standard violations cited during an inspection provide one measure of the degree to which a contractor's operations comply with OSHA standards.

The level to which contractors' operations comply with OHSA regulatory requirements on construction sites in the Western Cape Region of South Africa were determined by the Master Builders Association Western Cape (MBAWC) for Year 2007, 2008, 2009 and 2010 for annual Safety Competitions using an audit system which was designed by the association for grading the H&S regulation compliance of construction projects. The Master Builder South Africa Audit Tool (MBSAT) used in the H&S assessment is classified into 19 different elements and the associated points achievable, each element targeting different requirements of OHSA. While scaffolding, formwork and support accounts for 366 or 19.9%, administrative and legal; cranes; personal, protective equipment and clothing account for 316 or 17.2%, 166 or 9% and 102 or 5.5% respectively of the H&S requirement total points of an achievable 1840.

The H&S audit undertaken by MBAWC between 2007 and 2010 when averaged and distributed by compliance to the 19 H&S requirements revealed that the contractors attained acceptable standards in three elements – cranes, demolition, and transport and material handling; unacceptable standards in eleven elements; and very poor standards in five elements - site plant and machinery workplace environment, health and hygiene, personal protective health and clothing, plant and storage yards, and excavation (See Table 2; Warwick 2011; and MBAWC 2011). While compliance to statutory regulations should mandatorily be set at 100%, the level of contractor compliance was rated by MBSAT based on the view that scores above 95% are seen as acceptable standards, scores between 90 and 95% can be rated as unacceptable standards which require attention in the short term and below 90% is rated as very poor standard which require immediate attention (Bester 2011; and Warwick 2011).

Cost of Implementation/Compliance to OHSA Regulations

Lin and Mills (2001) acknowledge that the higher the investment in safety, the better the performance. According to McGarity and Ruttenberg (1997), the proper measure of the cost of complying with a regulation is the incremental cost above the "baseline" state of the world that would have existed in the absence of the regulations. The real resource cost of compliance according to McGarity and Ruttenberg (1997) consists of three elements: (1) costs associated with the purchase, installation, operation and maintenance of new equipment; (2) changes in the inputs or mixtures used in the production process; and (3) the capture of waste products that can either be disposed of, sold or reused. There are various strategies and systems that have to be put in place for H&S on construction sites and that have to be paid for by contractors in order to comply with the legislation (cidb 2009a). Smallwood (2004) estimates that the cost of implementing H&S systems within a construction company lies between 0.5% and 3% of total project costs. This study measures the cost of implementing an H&S system for a project, in line with the estimates provided by Smallwood (2004).

For all but a very few health, safety and environmental regulations, cost according to McGarity and Ruttenburg (2002) is an important consideration because decision makers want to decide whether a given option is 'worth it'. The cidb (2009a) posits that if potential losses relative to labour, materials, plant and equipment as a result of not implementing the requisite preventive measures, are cited, contractors will address H&S issues.

Safety/Performance Outcome and other Benefits of Compliance to OHSA Regulations

Occupational Health and Safety (OHS) Regulations impose on enterprises an expected penalty that is positively related to the presence of unsafe working conditions for firms not in compliance with the standards (Viscusi 1979). According to cidb (2009a), the benefits of compliance to the OHSA regulations include reduction in work related illnesses, ergonomics, and more primary health care methods. In addition, cidb (2009a) notes that factors such as increases in insurance premiums, direct costs associated with hospitalization, disability pay, lost workdays and a decrease in productivity are also changing the focus of contractors towards the management of H&S.

The Cost of Accident (CoA) is the final measure that can readily be related to by all stakeholders as it can be expressed as a percentage of organizational business volume or construction completed, it could also be categorized as being either direct or indirect which collectively constitutes the total CoA and it is noted that in South Africa, the estimated CoA is around 5% of the value of the completed construction (cidb 2009a). The direct costs of accidents which includes wages and medical expenses tend to be those associated with the treatment of injury and any unique compensation offered to workers as a consequence of being injured and are covered by workmen's compensation insurance premiums. Research by Pillay and Haupt (2008) indicates that direct costs contribute 27% to the CoA in South Africa. The indirect costs which are borne by the contractors and which are found to contribute 73% to the CoA according to Pillay and Haupt (2008) include incident investigation costs, production loss, reduced productivity and process delays, the cost of overtime, replacement worker orientation costs, administrative costs, consumables, legal fees, and funeral and compensation due to fatalities. Smallwood (2000) establishes that the indirect costs of accidents are 14.2 times the direct costs.

This study measures cost savings on construction projects by using the CoA established by cidb as proxy. The study argues that compliance to the OHSA requirements will bring about safe working conditions (Viscusi 1979) and therefore lower costs of accidents (<5% of the value of completed construction) on construction projects.

Degree and Level of Risks, which the OHSA Regulatory Requirements attempt to Prevent

The degree and level of risks which the OHSA regulatory requirements attempt to prevent include: fatal accidents which are caused by contact with moving vehicles; injurie by falling objects; transport; slips/falls on levels; falls from a height; being struck by something; drowning; fire or explosion; and contact with electricity (Dalton 2002; HSE 2004 and Eppenberger 2008). According to the cidb (2009a) report, accident frequency and severity rates are the two essential standards required for reviewing accident statistics. These rates show on average, how often disabling injuries occur in any particular industry; the accident frequency rate; the seriousness of the time loss involved and the accident severity rate. Construction project risks can be evaluated and ranked according to their probability of occurrence and severity of consequence within a project risk matrix (see McGeorge and Zou 2013; and Pinto 2002) as shown in Table 1.

Potential Severity of Injury	**Probability of Occurrence**			
	Frequent	**Occasional**	**Remote**	**Unlikely**
Catastrophic	Extreme Risk (9)	Extreme Risk (8)	High Risk (7)	High Risk (6)
Critical	Extreme Risk (8)	High Risk (7)	Medium Risk (5)	Medium Risk (5)
Moderate	High Risk (7)	Medium Risk (5)	Medium Risk (4)	Low Risk (3)
Minor	Medium Risk (4)	Low Risk (3)	Low Risk (2)	Low Risk (1)

Table 1 Project Risk Matrix After McGeorge and Zou (2013); and Pinto (2002)

The severity rates (SR) indicate to management the seriousness of the disabling injury and the amount of time that will be lost by the worker for every 1,000 hours worked (cidb 2009a). The degree and level of risk (significant risks faced by employees), which the regulatory requirements are trying to prevent/control, will be evaluated in the study in terms of their severity (human capital) and frequency of occurrence by the following classes of accidents (Mthalane et al., 2008):

- Very Severe/catastrophic injuries/fatalities – death results from work injuries on site;
- Severe/major accidents and critical injuries – non-fatal, permanent disablement. These are injuries, which result in amputations or fractures. Injured persons would be absent from work for about 30 days. Major accidents include slips, trips, falling from heights, machinery, transport or electrical accidents;
- Mildly severe/moderate accidents or injuries – non-fatal, temporary disablement; and
- Not severe/minor accidents or injuries – non-casualty, requiring only medical aid. These are accidents, which yield less severe injuries. The absence from work due to this type of injury would usually be two days or less. Most accidents fall into this category – stepping or striking against small objects, tools, sprains or strains.

Research Methods

The research process used for the study reported started with a literature review According to Yin (1994) the main use of documentation is to corroborate and argue evidence from other sources. In addition, effective design of the research instrument requires a good understanding of the theories underpinning the research area. The research therefore began with a holistic review of the research problems, the OHSA regulations, the degree of risk the prescribed regulation requirements attempts to prevent, and cost and benefits of compliance to regulations, in order to understand the variables used in the study. Compliance to regulations and risk prevention and management are areas that have been comprehensively studied and documented in project and construction management, knowledge management, design and technology management, and social studies. The study undertaken is supporting and is aligned to the work of McConnaughey (1978), Weil (2001) and Haupt (2003).

Research Positioning

The study employs a sequential mixed method research approach, which combines quantitative and qualitative approaches in research design and collection (Abowitz and Toole 2010). The sequential explanatory mixed method involves collection of both quantitative and qualitative data sequentially (Ivankova et al., 2006). The research design allows quantitative data to be collected first and analysed. Second in the sequence is the qualitative (text) data, which are also collected, analysed and used to explain, or expound on the quantitative results obtained in the first place.

Jaselskis (1996) acknowledges that an effective way of measuring the safety performance of a construction company is by using a combination of both qualitative and quantitative measures. While the quantitative approach formulates hypotheses and tests them, the qualitative approach seeks to describe and analyse the behaviour of humans and their groups from the point of view of those being studied (Amaratunga, et al. 2002). Ontologically, the research is premised on an objective view of reality that exists independently and can be uncovered and known through research (Runeson and Skitmore 2008). Following that knowledge is obtained through an understanding of how the degree and level of risk, which the OHSA regulatory requirements attempt to prevent, and the likelihood of its occurrence is constructed through information provided by the respondents. Epistemologically, the research assumes a constructivist paradigm where the nature of knowledge and meaning is a subjective human creation (Baxter and Jack 2008). Quantitative data on the level of compliance to OHSA regulatory requirements was obtained from past records of scores achieved by contractors in the H&S audit undertaken by MBAWC between 2007 and 2010 using an audit system, which was designed by the association for grading the H&S regulation compliance of construction projects (see Table 2).

The research methodological stance and positioning adopted a descriptive study approach, which engaged contractors based in the Western Cape province of South Africa between May and September 2011. A combination of convenience and snowball sampling was used to obtain data from 53 contractors who are registered in the MBAWC safety competition and who agreed to participate in the study. In addition, case studies of four construction sites chosen from among those surveyed by the MBAWC for their Occupational Health and Safety site audit were conducted. This allowed for survey data validation and for linking perceptions and costs of OHSA compliance, to the data collected by MBAWC.

The study examined the main sub-clauses of the OHSA legislation that govern H&S in the South African construction industry. The values for the perceived level of risk and the likelihood of occurrence of the risk which was used in computing the degree of risk, cost of compliance and cost savings shown in Table 2 were measured based on the respondents' rating of the 4-point Likert-scale of Not Severe (1), Mildly Severe (2), Severe (3) and Very Severe (4); and Unlikely (1), Remote (2), Occasional (3), and Frequent (4) respectively. In addition, the cost of compliance and cost savings were measured based on the respondents' rating of a 3-point Likert-scale – Less than 0.5% (1), Between 0.5% and 3% (2), and above 3% (3) of total project costs; and Low Effect or <5% (1), Medium Effect or 5% (2), and High Effect or >5% (3) cost savings on the project.

Methods of Data Collection and Analysis

The method of data collection involved a questionnaire survey where the same questions were asked to different respondents (general building contractors) registered in the MBAWC safety competition. This questionnaire survey enabled the consideration of stakeholders who were difficult to reach. In addition, case studies of four construction sites were carried out, involving direct observations and personal interviews. These observations provided useful primary additional information. The questions developed were piloted by way of an interview protocol with three construction industry experts. The pilot study revealed the need to reduce the 19 elements used in the MBSAT audit tool for rating H&S compliance on construction projects to all the three elements that have acceptable compliance standards; four that have very poor compliance standards minus the plant and storage yard requirement; and two other elements that have the highest achievable points (See Table 3). This was to facilitate the ease of filling in the questionnaires and to improve response rates.

The data collected from the survey and past records on level of compliance maintained by the MBAWC were analysed using frequency analysis, percentage scores, risk matrix analysis (following Table 1), composite risk index (see Ciobanu and Mazilu, n.d), cost of compliance and savings relative importance index, and Pearson Product Moment Correlation and Regression analysis, which was used to test the relationship between the variables.

Test for Reliability

The study sought to know whether the information supplied by the respondents was reliable. Cronbach's Alpha (α) an index used for objectively measuring the reliability of a questionnaire instrument was used in testing the data collected (Bland and Altman 1997). The acceptable values of α for consistency range between 0.70 and 0.95 (Tavakol and Dennick 2011; DeVellis 2003). The Cronbach's α reliability score of perceived severity of risk, frequency of risk occurrence, direct cost of compliance to OHSA requirements and cost savings due to OHSA compliance investigated is presented in Table 2.

Perceptions	Cronbach α values	No. of Items
Severity of risk that the OHSA requirement attempts to prevent/control	0.913	9
Frequency of occurrence of the risk	0.920	9
Direct cost of compliance to OHSA requirement	0.859	9
Cost savings as a result of OHSA requirement compliance	0.983	19
Overall Cronbach α values	0.940	46

Table 2 Results of Test for Reliability (Cronbach's α values) of Respondents' Perceptions

Table 2 shows that (1) the values of the Cronbach α are greater than 0.85 for each of the respondents' perceptions investigated, based on nine standardized OHSA requirement and (2) the overall Cronbach α value of 0.940. The results thus suggest that all the perception variables are of high reliability, that there is high interrelatedness between the perceptions, that the research tool developed is reliable because it shows that the respondents genuinely have different opinions and the differences are not because the questionnaire is confusing or has multiple interpretations.

Results and Discussion

The data collected from the study conducted will be discussed under the following headings:

Respondents' Distributions and Profiles

Construction companies surveyed had the following background – 45% are private liability companies, 38% are national and international companies, 40% have been in operation in the construction industry for more than 21 years and 41% are large companies with more than 100 employees. In addition, 61% of the respondents are owners or directors of the companies surveyed. The type of business, geographical spread, age and size of the companies surveyed and the status of the respondents indicates that the respondents have the necessary experience and exposure to provide credible and reliable information.

Breakdown of detailed Scores for Variables used in the study

The detailed breakdown of overall scores for the level of compliance obtained from MBAWC records; relative importance index for cost of compliance and cost savings as a result of compliance (benefits of compliance), probability of accident occurrence and severity of accident; and composite risk index (degree of risk), computed from the data collected through the questionnaire survey and case studies are shown in Table 3 and Figure 2.

OHSA Elements	Level of Compliance [1]	PI[2]	SI[3]	Degree of Risk[4]	Cost of Compliance Index	Cost Savings Index
Cranes (Tower, Mobile & Gantry)	0.99	0.44	0.52	0.23	0.52	0.52
Demolition Work (Safety Requirements)	0.97	0.49	0.53	0.26	0.58	0.54
Transport & Material Handling (safety in use)	0.97	0.48	0.66	0.32	0.62	0.56
Admin Structure & Legal Requirements for H&S	0.94	0.46	0.79	0.36	0.63	0.53
Scaffolding, Formwork & Support (safety in use)	0.90	0.56	0.53	0.30	0.65	0.58
Site Plant & Machinery	0.90	0.51	0.59	0.30	0.62	0.58
Work Place Environment Health & Hygiene	0.89	0.45	0.84	0.38	0.55	0.57
Personal Protective Health & Clothing	0.88	0.55	0.71	0.39	0.62	0.58
Excavation (adequacy of side support, etc.)	0.83	0.49	0.59	0.29	0.59	0.58

Table 3 Breakdown of detailed scores/indices for variables used in the study

[1] Source: MBAWC (2011); 2 Probability of Accident Occurrence Index (PI); 3 Severity Index (SI); and 4 Degree of Risk = PI x SI

Table 3 shows that from a ranking perspective, all the OHSA elements considered have fairly equal level of compliance, degree of risk, cost of compliance and cost saving indices.

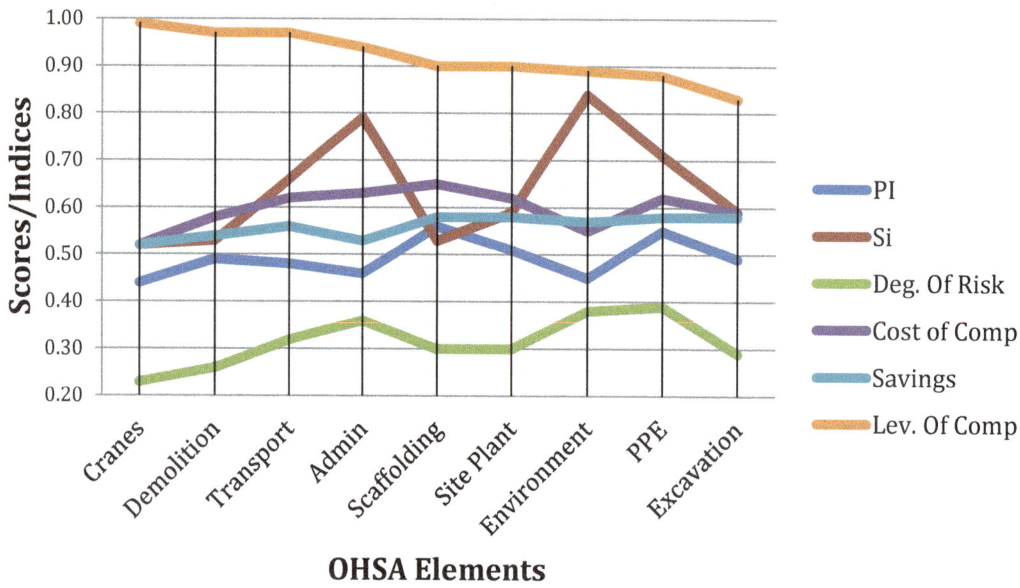

Figure 2 Breakdown of detailed scores/indices for variables used in the study

Relationship between the Degree of Risk, Level of Compliance, Cost of Compliance and Savings

The correlation coefficients between the degree of risk, overall level of compliance scores achieved for the nine OHSA elements, cost of compliance and savings is shown in Table 4.

	Level of Compliance	PI	SI	Degree of Risk	Cost of Compliance	Cost Savings
Level of Compliance	1					
Probability of accident occurring (PI)	-0.436	1				
Severity Index (SI)	-0.224	-0.284	1			
Degree of Risk	-0.430	0.189	0.885**	1		
Cost of Compliance	-0.277	0.720*	0.053	0.377	1	
Cost Savings	-0.811**	0.686*	0.075	0.406	0.474	1

Table 4 Correlation Matrix for degree of risk, level of compliance, cost of compliance and savings

** Correlation is significant at the 0.01 level (1-tailed); * Correlation is significant at the 0.05 level (1-tailed)

The results of the study presented in Figure 2 and Table 4 show that the probability of an accident occurring, which is a component of the degree of risk is significantly and positively correlated with the cost of compliance and the perceived cost savings for complying with the OHSA regulatory requirements. This result implies that the higher the probability of a risk event occurring, the more the contractor perceives the direct cost of compliance/implementation and the expected cost savings on accidents to be. The correlation coefficients presented in Table 4 also show a significant but negative relationship between the perceived cost savings on accidents and the level of compliance to OHSA regulatory requirements. This result suggests that the cost savings made by the contractor are perceived to decrease as compliance to OHSA regulatory requirements increases.

Given the rather poor correlation reported between the other relationships proposed and investigated and shown in Table 4, regression analysis was used to provide a better understanding as to what extent the level of compliance (LoC) to H&S regulatory requirements is affected by the independent variables of cost of compliance (CoC), degree of risk (DoR) and cost savings (CS). It should be noted however that the purpose of the regression analysis is not for developing a model for predicting a contractor's future level of compliance to H&S regulatory requirements, but for use in establishing causal relationships between the variables used in the study.

$$LoC = 1.845 - 0.154DoR + 0.220CoC - 1.802CS$$

Table 5 presents the summary of the regression analysis results.

Variables in the model		B	Standard error	Beta	t-stat	Sig
constant		1.845	0.312		5.911	0.002
Degree of Risk (DoR)		-0.154	0.269	-0.16	-0.573	0.592
Cost of Compliance (CoC)		0.220	0.361	0.177	0.61	0.568
Cost Savings (CS)		-1.802	0.631	-0.83	-2.83	0.037
R	R^2	Adjusted R^2	Standard error of the estimate		F	Sig
0.832	0.693	0.58	0.03651		3.756	0.094

Table 5 Regression analysis results for level of compliance (dependent variable), degree of risk, cost of compliance and savings

Table 5 reveals that cost savings make the greatest contribution to level of compliance followed by cost of compliance and degree of risk. Following that a change of one standard deviation on cost savings produces a change of 0.83 standard deviations in the level of compliance, while cost of compliance and degree of risk – neither of which were significant – indicate changes of 0.177 and 0.16 respectively. The total variation in the level of compliance is well explained by the independent variables with $R^2 = 0.693$.

This finding suggests that the level of compliance of contractors to H&S regulatory requirements is responsive to changes in perceived cost savings gained through compliance. Furthermore, the negative correlation coefficient computed between cost savings and level of compliance implies that the cost savings made by the contractor are perceived to decrease as compliance to OHSA regulatory requirements increases. The negative relationship between degree of risk and overall level of compliance indicates that the level of compliance to OHSA regulatory requirements decreases as the degree of risk increases and therefore suggests that the tendency for contractors to comply with OHSA regulatory requirements is not based on the degree of risk which the requirements attempts to prevent or control but on perceived cost savings or loss from non-compliance or compliance respectively. In addition, the positive relationship between the level of compliance and cost of compliance suggests that the level of compliance to H&S statutory regulatory requirements may increase (all else being equal) with money spent on the implementation of compliance requirements.

Case Study Results

The results of the questionnaire survey have given some indication of the relationship between the degree of risk, level of compliance, cost of compliance and cost savings. However, four case studies - referred to below as Site A, Site B, Site C and Site D in Table 5

- were undertaken sequentially within the study sample to reinforce and make the survey findings more meaningful. The type and size of construction projects ranged from housing projects with a budget of less than R20m (Site C and D), to a hospital project of R500m budget (Site B) as well as a multi-billion rand power station site (Site A).

Questions Posed	Site A	Site B	Site C	Site D
Is OHSA implemented on site?	Yes	Yes	Yes	Yes
Level of compliance with OHSA	High	High	High	High
Is the project running within time?	No	No	Yes	No
Is the project running within budget?	No	No	Yes	Yes
Have there been any reports of accidents/ injuries?	Yes – Fatalities, accidents and injuries	Yes - Accidents and injuries	Yes – Minor injuries	Yes – Minor injuries
Reasons for non-compliance	Lack of knowledge; Cost mitigation – profit maximisation; Time consuming	Cost mitigation	Lack of knowledge; Cost mitigation	Lack of knowledge; Cost mitigation; Attitude/negligence
Effects of OHSA compliance on performance	Less accidents	H&S is costly but cost is justified; overall performance is improved	OHSA compliance is unnecessarily costly and time consuming	Quality is improved; time and cost are compromised
Benefits of complying with OHSA legislation	Competing on international standards; Penalties & loss in production due to accidents are avoided;	Increased productivity	Safe working environment	Workplace is safe and thus more productive; reduces accidents
Any improvements to existing OHSA legislation?	None	Adapt to suit individual project requirements	Adapt to suit individual project requirements	Adapt to suit individual project requirements, which is not practicable

Table 6 Summaries of Case Study Interviews

The interviewees were asked questions pertaining to the requirements of the OHSA and the effects they have on level of compliance, accidents, injuries and benefits of compliance. A few project specific questions to establish whether the level of compliance with OHS requirements had an impact on health and safety performance and other performance parameters on a particular site were also examined. The interviewees were then requested to give their opinions on how the regulations in the OHSA Act could be improved. The interview results (Table 6) show that negligent attitudes, lack of knowledge (ignorance) and/or understanding of H&S regulations by construction site employees and contractors' profit motive are the main causes of non-compliance to H&S regulatory requirements. This is widely supported by the literature and the survey result analysis. The interviewees believed that non-compliance because of cost mitigation is due to small contractors not including allowances for H & S requirements in their tenders as a deliberate strategy to win tenders.

Two of the interviewees stated that they did not comply with some of the requirements of the OHSA Act because they perceived them as unnecessarily expensive, time-consuming and unjustified for their particular site (housing projects). They argued that most of the requirements of the OHSA regulations are more applicable to larger projects and would be better complied with if they were more project-specific. Other interviewees stated that non-compliance by site operatives could be significantly attributed to negligence/worker attitude, which is a product of risk-taking and each individual's cultural background. In addition to providing a safe work environment, the interviewees stated that compliance with H&S regulations also gives the firm a competitive advantage. This implies an increased level of compliance with the OHSA requirements is an investment by contractors into the pursuit of corporate growth and profitability.

The interview findings further validate the statistical inference drawn from the analysis of data obtained from the questionnaire and literature survey that the level of compliance by contractors to H&S regulatory requirements is essentially not influenced by the degree of risk which the H&S regulatory requirements attempt to prevent, but by perceived cost savings.

Discussion of Findings

The H&S regulations in the construction industry were enacted to safeguard lives and to improve the quality of construction products including processes (HSE 2004; Coglianese et al. 2003; and Wolski et al. 2000). However, contractors have been reported to be non-compliant with these regulations (Warwick 2011; Baxendale and Owain 2000; and Coble et al 2000). South Africa is not lacking in terms of H&S legislation (cidb 2009a) yet, the number of people that get injured or die on construction sites is still high (cidb 2009b; and DoL 2007). Windapo (2011), Bettesworth (2011), MBAWC (2011) and cidb (2009a) attribute factors responsible for injuries and fatalities on construction sites to non-compliance to H&S statutory regulatory requirements by contractors. This study sought to understand the mindset of contractors and establish whether the decisions made by contractors to comply with H&S regulations are influenced by the cost of compliance, cost savings and the degree of risk, which the regulation is trying to prevent.

The study established that the level of a contractor's compliance with H&S regulatory requirements is significantly related to perceived cost savings and unrelated to the degree of risk, which the regulation is trying to prevent. The study findings also reveal that OHSA compliance is considered by contractors to be unnecessarily costly and time consuming to implement and as a result, a contactor perceives that higher level of compliance requires more money. It also emerged in the study that cost savings are positively related with probability of accident occurrence which is a component of the degree of risk. These findings suggest that contractors have a fixation on cost saving where H&S is involved on projects and that the risk involved in certain construction activities are not given due consideration. This mindset of contractors may not be unconnected with injuries, fatalities and unsafe work conditions reported on construction sites (see Table 6).

According to Viscusi (1979), compliance to OHSA requirements will bring about safe working conditions on construction sites. The study findings are aligned to conclusions made by Emrath (2011) that contractors consider precautions as an unnecessary cost; McGarity and Ruttenburg (2002) that cost is an important consideration in compliance; Williams (1995) that regulatory authorities impose regulations that are in practice unattainable or prohibitively expensive; and McConnaughey (1978) that cost savings derived from the implementation of H&S regulatory requirements on site are related to the degree of risk which the regulation attempts to prevent – in this study, cost savings were related to probability of accident occurrence.

Conclusion

Based on the study findings, this paper concludes that the decision made by contractors to comply with H&S regulatory requirements is influenced by the perceived cost saving on account of compliance and that cost savings are influenced by the probability of accident occurrence which is an element of the degree of risk which the regulation is trying to prevent or control (see Figure 1). This paper also concludes that the tendency for a building contractor to comply with statutory H&S requirements, decreases with increase in cost of compliance, and does not increase with degree of risk or perceived cost savings. The survey results and case studies provided evidence of the profit maximisation motive of the contractors, driven by the competitive nature of the construction industry. In addition, the standards complied with by contractors in the study are not linked with perceived severity of the physical hazards that may occur as a result of non-compliance to H&S regulatory requirements and estimated degree of risks. This probably explains the fatalities, accidents and injuries that occur on site (see Table 6).

The main implications of the findings to the different stakeholders involved in construction are: non-compliance to H&S regulatory requirements by contractors because of cost implications will lead to unsafe work condition, injuries and fatalities on construction sites. Contractors will benefit from a safe work environment, reduction in CoA and improved productivity with increased levels of compliance with H&S regulatory requirements. Although complying with the H&S regulatory requirements involves upfront costs, the safety and health of construction operatives should take precedence. Secondly these findings suggest that some elements of the H&S regulations are considered unnecessary, costly and time consuming in implementation by contractors and regulatory authorities should therefore expect issues of non-compliance to H&S regulations. Thirdly, the study findings imply that accidents will continue to occur on construction sites due to the contractors cost saving mindset, and this should be of concern to both public and private clients.

One possible way to improve compliance would be that, in the same way that provisions are made in tenders for contingencies and insurances, statutory allowances of could be allowed in tenders as part of contract agreements, to enable the implementation of viable H&S systems by construction companies on site and ultimately the promotion of healthy and safe work environments at the project site. An H&S consultant could administer these allowances. Future research work can be based on the relationship between the degree of risk, level of compliance with OHSA regulatory requirements, cost of compliance and cost saving, using all nineteen OHSA elements as the basis for analysis, and developing a model which can be used for predicting the likelihood of contractor compliance to statutory H&S regulatory requirements, based on the parameters of cost of compliance, degree of risk (probability of accident occurrence and severity) and cost savings.

References

Abowitz, D. and T. Toole (2010) 'Mixed Method Research: Fundamental Issues of Design, Validity, and Reliability in Construction Research', *Journal of Construction Engineering and Management*, **136**, SPECIAL ISSUE: Research Methodologies in Construction Engineering and Management, 108-116

Amaratunga, D. Baldry, D. Sarshar, M. and R. Newton (2002) 'Quantitative and Qualitative Research in the Built Environment: Application of "Mixed" Research Approach', *Work Study*, **51** (1), 17-31

Ang, G. Groosman, M. and N. Scholten (2005) 'Dutch performance-based approach to building regulations and public procurement', *Building Research and Information* (March – April) **33** (2), 107-119

Ayininuola, G. M. and O. O. Olalusi (2004) 'Assessment of Building Failures in Nigeria: Lagos and Ibadan Case Study', *African Journal of Science and Technology*, **5** (1), 73–78

Baxter, P. and S. Jack (2008) 'Qualitative Case Study Methodology: Study Design and Implementation for Novice Researchers', *The Qualitative Report*, **13**, 544-559

Baxendale, T. and J. Owain (2000) 'Construction design and management safety regulations in practice-progress on implementation', *International Journal of Project Management*, **18** (1), 33-40

Benjamin, P. and J. Greef (1997) *Report of the Committee of Inquiry into a National Health and Safety Council in South Africa*, Department of Labour, Pretoria

Bester M. D. (2011) *Analysis of Occupational Health and Safety Audits Conducted in the Western Cape for period 2007 to 2010*, Unpublished Report. Master Builder Association Western Cape, Cape Town

Bettesworth, D. (2011) The compliance debate. [Online] Available at: http://dbtownplanner.co.za/2011/02/21/the-compliance-debate/ [Accessed: 15 April 2011]

Bland, J. and D. Altman (1997) Statistics notes Cronbach's alpha, *British Medical Journal*, **314**, 572

Ciobanu, M. and M. Mazilu (n. d.) 'Environmental Crisis Management through Risk Management', *Recent Researches in Tourism and Economic Development*, ISBN: 978-61804-043-5, 537-541

Coble, R. J. Hinze, J. and T. C. Haupt (2000) *Construction Safety and Health Management*, Prentice-Hall, New Jersey

Coglianese, C. Nash, J. and T. Olmstead (2003) 'Performance-based regulation: prospects and limitations in health, safety and environmental protection', *Administrative Law Review*, **55** (4), 705–730

Construction Industry Development Board (cidb) (2009a) *Construction Health and Safety in South Africa: Status and Recommendations*, cibd, Brooklyn Square, Pretoria. [Online] Available at: www.cibworld.nl/app/export/baPOuCHu/20102760/cidb01.pdf [Accessed 17 January 2010]

cidb (2009b) cidb Construction Industry Indicators Summary Results: 2009, captured by the Construction Industry Development Board in partnership with the Department of Quantity Surveying and Construction Management of the University of the Free State, Available at: http://www.cidb.org.za/knowledge/publications/industry_reports [Accessed 4 March 2011]

Dalton, M. (2002) *Fatal Accidents in the Irish Construction Industry 1991 – 2001: A Survey of Contributory Factors*, Health and Safety Authority Publication, November

De Vellis, R. F. (2003) *Scale development: theory and applications*, Sage, Thousand Oaks

Department of Labour (DoL) (2007) [Online Resource] Available at: http://www.labour.gov.za [Accessed: 17 October 2009]

Emrath, P. (2011) How Government Regulation Affects the Price of a New Home, National Association of Home Builders, [Online Resource] Available at: HousingEconomics.com, http://www.nbnnews.com/NBN/issues/2011-07-25/ [Accessed 26 July 2011]

Gann, D. M. Wang, Y. and R. Hawkins (1998) 'Do regulations encourage innovation?' *Building Research and Information*, **26** (5) 280–296

Haupt, T. C. (2003) 'A study of management attitudes to a performance approach to construction worker safety', *Journal of Engineering Design and Technology*, **1** (1), 42-53

Health and Safety Executive (HSE) (2004) *Improving health and safety in the construction industry*, The Stationery Office, London

Eppenberger, M. (2008) *Older construction workers: a study of related injuries, underlying causes and estimated costs*, Unpublished Masters Theses, Cape Peninsular University of Technology, South Africa

Ivankova, N. V. Creswell, J. W. and S. L. Stick (2006) 'Using mixed methods sequential explanatory design: From theory to practice', *Field Methods*, **18** (1), 3-20

Jaselskis E. (1996) Strategies for achieving excellence in construction safety performance, *Journal of Construction Engineering and Management*, **122** (1), 61-70

Lin, J. and A. Mills (2001) Measuring the occupational health and safety performance of construction companies in Australia, *Facilities*, **19** (3/4), 131-138

Master Builders Association Western Cape (MBAWC) (2011) *Construction Occupational Health Safety Environment Regional/National Competition Audit System and Star Grading Program*, MBAWC, Cape Town

McConnaughey, J. S. (1978) *An Economic Analysis of Building Code Impact: A Suggested Approach*, National Bureau of Standards, NBSIR 78 – 1528, Washington, DC

McGarity, T. O. and R. Ruttenberg (2001-2002), 'Counting the Cost of Health, Safety and Environmental Regulation', *Texas Law Review*, **80**, 1997-2058, downloaded from http://heinonline.org [Accessed 17 June 2010]

Mthalane, D. Othman, A. A. and R. G. Pearl (2008) *The economic and social impacts of site accidents on the South African society*, School of Civil Engineering, Surveying and Construction, Faculty of Engineering, University of KwaZulu-Natal, Durban, South Africa

McGeorge, D. and P. Zou (2013) *Construction Management: New Directions*, 3rd Ed., Wiley-Blackwell, United Kingdom

Odeyinka, H. Larkin, K. Cunningham, G. McKane, M. Bogle, G. and R. Weatherup (2011) 'Development of models for assessing risk impacts on the variability between contract sum and final account', *Proceedings* 27th *Annual ARCOM Conference*, 12-13 September, School of the Built Environment, University of Salford, London, 614-623

Pillay, K. and T. C. Haupt (2008) 'The Cost of Construction Accidents: A Pragmatic Study', Proceedings of the 3rd Build Environment Conference, T. C. Haupt (ed), Cape Town, 6-8 July, 268-283

Pinto, J. K. (2002) 'Project management 2002', *Research-Technology Management*, **45** (2), 22-37

Runeson, G. and M. Skitmore (2008) 'Scientific Theories', in Knight, A. and Ruddock, L. (eds), *Advanced Research Methods in the Built Environment*, Wiley-Blackwell, United Kingdom, 75-85

Sidumedi, K S. (2009) 'An investigation into the Relationship between the corporate culture of South African construction firms and performance' [Online], Available at: http://wiredspace.wits.ac.za/handle/10539/7942[Accessed: 28 June 2011]

Smallwood, J. J. (2000) *A study of the Relationship between Occupational Health and Safety, Labour Productivity and Quality in the South African Construction Industry*, Unpublished PhD Thesis, Department of Construction Management, University of Port Elizabeth, Port Elizabeth

Smallwood, J. (2004) 'The influence of engineering designers on health and safety during construction', *Journal of the South African Institution of Civil Engineering*, **46** (1), 217-226

Spence, R. (2004) 'Risk and regulation: can improved government action reduce the impact of natural disasters?' *Building Research and Information*, **32** (5), 391-402

Stigler, G. (1971) 'The Theory of Economic Regulation', *Bell. Journal of Economics and Management Science*, **2**, 3-21

Tavakol, M. and R. Dennick (2011) 'Editorial: Making sense of Cronbach's alpha', *International Journal of Medical Education*, **2**, 53-55

Ulang, N.M. Gibb, A.G.F. and C.J. Anumba (2010) 'The communication of health and safety information in construction', *Proceedings of CIB World Congress, [CIB W099 Special Track],* 1 May 2010, Manchester van Huyssteen, S. van Heerden, L. Perkins, P. and O. Gyimah (2002) 'The identification and measurement of performance indicators for the South African construction industry', [Online], Available at: http://buildnet.csir.co.za/cdcproc/docs/3rd/vanhuyssteen%20_vanheerden.pdf [Accessed: 28 June 2011]

Viscusi, W. K. (1979) 'The Impact of Occupational Safety and Health Regulation', The Bell *Journal of Economics*, **10** (1), 117-140, downloaded from http://www.jstor.org/stable/3003322 [Accessed 17 June 2010]

Warwick, J. (2011) 'Building industry safety needs to be of stellar standard' [Online]. Available at: http://www.crown.co.za/news/2011/July/building_industry_safety_needs_to_be_of_stellar_standard.htm [Accessed: 23 July 2011]

Weil, D. (2001) 'Assessing OSHA Performance: New Evidence from the Construction Industry', *Journal of Policy Analysis and Management*, **20** (4), 651-674, downloaded from: http://www.jstor.org/stable/3325777 [Accessed 17 June 2010]

Williams, P. (1995) 'A regulation evaluation system: a decision support system for the Building Code of Australia', *Construction Management and Economics*, **13**, 197–208

Winch, G., and E. Campagnac (1995) 'The Organization of Building Projects: an Anglo/French Comparison', *Construction Management and Economics* **13**, 3–14

Windapo, A. O. (2011) 'Investigation into the Level of Compliance to Construction Health and Safety Requirements within the South African Construction Industry', *Proceedings of the CIB W099 Conference*, 23-25 August, Washington D. C

Wolski, A. Dembsey, A. N. and B. J. Meacham (2000) 'Accommodating the perceptions of risk in performance-based building fire safety code development', *Fire Safety Journal*, **34**, 297-309

Yin, R. K. (1994) *Case Study Research: Design and Methods*, 2nd Ed., Sage Publications, London

Ideological Leaning and Praxis of Housing Co-operatives in South Africa: Matters Arising

11

Jimoh Richard, (Federal University of Technology Minna, Nigeria)

Jasper van Wyk, (Nelson Mandela Metropolitan University, South Africa)

Abstract

Social movements came about as a result of dissatisfaction felt by society from the *status quo*. The grievances are presented through demonstrations, strikes, riots, occupation of land, boycott of business or by the development of social and economic alternatives such as self-help schemes or saving clubs. In South Africa, people join housing co-operatives because it takes a long time to wait for the Reconstruction Development Project (RDP) houses promised by the Government. The leaning of 5 housing co-operatives was examined within the context of ideology and praxis that are components of triad model. This was done through a multi-case study whereby the chairpersons of the housing co-operatives were interviewed using structured interview format. Findings from the study indicated that the housing co-operatives were not all open and voluntary; also, the housing co-operatives were not administered the same way. In all, efforts were put in by the housing co-operatives in engendering the ideology of co-operatives on one hand and the government on the other hand to provide the necessary subsidy so that the gap in the housing deficit could be bridged.

Keywords: Housing, Co-operatives, Ideology, Praxis, Principles, South Africa

Housing Co-operatives in South Africa

In 1996, the Gauteng Provincial Board enabled the approximately 2000 tenants of seven apartment buildings in Hillbrow, Joubert Park and Berea to become owners of the flats they were living in (SHF, 2000; Cull, 2001; Rust, 2001; Fish, 2003; Crofton, 2006 & NDoH, 2009). This marked the beginning of housing co-operatives in Johannesburg's inner city and to a large extent South Africa, as there was no prior documented evidence of the use of co-operatives to access the institutional subsidy of government. The question is why has it taken this long for this approach to be used in housing delivery?

TYPE OF PROPERTY	GAUTENG	NORTH WEST	WESTERN CAPE	KWAZULU-NATAL	EASTERN CAPE
PHP & INFORMAL	1	1	0	9	11
HOSTEL UPGRADE	0	0	2	0	0
GREENFIELD	2	0	0	1	0
FLATS RENOVATION	20	10	1	0	0
TOTAL	23	11	3	10	11
TOTAL UNITS	105,000				
OWNERSHIP	70% BLACKS, 25% COLOUREDS, 5% INDIANS				
DEMOGRAPHICS	50% WOMEN, 40% YOUTH & 10% DISABLED				
EMPLOYMENT OPPORTUNITY	1,715 PERMANENT EMPLOYEES				

Table 1 National Co-operative Housing Membership Statistics in Five Provinces

Source: Matsela, 2010; presentation on housing co-operatives to the National Housing Portfolio Committee Parliament, Cape Town.

Table 1 above is an indication of the low number of housing co-operatives in South Africa. Four provinces, namely Mpumalanga, Northern Cape, Free State and Limpopo are not represented. Even those represented show low membership statistics. However, in the later part of the 1990s, with the slower rate in the delivery of housing by government, people started looking back at the rationale behind the communal approaches to addressing their housing needs.

The Developmental Phases in Co-operative Housing

The CCMH (Commission on Co-operative and Mutual Housing, 2009) stated that there are three stages or phases of evolution of housing co-operatives up to the present time:

- The first phase is characterised by grass roots bottom up initiatives by trade unionists or housing and community activists experimenting with the development of housing co-operatives as a means of meeting the housing needs of people who are not able to get a decent home that they can afford. In countries such as Germany, Austria, Sweden and Canada, as stated by CCMH (2009), that were the pioneers of the co-operative housing approach, this phase was characterised by not well defined legal and financial frameworks and limited professional and technical expertise.

- Phase two replicates, consolidates and adapts successful pioneering experiments, leading to the emergence of a recognisable co-operative housing subsector. Regional and national federations of co-operative and mutual housing start to form; support services able to facilitate the development of co-operative and mutual housing begin to develop and codes of conduct and best practice begin to emerge. CCMH (2009) emphasised that the success of this phase was attributable to two reasons:

    - The advocacy role of the national co-operative movements in countries such as Germany, Austria, Sweden, Norway and Canada. This was dependent on the belief exhibited by the various governments that the co-operative housing could help to solve the housing problem of the people; and

    - The formation of savings and loan schemes established for members to save towards their co-operative membership shares.

- Phase three, according to CCMH (2009), is the period when co-operative housing has been the entrenchment in the psyche of the public; a situation where specific provision is made for co-operative and mutual housing within the national legislation and housing systems. The level of support received by the housing co-operatives from their various governments may be connected with the percentage of housing stock provided through the co-operative housing approach.

CCMH (2009) stated that the phases do not have a distinct division among them; a phase does not start and finish before the next one can begin. The development of the phases differs from one country to another but the common denominator is that housing co-operatives pass through the three phases for a sustainable co-operative housing subsector to be developed.

From the above characterisation, the South African housing co-operatives are at the first phase, where the rate of failure is at its highest. If efforts are not put in place to ensure that the housing co-operatives develop past that phase, it is not likely to move to the next phase. This phase requires a lot of sacrifices on the part of all the stakeholders. The developmental phases in co-operative housing has shown that without beneficial legislative and policy frameworks in place, the growth, development and sustainability of housing co-operatives will be near impossible.

The Problems of Housing Co-operatives

In spite of the benefits that could be derived from co-operative housing as a delivery option, housing co-operatives around the world are beset by problems ranging from inadequate legislative frameworks, lack of understanding of the public and government officials on co-operative housing to inadequate finance as established by Nubi, 2009; Eglin, 2008; Nnkya, 2001; Byaruhanga, 2001; Fruet, 2005; Fall, 2009; Mubvami & Kamete, 2001; CMHN & VNC, 2004; CMHC, 2003 and McClean & Onyx, 2009. These problems are tabulated in Table 2:

Developing countries	Developed countries
Lack of awareness by the officials of government.	Lack of information and knowledge on co-operative housing.
Unfavourable legislation towards co-operative housing.	Inappropriate regulation framework.
Interference by agency responsible for housing delivery.	Lack of support structures.
Opposition to co-operative model.	Hostile environment.
Lack of support by all the spheres of government.	Restrictive regulation.
Lack of access to finance.	Lack of sustainable finance.
Weak internal control.	Inadequate management structure.
Non payment of fees by the members.	Default in the payment of rents.
Lack of understanding by the members on co-operatives.	Inadequate education and participation by the members.

Table 2 Problems Experienced by Housing Co-operatives in Developing and Developed Countries (Source: Authors own summary)

For the purpose of the summary in Table 2, the developing countries are Nigeria, South Africa, Tanzania, Uganda, Brazil, Senegal and Zimbabwe. Developed countries are United States, U.K, Canada and Australia.

The above tabulation brings a salient issue to the fore, irrespective of where the housing co-operatives are based, the challenges are the same.

Theoretical Framework

From the previous literature review, the problems that housing co-operatives experienced in different countries around the world were identified and presented in Table 2. The problems can be grouped into two categories: The first has to do with the mobilisation of resources such as land and finance and secondly, organising these resources based on the principles of co-operatives, which form the mainstay of any housing co-operative. These principles distinguish housing co-operatives from other co-operative-like initiatives; hence the need to locate the study within a theory or model for proper understanding of the research was imperative. The study was anchored on the Develtere model, called the “triad” model shown in Figure 1 below.

It was used to study social movements in co-operatives, since housing co-operatives is a microcosm of a co-operative movement; hence the model can be used. Gerard and Martens (cited by Develtere, n.d.) distinguished three components or forces in all social and co-operative movements. The ideology of the movements present the images of a desirable society based on specified values and the ways to achieve them. There is also the praxis or the action that is responsible for the mobilization and participation of the membership base.

To realize their objectives, as stated by Gerard and Martens (cited by Develtere, n.d.), they also develop at least a minimal organisational structure.

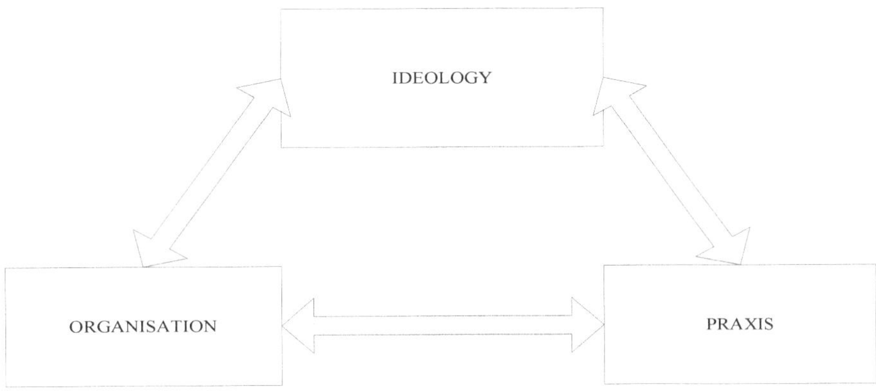

Figure 1: The three components of co-operative movements in developing countries
adapted from Develtere (n.d.), page 39.

Develtere (n.d.) established that there is a continuous interaction among the components (the ideology, the praxis and the organisation). Over the years, due to the resources deployed by the members of co-operatives through their participation in co-operative activities, the ideology is created and recreated. This ideology determines who can be members and the resources expected from the members. Based on the ideology and the praxis, according to Develtere (n.d.), the structure of the organisation (in this case the co-operative) is defined. The interaction of these components creates the identity of a social movement (Develtere, n.d.).

Social movements came about as a result of dissatisfaction felt by society from the status quo, as emphasized by Develtere (n.d.). They present their grievances through demonstrations, strikes, riots, occupation of land, boycott of business or by the development of social and economic alternatives such as self-help schemes or saving clubs. In South Africa, people join housing co-operatives because it takes a long time to wait for the Reconstruction Development Project (RDP) houses promised by the Government. They also want to have control of the outcome of their houses when completed. In other climes where there are no subsidies, as is being provided in South Africa, people form housing co-operatives where ordinarily they would have been incapacitated as individuals to get control of their destiny (Willie-Nwobu cited by Olusanya, 2000).

As stated at the outset, housing co-operative is a microcosm of the co-operative movement. Since the study was based on an in-depth understanding of housing co-operatives in South Africa, it has to do with the interplay of the principle of co-operatives (ideology); the praxis (the resources employed by both the members and external bodies such as donor agencies and government) and the way the organisation is structured. In order words, it investigated how the housing co-operatives adhered to these principles, the way resources were employed by both the government and the housing co-operatives members and finally, the way the members structured their various housing co-operatives. Develtere's "triad" model was suitable for the study. Only 2 (two) of the components of the model (ideology and praxis) were reported in this paper which covered open and voluntary membership; education, training and information; democracy; autonomy and finance.

Research Methodology

A multi-case study method was used. Gummesson (2007) defined a case study research as one where cases from real life are used as empirical data for research, especially when knowledge of an area is sparse or missing. Yin (2009) stated that a case study is used when a "how" or "why" question is being asked about a contemporary set of events over which the investigator has little or no control, and the boundaries between phenomenon and context are not clearly evident. For the purpose of this study, structured interviews were used with both closed and open ended questions asked from the interviewees.

In determining the number of housing co-operatives to be interviewed, purposive sampling was used. This sampling technique is a non-probability sampling procedure which is usually used in qualitative research that has to do with selecting the people to be interviewed based on the interviewer's knowledge on the appropriateness and typicality of the sample selected (David & Sutton, 2004; Cohen *et al*, 2005; Teddlie & Yu, 2007). Eisenhardt (cited by Meyer, 2001) stated that the logic of the sampling is different from statistical sampling because the idea is to select cases that are replicable or be able to further the emergent theory. Yin (2009) suggested that 2 or 3 cases could be selected for literal replication whereas 4 to 6 cases can be used to study theoretical replication (predicting contrasting results).

Initially, 6 cases were selected from the housing co-operatives in Johannesburg for this study based on the following parameters:

- Due to the large concentration of housing co-operatives in Johannesburg; there were twenty-nine housing co-operatives identified based on the list from the Registrar of Co-operatives Office;

- The cosmopolitan nature of the city that makes the poor, with limited or inadequate housing, continue to migrate there in search of opportunities. It is the belief of the National Government that co-operative housing could be explored to solve the housing problem of this group of people (NDoH, 2009); and

- The selection of housing co-operatives that have been in business of providing housing for their members for more than 3 years. It was the belief of the authors that this set of housing co-operatives will provide more insight than those that have just started out.

However, only 5 housing co-operative chairpersons were interviewed, due to the fact that one of the housing co-operatives identified requested an interpreter. The authors decided not to interview this chairperson because of fear of misinterpretation or inadequate interpretation from the interpreter. The first interview was conducted on 26 May 2011 and lasted for 30 minutes. It was recorded and notes were also taken. The permission of the interviewee was sought before recording the interview.

The second interview took place on 29 May 2011. It was difficult to interview the other 4 chairpersons individually because the convenient time for them was before their provincial meeting. As a result of this, it was necessary to change the format of interviewing adopted in the first interview to group interviewing. In this approach, the authors asked questions based on the research guide, while the chairpersons answered them in the interview guide given to each of them based on the peculiarity of their respective housing co-operatives. There was no need to sound record this session, since each individual housing co-operative answered in writing and enough time was usually given for all those present to complete a particular question before moving to the next one. This session lasted for 2 hours and 25 minutes.

Results and Discussions

Profiles of the Housing Co-operatives Interviewed

Profile	Co-op A	Co-op B	Co-op C	Co-op D	Co-op E
Year established	2000	1999	1999	1999	1999
Year registered at Companies & Intellectual Property Registration Office (CIPRO)	2001	2001	1999	2000	2001
Number of members	120	351	55	84	28
Number of houses	120	351	55	84	28

Table 3 Profiles of the Five (5) Housing Co-operatives Interviewed

The above table shows the years the housing co-operatives were established and when they were registered at CIPRO. As stated earlier, the year of establishment was a condition in the selection of the housing co-operatives to be interviewed. All the housing co-operatives have been in 'business' for more than twelve years which is an indication that the information supplied can be relied upon.

Open and Voluntary Membership

The first principle of all the co-operatives irrespective of t type is for the membership to be open and voluntary. The extent to which this principle was pursued by the co-operatives varied based on underlying parameters establishing such co-operatives. In South Africa the development of housing co-operatives hinged on the subsidy provision of the National Government from the beginning hence the open and voluntary nature of the membership was restricted. To be eligible for the subsidy, prospective beneficiaries must be in a household with a joint income of between $143-$700. In this respect none of the housing co-operatives could be said to be open and voluntary to new members.

Education, Training and Information

It was discovered that out of the four organisations listed [the Small Enterprises Development Agency (SEDA), Social Housing Foundation (SHF), South Africa Housing Co-operatives Association (SAHCA) and Non Governmental Organisations (NGOs)], the housing co-operatives only availed themselves of the training conducted by SAHCA. The training provided by SAHCA according to the interviewees ranged from adequate to fairly adequate; hence, they all canvassed for improvements to be made. The chairperson of housing co-operative A was aware of the existence of the SHF, but did not know the procedures to follow in order to benefit from its workshops and training opportunities. Education and information are important, but were lacking in this instance; probably not due to the fault of the chairpersons but as a result of limited promotion of SHF activities that would have informed housing co-operatives of how to approach the SHF for training.

SEDA, according to DTI (2010), was an agency of the Department of Trade and Industry which among other services provided social and skills enhancement for co-operative members and to support and promote co-operatives. The support provided was usually in the form of training on how business plans were prepared and how to run the co-operative. The SHF (the responsibilities has now been taken over by Social Housing Regulatory Authority [SHRA]) was also a government agency accountable to the Department of Human Settlements and was expected to develop a sustainable social housing sector for South Africa. This goal was approached by implementing rental strategy and providing training and workshops within the confines of the social housing sector (UN-Habitat, 2008 & SHF, 2009).

It was stated that the housing co-operatives offered education and training to both the members of the housing co-operatives and the employees. In addition to these, public enlightenment was also provided. A classic example was given by housing co-operative B to the effect that across the road, there was rental housing that charged more than the housing co-operative was charging on a monthly basis and that no better enlightenment could be found. Corroborating the responses of the interviewees, during the second interview, someone from Canada was introduced by the President of SAHCA to the members present. The national body facilitated the arrangement in order for the housing co-operatives to be taught the development in co-operatives in general and housing co-operatives in particular.

Democracy

In housing co-operative A, the Board members were usually elected by the members and were expected to serve for three (3) years. The committee members were also expected to serve for three (3) years and were selected by the Board members. This arrangement for selecting committee members is not in line with democratic tenets and is capable of being abused by the Board of Directors, especially when the Board of Directors' associates are selected to be members of the committee. One good thing here was that the housing co-operative had only one committee: the finance committee; hence the extent of abuse of power was reduced. The positive aspect of this selection of committee members was that the chairperson stated that the finance committee was often effective in discharging its functions.

In taking decisions, the following were usually adhered to by the housing co-operative:

- Major decisions were taken at the general meetings.
- Important and minor decisions were taken by the Board members.

In housing co-operative B, the situation was somehow different. The Board members were usually elected for more than four (4) years and due to the re-organisation going on in the housing co-operative, there was no committee in place. Electing Board members to run for more than four years could lead to vices such as lack of accountability, favouritism, corruption, apathy and perpetuation of self interest (CCMH, 2009). In taking decisions, the general secretary indicated that major and important decisions were normally taken by the Board members while minor decisions were left to be decided at the general meetings. This type of arrangement will only further perpetuate some of the vices earlier mentioned. Though running housing co-operatives by general meetings according to CCMH (2009) could lead to a situation where the housing co-operative may become difficult to administer it would have been more appropriate to have major and important decisions vested in the general meetings to ensure checks and balances. However, annual general meetings have not been held for sometime due to the re-organisation going on in the housing co-operative.

In housing co-operative C, the situation was that both the Board members and committee members were elected and are expected to run for four (4) years and one (1) year respectively. There were maintenance and social committees which were mostly effective in the discharge of their responsibilities. This may be connected with the limited time allowable for the committees; the idea of positively impacting the housing co-operative cannot be overemphasised. Important and minor decisions were usually taken by the Board members while major decisions were taken at the general meetings. Annual general meetings were held annually. In housing co-operative D, both the Board members and committee members were elected for a period of three (3) years and there were maintenance, finance and social committees in place. Major decisions were taken at the general meetings while Board members take important and minor decisions. Annual meetings took place every year.

Lastly, in housing co-operative E, Board members and committee members were elected for a period of four (4) years and one (1) year respectively and there were maintenance, finance and social committees in place. Major decisions were usually taken at the general meetings while Board members took important and minor decisions.

Of note is that in all the housing co-operatives, women have been very active in their participation in the activities of the co-operatives. For housing co-operatives to be sustainable, the interviewee stressed that women should be given equal opportunities and should be supported to achieve this.

The strength of any housing co-operative is hinged on the members' participation and as such the housing co-operatives embarked on the following measures to ensure adequate members participation in addition to the annual general meetings:

- New members were informed of the need to participate actively in the co-operative activities. This measure was sometimes effective as observed by the chairperson.
- Providing training for all the committee members. This measure was important and has been found to be very effective.
- Regular house to house visits was also an effective measure usually adopted by the housing co-operative in ensuring members' participation.
- In addition to the above measures, housing co-operatives D and E imposed fines when members fall due in their participation.

Autonomy

One of the principles of co-operatives is that they should be autonomous; hence they should not be subjected to external control from bodies such as government and donor agencies. Housing co-operatives A and B claimed that there has never been a time when any agency of the government, SHF, SAHCA and the Board of Directors tried to impose their will on the housing co-operatives. However, when looked at from the perspective of monitoring and evaluation, this is not a positive development because it implies that the housing co-operative was left to either 'swim or sink'. A modicum of control was expected especially when the subsidy of government was involved, as it was in these cases. However, the story was different in housing co-operatives C, D and E that have been subjected to some form of control.

Finance

The chairpersons were asked about the various sources of finance available to the housing co-operatives and the level of adequacy. It was discovered that the housing co-operative A had used members' contribution, government institutional housing subsidy and a loan from the National Housing Finance Corporation (NHFC) to finance the day to day activities of the housing co-operative. The NHFC, according to UN-Habitat (2008), was an agency of the Human Settlements Department, whose objective among others was the development and appropriate funding of institutions such as housing co-operatives, offering a variety of tenure arrangements for residential purposes in the areas of the housing market that were not well covered. As to the adequacy, members' contributions and the government institutional subsidy were very adequate while the loan from NHFC was adequate.

However, the other housing co-operatives (B, C, D and E) had only used members' contribution and government institutional housing subsidy. As to their adequacy, they claimed that only the subsidy was adequate except housing co-operative E where both the subsidy and the members' contribution were adequate. None of the housing co-operatives was registered as a Social Housing Institution (SHI) in order to be able to benefit from government grants apart from the subsidy. Efforts of policy makers were not encouraging in creating sustainable co-operative housing. This statement was borne out of the statement that appeared on page 9 of the 2005 Social Housing Policy and page 18 of the 2009 Social Housing Policy as shown below:

> *"Primary and secondary housing co-operatives registered under the Co-operatives Act of 1981, and accessing funding through this programme will be considered together with the social housing institutions and will have to be accredited as social housing institutions. Separate guidelines, however, will be drafted to accommodate the specific nature, operations and regulatory requirements of the housing co-operatives".*

Four years after the first social housing policy was published, separate guidelines for housing co-operatives have not yet been drafted. Hence, it will become difficult for the housing co-operatives to access the social housing grant. Interestingly, the Co-operative Act of 1981 does not distinguish housing co-operatives from agricultural based co-operatives, which were predominant at that time; it was the Co-operatives Act 14 of 2005 that recognised other specialised co-operatives such as housing co-operatives. This further showed that there is a minimal understanding by the agency responsible for the formulation of housing policies and legislation vis-à-vis co-operatives policies and legislation. This limited understanding may be connected to the low level of growth and development experienced so far in the co-operative housing subsector. Apart from this, it was going to be difficult for housing co-operatives to benefit from the current grants, except if some aspects of the Social Housing Act 16 of 2008 were changed. This issue came up during the presentation of the Social Housing Act Regulations by the SHRA to the Portfolio Committee on Human Settlements on 7th September 2011. The Committee remarked that no funding framework existed for housing co-operatives at the national level. This can be validated from page 9 of the 2005 Social Housing Policy and page 18 of the 2009 Social Housing Policy highlighted above.

The institutional subsidy that formed the main subsidy that housing co-operatives use, was not exclusively for housing co-operatives but also for other housing institutions such as rental, share block and instalment sale. The main policy intent of the subsidy was to provide capital for the provision and maintenance of affordable rental housing (NDoH, 2009a). In essence, it amounted to the survival of the fittest in getting the subsidy.

In spending the funds of housing co-operative A, it was the responsibility of the members and the Board members to decide on how the money of the co-operative was spent. This was a good method as it checked the unilateral spending by either the chairperson or the person in charge of finance. Also, this housing co-operative as a matter of statutory regulation ensured that its accounts were audited as enshrined in the 2005 co-operatives Act 14 in Section 47 Subsections 1d, although a co-operative may apply for exemption to the Registrar of Co-operatives if for reasons of finance the co-operative could not meet the requirements. Housing co-operatives B and C accounts were (as at the time of the interview) administered in a trust account. This may be indication of a problem in the housing co-operatives. In co-operatives D and E, the chairperson and the person in-charge of finance authorized the spending of the co-operative's funds.

To ensure that members pay their fees, the following measures were put in place by all the housing co-operatives:

- Initially, electricity to defaulters were disconnected but this policy resulted in the defaulters seeking redress in court. The policy was later terminated as it was not in the statutes of the housing co-operative;
- Presentation of letter of demand to the defaulter and
- Evicting defaulters. Over the years, members have been evicted.

The eviction of members of the housing co-operatives was something that should have been looked into as it was not a positive development. In a study of the co-operative housing subsector conducted by the Community and Neighbourhood Services Department, City of Toronto (2004), the following eviction prevention strategies were suggested:

- Eviction should be the last resort;
- Provision of late notices;
- A progressive approach to arrears management and conflict resolution;
- Repayment schedules;
- Provision of clear information on the eviction process, options and rights; and
- Allowance of time and opportunity to explain and resolve problems such as arrears and to contest eviction.

Matters Arising

All co-operatives, according to Johnston (1999), are in several ways created with ideology that is predetermined. In essence, there is a mission that each co-operative wants to achieve that is encapsulated in the co-operative principles. The implementation of these principles as stated by Johnston (1999) differs from one country to the other and also from one co-operative to the other. Hence, the essence of the principles is to provide a framework for co-operatives to carry out their businesses. Co-operative business is just a means to an end, a method of doing business which results in advantages to the society not otherwise possible in a capitalist system.

In a collection of 21 cases of housing co-operatives from 12 countries reported in Munkner (2009), none of the housing co-operatives was able to apply all the principles of co-operatives but became distinctive by one form of the principle or the other. Munkner (2009) concluded that what is important in any housing co-operative is the promotion of the interest of the members and that the existence of housing co-operatives hinges on this. Of note is that all over the world, there are different models of housing co-operatives available because of their flexibility to suit the environment in which they operate. In the words of Munkner and Trodin (1999), there is no single model for the development of self-help organisations, of which housing co-operatives is one that can be transferred from one country or region to another. Munkner and Trodin (ibid) are of the view that experience has shown that housing co-operatives differ from one community to another. For housing co-operatives to be successful, they have to be structured in line with the local needs of the members. This is not to say that best practice cannot be copied but with adaptations to suit the local needs of the members. Irrespective of the model chosen, the principles of co-operatives have to be embedded because without that, there is no housing co-operative.

The praxis may be broken down into internal and external actions. The internal praxis consists of members participating in the decision making in areas such as electing/selecting members to Boards of Directors, attending meetings or paying all levies. While external praxis refers to the relationship between housing co-operatives and other bodies such as government and donor agencies in terms of resources provision. Where the State provides the bulk of the finances through institutional subsidies the issue of autonomy may in the long run become challenging as government may control operational decisions hitherto taken by the Board of Directors. This will make the housing co-operatives unable to function as member owned and democratically controlled organisations. This scenario was seen in the membership structures of the housing co-operatives studied, as prospective members have to be in an income bracket to qualify for the subsidy. To curb this dependence on the government, members have to demonstrate commitment to internal practices.

Conclusions

Five housing co-operatives based in Johannesburg, South Africa have been examined using the triad model for their ideology and practises. The housing co-operatives examined were not open since membership depended on being a citizen within certain income bracket due to the institutional subsidy from the government that is involved. The test showed a mixed result on the matter of democracy as reflected in the ways the housing co-operatives chose their executives and committee members (internal praxis). The tests showed the same result for finance. In all, efforts were put in by the housing co-operatives in engendering the ideology of co-operatives on one hand and the government on the other hand to gain the necessary subsidy (external praxis) designed to reduce the housing deficit.

References

Byaruhanga, E.M. (2001) *Shelter Co-operatives in Uganda: Contributions of the Co-operative Sector to Shelter Development*. Available from: www.UN-Habitat.org/publications/ (Accessed 25 January 2010)

CCMH (2009) *Bringing democracy home*, Edited by Bliss, N. Available from: www.cch.coop/.../bdh-commission-report.pdf (Accessed 17 May 2010)

CMHC (2003) Crisis situations in co-operatives: Better interventions hinge on a better understanding. *Research Highlight: Socio-economic Series 03-003*, April

CMHN & VNC (2004) *Affordable housing co-operatives conditions & prospects in Chicago*, Available from: www.uic.wisc.edu/cuppa/voorheestr/ (Accessed 26 April 2010)

Cohen, L., Manion, L. & Morrison, K. (2005) *Research methods in education.* 5th edition, Taylor & Francis e-library, London

Community & Neighbourhood Services Department (2004) *Analysis of evictions in the City of Toronto: Co-operative housing sector*. Available from: www.toronto.ca/housing/... (Accessed 27 May 2011)

Crofton, O. (2006) Housing co-operatives in South Africa: Experiences and lessons. *Housing in Southern Africa, February*

Cull, T. (2001) The seven buildings project: Is co-operative housing the answer to South Africa's housing crisis? *Africa Insight,* **31** (1), 43-49

David, M. & Sutton, C.D. (2004) *Social research: The basics.* 1st edition, Sage Publications Inc., California

Develtere, P. (n.d.) *Co-operative and Development: Towards a Social Movement Perspective*. Available from: www.usaskstudies.coop/pdf-files/ (Accessed 10 February 2011)

Department of Trade & Industry (2005) *Co-operatives Act (No 14 of 2005)*. Available from: www.dti.gov.za/DownloadsFileAction/ (Accessed 23 January 2010)

Department of Trade & Industry (2010) *The DTI co-operatives handbook 2010*. Available from: www.dti.gov.za/DownloadsFileAction/ (Accessed 19 September 2011)

Eglin, R. (2008) *Housing co-operatives: Lessons from Afesis-Corplan's co-operative housing experience*. Available from: www.afesis.org.za/publications (Accessed 24 March 2010)

Fall, A.S. (2008) 'The Senegalese co-operative movement: Embedded in the social economy', In *Co-operating out of poverty: The renaissance of the African co-operative movement*. Edited by Develtere, P, Pollet, I & Wanyama, F. Available from: www.ilo.org/coop_out_of_poverty/ (Accessed 11 March 2009)

Fish, M. (2003) 'Social housing', In *Housing policy and practice in post-apartheid South Africa*. Edited by Khan, F & Thring, P. Heinemann publishers limited, South Africa, 409-410

Fruet, G.M. (2005) The low-income housing co-operatives in Porto Alegre, Brazil: A state/community partnership, *Habitat International,* **29**, 303-324

Gummesson, E. (2007) 'Case study Research', In *The Principles of Knowledge Creation: Research Methods in the Social Sciences.* Edited by Gustavsson, B. Edward Elgar, UK, 87

Johnston, H. (1999) Social movement identity: an application of theory to the co-operative housing movement in Toronto. Available from www.collectionscanada.gc.ca/.../MQ46756.pdf (Accessed 10 February 2011)

Matsela, Z. (2010) Presentation on housing co-operatives by SAHCA to National Housing Portfolio Committee Parliament, Cape Town, 24 February

McClean, S. & Onyx, J. (2009) Institutions & social change: Implementing co-operative housing & eventually sustainable development at Christie Walk, *Cosmopolitan Civil Societies Journal,* **1** (3), 109-131

Meyer, C.B. (2001) A case in case study methodology, *Field Methods,* **13** (4), 329-352

Mubvami, T. & Kamete, A. (2001) *Shelter Co-operatives in Zimbabwe: Contributions of the Co-operative Sector to Shelter Development.* Available from: www.UN-Habitat.org/publications/ (Accessed 25 January 2010)

Munkner, H.H. (Editor) (2009) *Application of Co-operative Principle: 21 cases of housing co-operatives from 21 countries*, Marburg Consult, Germany

Munkner, H.H. & Trodin, R. (1999) *Organised self-help to solve housing problems: A manual for organisers and local leaders of housing co-operatives among the urban poor*, Marburg Consult, Germany

National Department of Housing (2005) *Social Housing Policy for South Africa: Towards an enabling environment for social housing development.* Available from: www.housing.gov.za/ (Accessed 3 February 2010)

National Department of Housing (2009) *The National Housing Code, Part 3: Social Housing Policy*. Available from: www.housing.gov.za/ (Accessed 6 May 2010)

National Department of Housing (2009a) *The National Housing Code, Part 3: Institutional Subsidies*. Available from: www.housing.gov.za/ (Accessed 6 May 2010)

Nnkya, T. (2001) *Shelter Co-operatives in Tanzania: Contributions of the Co-operative Sector to Shelter Development.* Available from: www.UN-Habitat.org/publications/ (Accessed 25 January 2010)

Nubi, T.O. (2009) Housing co-operatives as tools for housing affordability and availability, *Housing Today,* **1** (11), 14-25

Olusanya, O. (2000) *Capacity Building in Housing Delivery: Strategy and Tactics for a Developing Economy*. Paper presented at the 4th International Conference on Housing, Abuja, Nigeria

Rust, K. (2001) *Update on the status of Housing Co-operatives in South Africa*; Draft report prepared for the Social Housing Foundation (SHF), October

SHF (2000) *Guidelines for social housing design*, SHF, Johannesburg

SHF (2009) *The Social Housing Foundation annual report for 2008/2009*. Available from: www.pmg.org.za/ (Accessed 22 March 2011).

South African Portfolio Committee on Human Settlements (2011) *Social Housing Act regulations: Department of Human Settlements & Social Housing Regulatory Authority Briefing*, Parliamentary Monitoring Group, Cape Town

Teddlie, C. & Yu, F. (2007) Mixed methods sampling: A typology with examples. *Journal of Mixed Methods Research,* **1** (1), 77-100

UN-Habitat (2008) *Housing Finance System in South Africa: The Human Settlements Finance Systems Series*. Available from: www.unhabitat.org/publications/ (Accessed 25 January 2010)

Yin, R.K. (2009) *Case Study Research: Design & Methods*, 4th edition, Sage Publications, California

Employee Preferences for Work-life benefits in a Large New Zealand Construction Company

12

Emily Morrison, (NZIQS, New Zealand)

Derek Thurnell, (Unitec Institute of Technology, New Zealand)

Abstract

The construction industry is a challenging environment in which to work, with job characteristics that are linked to work-life conflict, and it is important for its future sustainability that initiatives to support employee work-life balance are undertaken. 121 head office and site-based employees within a large New Zealand construction company rated their preferences for work-life benefits, and the results were compared with those of a similar previous (Australian) study. The most preferred work-life benefit factor (as for the previous Australian study) was found to be 'wellness and personal development'. Results suggest that company provision of a wide variety of work-life benefits from which employees can choose during different stages in their life and career is ideal. Qualitative results suggest some work-life conflict associated with working long hours and weekend work exists. It is proposed that to attract and retain valuable employees, the New Zealand construction industry must provide useful work-life benefits, reasonable working hours, and supportive workplace cultures in line with such initiatives.

Keywords: Work-life balance, New Zealand, Construction company, Human resource management

Introduction

Due to its project based nature, the construction industry can be a particularly challenging environment in which to work. Jobs can be highly stressful, with pressures such as achieving tight programme deadlines, meeting the needs of multiple stakeholders, dealing with disputes, and maintaining the safety of all individuals on site (Lingard & Francis 2004a). Furthermore, long working hours and weekend working are an integral part of many jobs in the industry. Many of these job characteristics are linked to work-life conflict, considered to be a lack of 'fit' between employees and their life responsibilities, and the goals of their organisation (Becker & Huselid 1998). Certain workplace attributes such as work-life balance initiatives (amongst others) are related to reduced work-life conflict (Eby et al. 2005). As such, it is posited that it is important for the future sustainability of the construction industry to undertake initiatives to support employee work-life balance. Lingard and Francis (2005) assessed employee preferences for work-life balance benefits in two Australian construction companies and suggested that further study be carried out in other countries. This research surveyed managerial and professional employees' preferences for work-life benefits in a large construction company in New Zealand, and the results were analysed and compared with Lingard and Francis' (2005) findings in Australia.

Work-life balance

Work-life balance is defined as the 'the ability of individuals to pursue successfully their work and non-work lives, without undue pressures from one undermining the satisfactory experience of the other' (Noon & Blyton 2007, p.356). The concept of work-life balance has evolved in response to a changing culture where work has become more demanding of both sexes and free time has become less available, in turn increasing pressures in both domains causing endemic work-life imbalance (Dallimore & Mickel 2006; Lewis, Gambles & Rapoport 2007; Families and Work Institute 2008). Rates of employment have increased for women and remained constant for men resulting in less time available outside of work. For some time society has been shifting away from the traditional male breadwinner model of work (Brough et al. 2008) and dual-earner households have become the norm, where both partners work and contribute to family responsibilities (Eby et al. 2005; Families and Work Institute 2008). Increasingly, the younger generation of working parents expect more flexible working practices and are more mindful of not over-committing themselves and achieving work-life balance (Loosemore, Dainty & Lingard 2003; Lingard & Francis 2004b). In addition, the percentage of single parents in the workforce has also risen, and an ageing population has resulted in an increase in elderly care demands (De Cieri et al. 2005; Families and Work Institute 2008). At the same time, work intensification has increased the pressures within the workplace. Due to these societal shifts, pressures in both work and life outside of work are at an all time high (Bardoel et al. 1999; McPherson & Reed 2007).

The right work-life balance is different for everybody depending on their circumstances, personality, desires, and stage in life (McPherson & Reed 2007). Moreover, work-life balance is multi-faceted and can be addressed on multiple levels (De Bruin & Dupuis 2004). Whilst some think it is an individual issue, others believe it is a responsibility for organisations to address, or believe it should be addressed and/or regulated by government, while some believe it is a shift in societal norms that is needed (Dallimore & Mickel 2006; Lewis, Gambles & Rapoport 2007; Kossek, Lewis & Hammer 2010). Purportedly, work-life research has not significantly helped improve the array of work-life balance issues that are faced by employees and employers today, and there is arguably a lack of theory and model development, a lack of exploratory research, and hence a lack of understanding of why certain relationships exist (Eby et al. 2005).

In response to the above-mentioned societal changes and increasing work-life conflict, employers have begun implementing work-life benefits; these initiatives are aimed at helping improve work-life balance. Work-life benefits can be broken down into subsets; four work-life balance categories have been identified: flexible work arrangements; leave arrangements; dependent care assistance, and general services (Frone 2003).

Implementing work-life balance policies which accommodate a diverse range of employees purportedly leads to having higher quality employees, which in turn increases company performance and competitive advantage (Kossek et al. 2010). Work-life balance policies attract and maintain employees, which leads to cost savings through decreased turnover, and if work-life benefits contribute to reduced stress, reduced work-life conflict, and increased well-being, the whole organisation experiences improved performance (Casper & Buffardi 2004; Eby et al. 2005; Kossek et al. 2010). However, 'there is a lack of consensus about whether the positive effect of work-life benefits is universal (i.e. experienced by all employees, irrespective of their individual characteristics or circumstances) or whether the effect of work-life benefits differs for particular sub-populations of employees' (Lingard & Francis 2005, p.1047).

What employees say they prefer, and what employees actually do once the options are available, can be different, and evidence of a gap between the organisational provision of work-life balance initiatives and actual employee usage of the initiatives has recently emerged. Utilisation is dependent not only on availability, but also on other issues, such as organisational support. The difference in uptake is possibly due to the perceived level of supportiveness of the workplace culture for using such policies (De Cieri et al. 2005; Dallimore & Mickel 2006; Brough et al. 2008). People often do not use benefits because they feel that usage may have a negative effect on their career path; therefore it is key that management communicate and make clear their objectives and expectations (Thornthwaite 2004).

Work-life balance in New Zealand

Work-life balance issues in New Zealand are similar to those reported internationally. A New Zealand national work-life balance survey found that just over half of employees experienced work-life balance, but a significant number of employees experienced work-life conflict, and found it difficult to achieve work-life balance for a variety of reasons. A major source of conflict was that many people worked long hours and unpaid additional hours, and that many New Zealanders would like more time away from work and flexible start and finish times (Department of Labour 2006).

Most of the research on work-life balance focuses on alternative work arrangements with far less research on wellness, care-giving support, and crisis assistance. In New Zealand, flexible work hours and location are the work-life benefits most often offered, followed by work and life information, while child/elderly care is offered significantly less often (Kossek et al. 2010). Findings from a 2008 survey of New Zealand employers showed an increase in the uptake and awareness of work-life benefits overall since the previous 2006 survey (Equal Employment Opportunities Trust 2008). However, there are perceived barriers to implementation of work-life benefits in New Zealand, the largest being 'needing everyone in the workplace at the same time' and being 'too complicated' (Department of Labour 2006, p. 36).

The results from a survey of New Zealand companies showed that benefits were able to be implemented over a diverse range of businesses, that organisational culture was important for their success, and that communication from managers was important for employee awareness and use of the benefits. From the organisation's perspective, the policies had to be economically sustainable and were important for retaining dedicated employees (New Zealand Business Council for Sustainable Development 2006). These findings reinforce those of other work-life balance studies which suggest that there is no one-size-fits-all policy that works across all business (Yasbek 2004; Lingard & Francis 2005).

Work-life balance in the Construction Industry

Interestingly, much of the work-life balance research has occurred within the Australian construction industry, due to the substantial amount of works from the same authors (see Lingard & Sublet 2002; Francis 2004; Lingard & Francis 2004a, 2004b, 2005, 2009; Lingard 2008; Lingard et al. 2008; Mackenzie 2008; Brown et al. 2009; Lingard, Francis & Turner 2010a, 2010b).

A survey of employees in professional and managerial roles within the Australian construction industry indicated that workplaces were not perceived to be supportive of the need to balance work with life outside of work, and it is suggested that work-life balance policies are virtually ineffective if the organisational culture does not strongly value such initiatives (Francis 2004). Furthermore, Townsend et al. (2006) found that employees think work-life balance problems do exist in the construction industry. They suggested that reform should happen in the following areas: government and/or organisation regulation of working hours; formal work-life balance policies and informal support by supervisors; and employee control over working time.

Turner, Lingard and Francis (2009) identified workplace culture, project resourcing, and schedule demands as barriers to achieving work-life balance in the construction industry, and identified collaborative project alliance contracts, flexible working hours, and project management support as supports.

A respondent in research carried out in Australia with a focus on wellbeing in commercial construction appositely summarises the construction industry working environment with regard to work-life imbalance: 'people in the industry...are collectively weary and generally sad about the personal cost they endure as a consequence of their work... ongoing pressure that has a detrimental impact on physical and mental wellbeing, are taking a serious toll...unrealistic programming appears to be the root cause of work/life conflict' (MacKenzie 2008, p. 53).

Judging by the Australian research, there is a significant amount of work-life imbalance in the construction industry. However, extensive research on work-life balance in the construction industry in New Zealand does not exist. As Lingard and Francis (2005) asserted, knowledge about employees' preferences for work-life benefits can aid in the development of effective work-life balance initiatives, and this research aimed to investigate this, so that the needs of employees in the New Zealand construction industry are better understood.

Research Objectives

This research aimed to identify employee preferences for work-life benefits within a large New Zealand construction company, and compare the results with those of a similar study undertaken within two large Australian construction companies (Lingard & Francis 2005). Therefore, the research design was based on that of Lingard and Francis' (2005) study, and was carried out in a broadly similar manner in order to ensure that the data was comparable. There is a dearth of research regarding which types of work-life benefits employees in the New Zealand construction industry prefer, and knowledge about employees' preferences may help people within the construction industry better understand the wants and needs of employees.

Research Context: Lingard and Francis' (2005) Research

Lingard and Francis (2005) collected data from professional and managerial employees of two different organisations (a large private civil engineering and building construction company, and a public sector organisation involved in large infrastructure construction) to measure employees' preferences for work-life benefits. Their results were compared according to gender, age, and family structure; they concluded that 'distinct sub-populations of employees differ in terms of their preferences for work-life benefits. In particular, employees' preferences differ according to their age and stage of family development' (Lingard & Francis 2005, p.1055).

Research Methods

Data Collection

Due to constraints regarding accessibility and permission, an opportunity sample of employees within one large New Zealand construction company was used. The company is relatively large, and is involved across all construction industry sectors and regions of New Zealand. By using an opportunity sample from one construction company, generalisations may not be valid; the generalisability of the findings to other large New Zealand construction companies would depend on how similar the companies are. However, collecting data from one company was considered appropriate, as this has advantages, for example 'that context variables such as organisational culture, policies and procedures are controlled' (Lingard & Francis 2004b, p. 21).

All the company's professional and managerial staff in the Auckland region were emailed with the questionnaire instrument. The email included a description of, and reason for, the survey. The first page of the survey contained a consent clause to which the respondent was required to agree before carrying out the survey.

Questionnaire Design

Section 1 was made up of closed questions which requested demographic information. No questions were mandatory; that is, there were no forced responses. Section 2 comprised twenty closed questions where respondents were asked to rate their preference for a variety of work-life benefits using a 5 point Likert scale where: 1=not at all useful, 2=of limited use, 3=moderately useful, 4=useful, 5=very useful. Based on Lingard and Francis' (2005) study, the work-life benefit items were drawn from a wide variety of sources (Biggs 1998; Scheibi & Dex 1998; Moen et al. 1999; Grandey 2001; Hill et al. 2001; Saltzein et al. 2001; Thornwaite

2002, all as cited in Lingard & Francis 2005). As with Lingard and Francis' (2005) study, participants were asked to ignore the likelihood of future usage when indicating their preferences. Two open questions were included at the end of the questionnaire, where employees could provide additional feedback and suggest additional work-life balance initiatives.

To avoid the need for additional ethics approvals, not all of the demographic questions from Lingard and Francis' (2005) study were used; to avoid asking respondents to reveal information that was considered to be too personal, respondents were not asked to reveal family status and birthplace. Demographic information collected from respondents included: age; gender; number of years worked in the construction industry; type of employment (full, part time or contract); job role, and work location (site-based or head office).

Data Analysis

Employees' perceptions of the utility of each of the twenty work-life benefits rated in Section 2 were analysed by determining their mean rating scores and standard deviations. Mean rating scores were also calculated for the four different categories of work-life benefit, based on the factors established by Lingard and Francis (2005) – 'childcare support', 'alternative work arrangements', 'crisis assistance/support', and 'wellness and personal development'. The mean scores of the four factors, firstly overall, then by age and gender, were then compared with Lingard and Francis' (2005) results. Mean scores were also analysed by work location (i.e. head office and site-based), which was not done in Lingard and Francis' (2005) study. A content analysis was carried out on the qualitative data; the responses to the open questions were coded by topic, and were considered in the discussion to assist in drawing inferences regarding the results from section 2 of the questionnaire.

Findings

Demographic Information

Of a possible 174 respondents, 121 complete and usable questionnaires were returned (a 70% response rate). Table 1 shows the demographic characteristics of the sample.

Age	n	%	**Type of employment**	n	%
Under 30 years	22	19	Full-time work	104	92
30-39 years	30	25	Part-time work	2	2
40-49 years	31	26	Contract work	7	6
50-59 years	26	22			
60 years +	10	8	**Work location**		
			On site	85	71
Gender			Head office	35	29
Male	108	90			
Female	12	10	**Job description**		
			Project managers*	40	33
Years worked in construction			Foremen, site managers	8	7
0-9	35	30	Site administration	8	6
10-19	27	23	Commercial managers**	41	34
20-29	28	23	Engineering services	7	6
30 years +	29	24	Upper management	6	5
			Other	11	9

Table 1 Demographic characteristics of sample (n=121)

* includes project, construction, site, and package managers and project leaders

** includes commercial managers, quantity surveyors and estimators

The mean age of the sample was undeterminable as responses were only given according to age bracket. The modal age was 40-49 years. Of the 120 responses to the question on gender, 108 (90%) were male while only 12 (10%) were female. Only 17% of females were over 50 years of age, whereas 32% of males were over 50 years of age, a gender distribution which indicated a younger cohort of females in general, similar to the gender/age profile in Lingard and Francis' (2005) study. There was also a similar gender mix to that of Lingard and Francis' (2005) study (which had 88% males and 12% females). The low proportion, yet younger cohort of females suggests that these large construction companies are still largely male dominated, but that more females may be starting to enter the industry.

Considering the number of years worked in construction, a relatively even distribution of respondents existed. Of the 113 responses regarding type of employment, the vast majority (104 respondents) had full-time working arrangements. Site-based employees made up 71% of respondents, while 29% were based off-site (in the head office). The largest groups of employees were quantity surveyors, estimators, and commercial managers (34% of respondents), and project managers, construction managers, site managers, package managers, and project leaders (33% of respondents).

Work-life benefit Preference Mean Scores

The mean preference scores and standard deviations of each work-life benefit (based on respondents ratings of how useful they thought each benefit would be for them personally) are shown in Table 2. The 4 factor groupings based on those identified by Lingard and Francis (2005) are also indicated.

The five benefits perceived to be the most useful overall had the smallest standard deviations, which indicated that they were not only the most preferred, but also the most agreed upon by employees. Four of the five lowest scoring benefits had relatively large standard deviations, which indicated that there was less agreement regarding employee preference for these benefits; in other words they possessed a greater variation in perceived utility.

That the highest overall mean score was for 'allowing for flexible work e.g. starting and finishing half an hour earlier or later' is not surprising, as flexible working arrangements relating to hours have been reported as by far the most desired (for 91% of New Zealanders) work-life benefit (Department of Labour 2006; McPherson & Reed 2007). The qualitative data reinforces this, as a relatively large number of comments regarding flexible working options were made by respondents. There is evidence to suggest that providing such work-life benefits could be beneficial on many levels. Having more time-flexible working options has been linked to employee loyalty, fewer missed deadlines, and decreased stress and time off work (Halpern 2005). Furthermore, flexible workplaces have higher rates of employee job engagement, job satisfaction, intention to stay with the company and mental wellbeing (Families and Work Institute 2008).

The high mean scores for 'allowing for special family leave e.g. to care for a sick dependent' (4th highest mean score), and 'offering temporary part time work options during a family crisis' (5th highest) lend support to other findings from New Zealand which describe a demand for a wider scope of flexible work options including 'career breaks, extended maternity and paternity leave, paid dependency leave, job share, subsidised health-care and leave for community, sports and volunteer work' (New Zealand Business Council for Sustainable Development 2006, p.3). One respondent from the present study recognised some of the adverse effects of a successful career in the industry, and pointed out that the changing workforce and subsequent provision of more flexible working arrangements may have positive effects for all employees:

'The key issue in construction is that our personalities are generally "goal driven". To achieve success... requires intense focus. This tends to fill our industry with [people] who leave behind them a trail of broken relationships whether with spouse or children or family. ...the industry has been dominated by men over most of [my career] and the "absent" workaholic father has been extremely common. As more women have entered the workforce, child care issues have become more prevalent. These may force more flexible working habits that are useful for all employees.'

Factor	Work-life benefit	Mean value	Standard deviation
W	Allowing for flexible work hours e.g. starting and finishing half an hour earlier or later	4.23	1.00
W	Reimbursing the costs of work related courses and further study	4.13	1.04
W	Providing a wellness programme i.e. health checks, guest speakers, fitness challenges	4.05	0.87
Cr	Allowing for special family leave e.g. to care for a sick dependent	4.02	1.03
A	Offering temporary part time work options during a family crisis	3.91	1.04
Ch	Providing scholarships for employees' children	3.85	1.41
W	Offering a fitness programme e.g. discounted gym memberships	3.6	1.14
Cr	Providing an employee assistance programme for employees with family problems	3.5	1.20
Cr	Ensuring people take their annual leave regularly	3.36	1.13
Cr	Providing a legal information service	3.23	1.07
Ch	Providing child care facilities	3.12	1.47
Ch	Offering extended parental leave	3.1	1.49
Ch	Providing care for children during school holidays	3.08	1.48
Ch	Providing assistance with child care costs	3.08	1.54
Ch	Offering care on short notice for a child or other dependents	3.08	1.39
A	Increasing flexibility in work location e.g. working from home/telecommuting	3.07	1.39
Cr	Providing an information and referral service to assist with care of dependent children	2.88	1.35
Cr	Providing an information and referral service to assist with care of elderly parents	2.61	1.36
A	Offering part time work options e.g. approximately 25hrs/week	2.59	1.35
A	Introducing job sharing where one job is split between 2 people working fewer hours	2.13	1.15

Table 2 Work-life benefit preference scores (with factors indicated)

Key:

W	= Wellness and personal development
Cr	= Crisis assistance / support
A	= Alternative work arrangements
Ch	= Childcare support

Considering the wider scope of flexible work options, it might then be expected that the following benefits would have received higher ratings as well: 'offering extended parental leave' (12th highest mean score), 'increasing flexibility in work location e.g. working from home/telecommuting' (16th), 'offering part time work options e.g. approximately 25hrs/week' (19th), and 'introducing job sharing' (lowest mean score). There are likely to be different reasons for these unexpectedly low scores, although it is impossible to decipher the reasons for these due to the nature of survey. The fact that 90% of respondents were men may have influenced the scores for 'offering extended parental leave', 'offering part time work options', and 'introducing job sharing', since traditionally and typically it tends to be women who reduce their working hours to care for dependents, by, for example, 'taking career breaks or seeking part-time employment while children are young' (Lingard & Francis 2009, p.129). Hence a large proportion of men may not find these benefits very useful.

The second highest mean score of the benefit 'reimbursing the costs of work related courses and further study' is not unexpected, as the provision of this benefit is highly valuable. Company funding of work-related courses and continuing education is likely to be useful to a wide-ranging, large number of employees, as the high mean rating score and low standard deviation indicate. Similarly the third highest mean score for 'providing a wellness programme i.e. health checks, guest speakers, fitness challenges' is also likely to appeal to a large number of employees from various demographic groups. In support of the quantitative data, there were a relatively large number of qualitative responses (9 out of 41) expressing interest in health and fitness initiatives. As one respondent said: 'fitness incentives can create good team bonding and are a very worthwhile method of making people aware of what they are missing out on and also help a greater work / life balance.'

Work-life benefit Factors

Table 2 earlier showed the mean preference scores for the 20 benefits, together with the 4 factor groupings. The benefits were grouped into factors in order to gain insight into whether certain categories of benefits might be more useful to employees overall, and by different demographic groupings. The factors as defined by Lingard and Francis' (2005) study were relied upon for the research. Table 3 provides a direct comparison of the mean values and standard deviations for the four factors.

Factor	Mean value		Standard deviation	
	L&F (2005)	Current (2011)	L&F (2005)	Current (2011)
Wellness and personal development	3.39	4.00	0.92	1.04
Crisis assistance/ support	2.51	3.44	0.93	1.23
Alternative work arrangements	2.41	3.17	1.06	1.44
Childcare support	1.95	3.17	1.12	1.47

Table 3 Comparison of work-life benefit factor preferences

(after Lingard and Francis 2005)

'Wellness and personal development' rated as the most useful overall factor, with the lowest standard deviation, followed by 'crisis assistance/support', then 'childcare support' and 'alternative work arrangements' (which were both jointly least preferred factors). It is important to note that, although these were least preferred, the value of the mean rating score still indicates that respondents perceived those benefits to be slightly more than 'moderately useful'. 'Wellness and personal development' having the highest mean score is not surprising; benefits that fall into this category are likely to be useful to a wide range of employees, and their family structure, age, or gender may not significantly limit how useful they are. There were many qualitative comments supporting the usefulness of benefits in this category.

As shown in Table 3 above, the results regarding the factors' relative importance and variation in standard deviations are strikingly similar to Lingard and Francis' (2005) results. Interestingly, judging from the mean values, all of the factors showed a slightly lower score in Lingard and Francis' (2005) Australian study, which, on face value, suggests that the Australian respondents found all of the factors slightly less useful than the respondents in the current (2011) New Zealand study. This difference may be real, but also may be due to other factors, such as cultural differences in terms of interpreting or expressing preference using the rating scale, as well as perhaps that employees' awareness in general of work-life benefits has been raised in the interim period between Lingard and Francis' (2005) study and the current (2011) study. Notably, the standard deviations in both studies got larger as the mean values decreased. This indicates that, in both studies, 'wellness and personal development' was not only rated as most useful, but respondents were also in the highest amount of agreement regarding the utility of this factor compared to the others.

Comparisons by Age

A comparison of both studies' mean values of respondent preferences for work-life benefit factors by age is presented in Table 4.

	Mean value									
	<30 years		30-39 years		40-49 years		50-59 years		60+ years	
	L&F 2005	Current 2011	L&F 2005	Current 2011	L&F 2005	Current 2011	L&F 2005	Current 2011	L&F 2005	Current 2011
W	3.68	4.43	3.49	4.03	3.37	4.05	3.09	3.67	2.82	3.75
Cr	2.44	3.2	2.39	3.7	2.55	3.45	2.82	3.5	2.05	3.1
A	2.49	3.34	2.51	3.46	2.06	3.02	2.57	2.98	1.83	2.95
Ch	1.76	2.72	2.16	3.85	2.22	3.44	1.62	2.79	1.07	2.34

Table 4 Comparison of work-life benefit preferences for different age cohorts

(after Lingard and Francis 2005)

Key:

W	= Wellness and personal development
Cr	= Crisis assistance / support
A	= Alternative work arrangements
Ch	= Childcare support

'Wellness and personal development' was by far the most preferred factor for all age cohorts in both studies. Although relative importance of the 'wellness and personal development' factor among age cohorts is not identical between the two studies, they are very similar. Employees in the 'under 30 years' age group expressed the strongest preference for the factor 'wellness and personal development' in both studies. Age cohorts 30-39 and 40-49 preferred the factor slightly less, and the age groups 50-59 and 60+ preferred the factor less than those younger than 50 years of age. This indicates a general diminishment of importance for this factor with age. In both studies, the 60+ age cohort expressed the least interest in 'alternative work arrangements', 'crisis assistance support', and 'childcare support'. The 30-39 year age cohort expressed the greatest interest in these three factors, but this was not the case in Lingard and Francis' (2005) study (although compared with the other age cohorts, the 30-39 age group did show a relatively strong preference for these factors in their study).

That the order of preference for various benefits varied by age group indicates that the utility of different benefits varied with age. For example, 'childcare support' was found to be the least preferred factor for the 50-59 year and 60+ age brackets, whereas it was found to be the second most useful factor for the 30-39 year age cohort, behind 'wellness and personal development'. This was to be expected, as respondents were asked to rate the benefits according to how useful they thought that they would be to them personally, and respondents over 50 years of age are likely to have far fewer needs for childcare support than the 30-39 year age group.

Comparisons by Gender

A comparison of both studies' mean values of respondent preferences for work-life benefit factors by gender is presented in Table 5.

Work-life benefit Factor	Mean value			
	Female		Male	
	L&F 2005	Current 2011	L&F 2005	Current 2011
Childcare support	1.74	3.71	1.58	3.09
Alternative work arrangements	2.89	3.95	2.33	3.08
Crisis assistance/ support	2.4	3.82	2.51	3.39
Wellness and personal development	3.84	4.29	3.84	3.97

Table 5 Comparison of preferences for work-life benefit by gender
(after Lingard and Francis 2005)

In both studies, both genders expressed their highest preference for the factor 'wellness and personal development'. Females expressed a higher preference for all factors, whereas this was not the case in Lingard and Francis' (2005) study. This difference may be due to a different make-up of respondents' family structures between the two studies (although this cannot be explored as this research did not request information regarding respondents' family structure, due to ethics approval constraints). Employees with partners and dependent children in Lingard and Francis (2005) preferred the childcare support benefits the most, whereas those without dependent children preferred them the least. This may explain why females in the current research rated 'childcare support' the lowest; as this female cohort were generally younger than the males, they are less likely to have children. The relative importance of factors was the same in each study. This once again shows that the perceived usefulness of different types of benefits varied amongst the different demographic groups.

Comparisons by Work location

Table 6 shows the mean preference scores for each of the work-life benefit factors by work location. No comparison can be made with Lingard and Francis' (2005) study, as they did not analyse their results by work location.

Factor	Mean value	
	Site based (n=86)	Head office (n=35)
Wellness and personal development	3.97	4.08
Crisis assistance/ support	3.46	3.38
Alternative work arrangements	3.16	3.19
Childcare support	3.14	3.18

Table 6 Work-life benefit factor preferences mean scores for different work locations

The relative importance of the 4 factors was identical for site (n=86) and head office (n=35) based employees. Although the head office employees' scores were slightly higher (i.e. they found the factors slightly more useful) than those of site-based employees, this finding suggests that the benefits are apparently valued equally by employees in both work locations, and that differences in preference for the various benefits does not stem from the type of working environment, but may be more associated with respondents' demographic characteristics such as age, gender, and family structure. Since site-based employees tended to work longer hours than head office employees, it might have been expected that site-based employees might show a stronger preference for work-life benefits, but the results did not support such a presumption.

Open Questions

Answers to one or both of the optional open ended questions were provided by 41 out of 121 respondents. Having more social events including sport, recreation, and entertainment (with and without family), was suggested by twelve respondents. More fitness initiatives such as weight loss programmes, sponsored events, and incentives were suggested by nine respondents. The third most discussed topic (mentioned by seven people) regarding work-life balance initiatives was the need for reduced working hours and/or reduced weekend work. Flexible work options such as flexible working hours were considered desirable by six respondents.

Work-life benefits – Implications of Findings

The quantitative results indicate that employee preferences for different types of work-life benefits varied among different demographic groups. These findings are in agreement with a conclusion from Lingard and Francis' (2005) study which stated that 'there is no "one size fits all" solution to work-life balance in the construction industry' (p. 1053). Although the provision of a multitude of work-life benefits is desirable, initiatives can be expensive for companies to implement. Because of this, organisations tend to offer different work-life benefits to reflect the particular needs of employees (Bardoel et al. 1999).

Qualitative responses showed employee interest in a wide variety of different work-life benefits, the most prevalent being: social events; fitness and health initiatives; reduced working hours/weekend work, and flexible work options. As suggested by Townsend et al. (2006) 'employee voice presents as an important resource for managers and policy-makers in determining work-life balance policies for industry, workplaces and individual employees' (p. 2). Information retrieved from employees can provide valuable insight for companies who wish to understand the needs and preferences of their workforce.

It is important to note that the provision of work-life benefits alone is not effective without other supports in place. A large body of research (Eaton 2003; De Cieri et al. 2005; Dallimore & Mickel 2006; Smith & Gardner 2007; Kossek et al. 2010) shows that if an organisation wants to provide effective and usable work-life benefits, it is extremely important to ensure that the organisational culture, managers and supervisors show support for such initiatives. Employee perception of organisational support for an individual's work-life balance has been related to increased job satisfaction and reduced work pressure, both of which reduce leaving intention (Forsyth & Polzer-Debruyne 2007).

Conclusions

Many of the findings are consistent with those of Lingard and Francis (2005), the most significant being that the strongest preference by work-life benefit factor, both overall and within the various demographic groups, was for 'wellness and personal development'. It is likely that this is the case because these work-life benefits are appropriate for a wide variety of demographic groups. Furthermore, as with Lingard and Francis (2005) findings, employees' preferences for work-life benefits vary among different demographic groups. Consistent with other studies, these findings support the notion that there is no 'one-size-fits-all' policy appropriate for all companies or group of employees, and that the provision of a wide variety of work-life benefits from which employees can choose during different stages in their life and career is ideal.

The results suggest that there is some work-life conflict associated with working long hours and weekend work. It is important that the New Zealand construction industry continues to address these issues in order to find a balance that is suitable to employees, companies, and the industry as a whole, so as to provide and maintain sustainable working environments.

As with Lingard and Francis' (2005) study, this research relied upon respondents' subjective reports of their preferences for different work-life benefits. It did not attempt to ascertain whether or not such benefits would actually be used, and, if so, what impact their use would have. Further research could address whether or not employees feel the organisational culture, supervisors and managers support the work-life benefits provided by the company, and investigate whether employees are aware of all available benefits, and how they are used. Furthermore, the study did not explore employee perceptions of their own work-life balance or conflict. Future research could investigate the prevalence of work-life balance or conflict according to job role and location.

References

Bardoel, E.A., Moss, S.A., Smyrnios, K. and Tharenou, P. (1999) 'Employee characteristics associated with the provision of work-family policies and programs', *International Journal of Manpower,* **20**, 563-576

Becker, B. and Huselid, M. (1998) 'High performance work systems and firm performance: a synthesis of research and managerial implications', *Research in Personnel and Human Resources Management,* **16**, 53-101

Brough, P., Holt, J., Bauld, R., Biggs, A. and Ryan, C. (2008) 'The ability of work-life balance policies to influence key social/organisational issues', *Asia Pacific Journal of Human Resources,* **46**, 261-274

Brown, K.A., Ling, S., Bradley, L.M., Lingard, H. and Townsend, K. J. (2009) 'What about me? Avoiding fatigue and gaining personal time in the work to leisure transition in work-life balance initiatives', *In proceedings of the 23rd Annual Australia and New Zealand Academy of Management Conference (ANZAM 2009)*, 1-4 December, Melbourne

Casper, W.J. and Buffardi, L.C. (2004) 'Work-life benefits and job pursuit intentions: the role of anticipated organizational support', *Journal of Vocational Behavior,* **65**, 391-410

Dallimore, E. and Mickel, A. (2006) 'Quality of life: obstacles, advice, and employer assistance', *Human Relations,* **59** (1), 61-103

De Bruin, A. and Dupuis, A. (2004) 'Work-life balance?: insights from non-standard work', *New Zealand Journal of Employment Relations*, **29** (1), 21-37

De Cieri, H., Holmes, B., Abbott, J. and Pettit, T. (2005) 'Achievements and challenges for work/life balance strategies in Australian organizations', *International Journal of Human Resource Management,* **16** (1), 90-103

Department of Labour (2006) *Work-life balance in New Zealand: A snapshot of employee and employer attitudes and experiences,* viewed 20th May 2011 http://www.dol.govt.nz/worklife/research/work-life-balance.asp

Eaton, S.C. (2003) 'If you can use them: flexibility policies, organizational commitment and perceived performance', *Industrial Relations,* **42**, 145-167

Eby, L.T., Casper, W.J., Lockwood, A., Bordeaux, C. and Brinley, A. (2005) 'Work and family research in IO/OB: content analysis and review of the literature (1980–2002)', *Journal of Vocational Behavior,* **66**, 124–197

Equal Employment Opportunities Trust (2008) *EEO Trust work-life survey,* viewed 13th May 2011 http://www.eeotrust.org.nz/research/worklife.cfm

Families and Work Institute (2008) *2008 National study of the changing workforce,* viewed 18th March 2011 www.familiesandwork.org/site/work/workforce/2008nscw.html

Forsyth, S. and Polzer-Debruyne, A. (2007) 'The organisational pay-offs for perceived work-life balance support', *Asia Pacific Journal of Human Resources,* **45**, 113-123

Francis, V. (2004) 'Supportive organisational cultures and their effect on male engineers', *The Australian Journal of Construction Economics and Building,* **4**, 1-9

Frone, M.R. (2003) 'Work-family balance', in Quick, J.C. and Tetrick, L.E. (eds), *Handbook of Occupational Health Psychology,* American Psychological Association, Washington, D.C., 143-162

Halpern, D. F. (2005) 'How time-flexible work policies can reduce stress, improve health, and save money', *Stress & Health: Journal of the International Society for the Investigation of Stress,* **21**, 157-168

Kossek, E.E., Lewis, S. and Hammer, L.B. (2010) 'Work-life initiatives and organizational change: overcoming mixed messages to move from the margin to the mainstream', *Human Relations,* **63** (1), 3-19

Lewis, S., Gambles, R. and Rapoport, R. (2007) 'The constraints of a 'work-life balance' approach: an international perspective', *International Journal of Human Resource Management,* ***18***, 360-373

Lingard, H. (2008) 'An exploration of the adaptive strategies of working families in the Australian construction industry', *Engineering Construction & Architectural Management,* **15**, 562-579

Lingard, H. and Francis, V. (2004a) *A Quantitative Study of Work-life Experiences in the Public and Private Sectors of the Australian Construction Industry,* Queensland University of Technology, Brisbane

Lingard, H. and Francis, V. (2004b) 'The work-life experiences of office and site-based employees in the Australian construction industry', *Construction Management & Economics,* **22**, 991-1002

Lingard, H. and Francis, V. (2005) 'The decline of the 'traditional' family: work-life benefits as a means of promoting a diverse workforce in the construction industry of Australia', *Construction Management & Economics,* **23**, 1045-1057

Lingard, H. and Francis, V. (2009) *Managing Work-Life Balance in Construction,* Spon, London

Lingard, H., Francis, V. and Turner, M. (2010a) 'Work-family enrichment in the Australian construction industry: implications for job design', *Construction Management & Economics,* **28**, 467 –480

Lingard, H., Francis, V. and Turner, M. (2010b) 'The rhythms of project life: a longitudinal analysis of work hours and work-life experiences in construction', *Construction Management & Economics,* **28**, 1085 –1098

Lingard, H. And Sublet, A. (2002) 'The impact of job and organizational demands on marital or relationship satisfaction and conflict among Australian civil engineers', *Construction Management & Economics,* **20**, 507 –521

Lingard, H. C., Townsend, K., Bradley, L. and Brown, K. (2008) 'Alternative work schedule interventions in the Australian construction industry: a comparative case study analysis', *Construction Management & Economics,* **26**, 1101-1112

Loosemore, M., Dainty, A. and Lingard, H. (2003) *Human Resource Management in Construction Projects,* Spon, London

MacKenzie, S. (2008) *A Close Look at Work and Life Balance/Wellbeing in the Victorian Construction Sector*, Report for the Building Industry Consultative Council, Reinventing the Image Steering Committee, Equilibrium Work-life Solutions, Melbourne

McPherson, M. and Reed, P. (2007) 'Work-life balance in the New Zealand context', in Waring, M. and Fouche, C. (eds.) *Managing Mayhem: Work-life Balance in New Zealand,* Dunmore Publishing, Wellington

New Zealand Business Council for Sustainable Development (2006) *Work-life Balance Report: How Member Companies are Addressing Work-life Balance,* Viewed 17th December 2010 http://www.dol.govt.nz/worklife/research/work-life-balance.asp

Noon, M. and Blyton, P. (2007) *The Realities of Work,* Palgrave, Basingstoke

Smith, J. and Gardner, D. (2007) 'Factors affecting employee use of work-life balance initiatives', *New Zealand Journal of Psychology,* **36** (1), 3-12

Thornthwaite, L. (2004) 'Working time and work-family balance: a review of employees' preferences', *Asia Pacific Journal of Human Resources,* **42** (2), 166-184

Townsend, K., Bailey, C., Brown, K., Bradley, L. and Lingard, H. (2006) 'How would employees 'construct' a balance between work and non-work life? Work-life balance in the construction industry', *in Procs 14th International Employment Relations Association Conference*, Hong Kong. Viewed 7th May 2011 http://eprints.qut.edu.au/9137/

Turner, M., Lingard, H. and Francis, V. (2009) 'Work-life balance: an exploratory study of supports and barriers in a construction project', *International Journal of Managing Projects in Business* **2** (1), 94 - 111

Yasbek, P. (2004). *The Business Case For Firm-level Work-life Balance Policies: a Review of the Literature,* Report for the Department of Labour. Viewed 17th November 2010 http://www.dol.govt.nz/PDFs/FirmLevelWLB.pdf

Improvement of Job Satisfaction and Organisational Commitment through Work Group Identification: An Examination of the Quantity Surveyors in Hong Kong

13

Wai Yee Betty Chiu and Fung Fai Ng, (The University of Hong Kong, Hong Kong)

Abstract

Though extant literatures in other sectors point to the fact that job satisfaction and organisational commitment are important for determining individual and organisational outcomes, limited related research has been conducted amongst Hong Kong quantity surveyors. Given cooperative working arrangement in the profession, work group identification is regarded as an important antecedent for determining job satisfaction and organisational commitment. The aim of this study is to examine whether work group identification improves job satisfaction and organisational commitment. A questionnaire survey is conducted to collect data from quantity surveyors working in private sector. A total of 71 valid responses are obtained from 509 contacted quantity surveyors in Hong Kong and used for the study. Bivariate correlation and multiple regression analyses are performed for indicating the significance of relationships amongst the variables. Data analysis results support most hypotheses. Work group identification is found to have significant positive influence on job satisfaction, affective and normative commitment. The finding is a step for surveying companies towards improving quantity surveyor's job satisfaction and commitment level. The roles of other contextual and organisational factors on job satisfaction and organisational commitment need to be subject to future research.

Keywords: Work group identification, Job satisfaction, Organisational commitment, Social exchange theory, Quantity surveyors

Introduction

Job satisfaction and organisational commitment have drawn great attention from scholars and researchers in different disciplines for several decades (e.g. Lok and Crawford, 2001; Meyer and Allen, 1997; Poon et al. 2007; Robbins, 2005; Spector, 1997). The two constructs are also an area of concern for most organisations, primarily for the reason that they significantly affect major organisational outcomes such as staff turnover intentions, absenteeism, organisational citizenship behaviour, and effectiveness etc. (e.g. Cheng and Francesco, 2000; Laschinger, 2001; Lu et al. 2005; Martin and Roodt, 2008; Moorman et al. 1993; Rafferty and Griffin, 2009; Sagie, 1998).

Job satisfaction is commonly regarded as a constellation of employee's feelings about various job elements. Organisational commitment, is about individual's attachment to the working organisation. The importance of job satisfaction and organisational commitment in facilitating the well-being of the construction industry has also been highlighted (Lingard and Francis, 2004; Loosemore et al. 2003). However, despite the wealth research into job satisfaction and organisational commitment, little research has been undertaken for the improvement of job satisfaction and organisational commitment in the quantity surveying context.

Social exchange theory (SET) is applied to provide conceptual underpinnings for the study. Sarantinos (2007) asserts that commitment is directly tied to the psychological contract, which describes the implicit agreement between employees and the organisation with regards to their reciprocal obligations and perceived expectations. Work group identification is about the identification with the work unit in which individuals conduct most of their day-to-day activities (Van Dick et al. 2008). Within quantity surveying consultant firms, a quantity surveyor usually has two identities: an employee of his working organisation and a member of his work group. Usually, a quantity surveyor is assigned into work group for handling construction projects and project tasks are divided among work group members. Generally, quantity surveyors would have better communication and information attainments among work group members, which also improves quantity surveyor's feeling to the job task and working environment. Because the work group is also part of the organisation, the employee who has a positive feeling about working experience in the work group may extend this belief to the organisation and develop a reciprocal behaviour that benefits the organisation. This research postulates that job satisfaction and organisational commitment are social exchange outcomes because the two constructs reflect a perception of the exchange quality (Van Knippenberg and Sleebos, 2006) of which the employee and the working organisation are required to fulfill their obligations to each other and establish ongoing reciprocity. Through this study, it is proposed that work group identification is likely to have a positive impact on job satisfaction and organisational commitment.

The purpose of this paper is to report on a study examining the role of work group identification in the development of job satisfaction and organisational commitment of quantity surveyors working in the Hong Kong construction industry. Quantity surveyors are one of the important players in the construction industry as they provide a wide range of professional services from project feasibility till completion (e.g. preliminary cost advice, cost planning, tender evaluation, cost monitoring and financial management) to the client for the decision making. The objectives of this study are (1) to investigate the perception work group identification of quantity surveyors and (2) to examine if work group identification enhances job satisfaction and organisational commitment level of quantity surveyors. This study hypothesizes that work group identification would be positively associated with two factors of job satisfaction (i.e. intrinsic and extrinsic job satisfaction) and three factors of organisational commitment (affective, normative and continuance commitment). The approach in this study is to test the association of work group identification with the two job satisfaction and three organisational commitment levels of quantity surveyors.

Literature Review

Social Exchange Theory (SET)

Social exchange theory (SET) is used to provide the theoretical basis for this study. According to Blau (1964, p. 91), social exchange is defined as '*voluntary actions of individuals that are motivated by the returns they are expected to bring and typically in fact bring from others.*' The theory suggests that when a person gives another person a reward, resource or other perceived commodity, there is an expectation of future return from the other party (Bernerth and Walker, 2009). In recent years, SET has received increased attention in organisational research context as it provides the conceptual underpinnings for understanding employee's workplace attitude (e.g. Setton et al. 1996; Tse and Dasborough, 2008; Wayne et al. 2002) and is able to provide explanations of employee's positive outcomes (e.g. Barlett, 2001; Kang and Stewart, 2007).

Within quantity surveying consultant firms, quantity surveyors in different position hierarchy are assigned to work group for handling different construction project of which project tasks are divided among group members. Through the cooperative and collaborative working practice, concurrent working relationship is built among work group members. Quantity surveyors in different position hierarchy are used to help each other within the work group for project tasks completion (goal attainment within the work group). Exchange relationships are built among work group members in undertaking project tasks to attain the goals. The content of exchange is not limited to the requirement of project task completion, but also extends to the exchange of valued resource and support that goes beyond what is required for task completion (e.g. sharing of knowledge / past experience learnt from other projects, information attainment). To the quantity surveyor, goal attainment within the work group creates a positive influence to be a member of the group. Hence, quantity surveyors are likely to gain or maintain their membership in the group.

Because a work group is also part of the organisation, the employee who has positive perception of the work group may extend this belief to the organisation and develop a reciprocal behaviour that benefits the organisation. Previous studies (e.g. Coble, 2002; Van Dick et al., 2004) have shown that employee's positive perception of being members of the work group has positive influence on individual and organisational outcomes such as job satisfaction and organisational commitment. Several studies (e.g. (Moreland and Levine, 2001; Van Knippenberg and Sleebos, 2006) have established that job satisfaction and organisational commitment as social exchange outcomes because the two constructs reflect a perception of the exchange quality of which the employee and the working organisation are required to fulfill their obligation to each other and establish ongoing reciprocity. Thus, it is considered that work group identification, job satisfaction and organisational commitment can hence be linked through the framework of social exchange.

Work Group Identification, Job Satisfaction and Organisational Commitment

Since 1990s, social and psychological researchers have expressed extensive interest in organisational identification as the construct has been shown to provide positive influence to various work outcomes such as increased job satisfaction, commitment and lowered turnover intention (e.g. Riketta and Van Dick, 2005; Tyler and Blader, 2001; Van Knippenberg et al., 2000). Ashworth and Mael (1989) suggest that as individuals often have multiple identities within work organisations, research should focus on identification with salient sub-groups such as work group, instead of simply examining organisational identification. It is because work group identification plays a crucial role in affecting employee's attitude and behaviour in organisations.

The concept of work group identification was initially defined by Tolman (1943) as '*a personal cognitive connection between an individual and the work group. It is the individual's perception of oneness with the group and the tendency to experience the group's successes and failures as one's own.*' Foote (1951), Mael and Tetrick (1992), Riordan and Weatherly (1999, p. 311) support this definition and declare that '*when an individual identifies with a group, that individual defines himself or herself by the same attributes (e.g. qualities, faults, common destinies) that define the work group.*' Riordan and Weatherly (1999) point out that there are three perceptions included in work group identification. Firstly, individuals within the group perceive themselves to be psychologically intertwined with the fate of the work group (Foote, 1951; Gould, 1975; Riordan and Weatherly, 1999). Secondly, when individuals identify themselves within the group, they can personally experience the successes and failures of the group (Foote, 1951; Tolman, 1943). Thirdly, when individuals identify themselves within the group, the individual's self-definition are strongly influenced by the work group attributes (Riordan and Weatherly, 1999). In this study, the Van Dick et al. (2008) definition about work group identification is employed, i.e. – 'work unit in which individuals conduct most of their day-to-day activities.' It provides a consistent and clear meaning about the term 'work group identification', which is easy for people to understand.

The concept of job satisfaction is initiated by Hoppock (1935) study, which is defined as '*any combination of psychological, physiological and environmental circumstances that causes an employee to be satisfied with his / her jobs.*' Broadly speaking, job satisfaction is regarded as a constellation of employee's attitudes about various job aspects (Lu et al. 2005). It is found that the facets included are generally related to the employee's feeling about the intrinsic and extrinsic job elements. In addition, several approaches are adopted by a number of studies (e.g. Cranny et al. 1992; Moorman, 1993; Weiss et al. 1967) to classify job satisfaction dimensions. In this research, the intrinsic-extrinsic dimension suggested by Weiss et al. (1967) is taken to classify job satisfaction. It is because both psychological and physical issues are included in the two dimensions for classifying employee's job satisfaction. According to Martin and Roodt (2008), intrinsic satisfaction is derived from performing the work and consequently experiencing the feelings of accomplishment, self-actualization and identity with the task. On the other hand, extrinsic satisfaction is derived from the rewards bestowed upon an individual by peers, supervisors or the organisation, and can take the form of recognition, compensation and advancement (Martin and Roodt, 2008). In other words, intrinsic job satisfaction is about how individuals feel about the nature of job task; whereas extrinsic job satisfaction is about how individuals feel about the aspects of work situation (Hirschfield, 2000). Generally, previous findings (e.g. Fellows et al. 2003, Rad and Yarmohammadian, 2006) have shown that the two dimensions of job satisfaction are significantly affected by work-related issues such as leadership style and power relations, this study tends to examine whether they are influenced by work group identification.

The concept of organisational commitment is initially explored by Porter et al. (1974) study, who noted that organisational commitment was characterized by '*...a strong belief in and acceptance of organisational goals and values; willingness to exert considerable effort on behalf of the organisation, and define desire to maintain organisational membership*' (Porter et al., 1974, p. 604). Generally speaking, organisational commitment is about individual's attachment to the working organisation. Researchers have variations in defining the concept because they look into the concept from three approaches. Some researchers (e.g. Becker, 1960; Farrell and Rusbult, 1981 and Sheldon, 1971) look into the commitment concept from a cost approach and regard it as a side bet i.e. non-transferable investments such as time, effort and money. To other researchers such as Kanter (1968), Mowday et al. (1982) and Porter et al. (1974), organisational commitment represents an individual's emotional response to the organisation, as well as its goals and values. On the other hand, researchers such as Wiener and Vardi (1980) focus on the obligation that an individual believes he has to stay in the organisation.

Based on these three approaches, Meyer and Allen (1991) categorize organisational commitment into three components: affective, normative and continuance commitment. Affective commitment is used to reflect the employee's emotional attachment, identification with and involvement in the organisation, which is also similar to Mowday et al. (1982) definition of attitudinal commitment. Normative commitment is about an employee's feeling of obligation to stay in the organisation. Continuance commitment is about the cost associated with the employee leaving the organisation. In this research, this three-component model is taken for studying organisational commitment as it is the leading multidimensional model used to study the concept in the past two decades. Moreover, the model has been widely adopted to account for multiple foci of workplace commitment (Meyer et al. 2004). Therefore, this study adapts the three-component model to further examine the ways in which it is influenced by work group identification.

Overview of research about Work Group Identification and Hypotheses Development

Although work group identification has been defined by scholars for more than 60 years, it has just generated interest from researchers in the past three decades. Since 1980s, researchers (e.g. Guzzo and Shea, 1992; Park et al., 2010; Van Knippenberg and Van Schie, 2000) have expressed the need to understand work group identification because it can provide positive influence on interpersonal relations and foster desirable outcomes to organisations. Empirical research conducted by a number of researchers (e.g. Ollkonen and Lipponen, 2006; Van Dick et al. 2008; Van Knippenberg and Van Schie, 2000) show that work group identification has strong influence on employee's job-related attitudes and behaviour towards the organisation such as job involvement, in-role and extra-role behaviour, organisational citizenship behaviour, turnover intention and absenteeism etc.

Regarding job satisfaction and organisational commitment, a few empirical studies show that there is a positive relationship between work group identification and the two constructs. Van Dick et al. (2004) studied the relationship of work group identification and job satisfaction in the education and banking sectors in Germany. Social identity and social categorisation theories are adopted to explain the relationship between work group identification and job satisfaction. In the study, it is observed that university teachers and bank accountants' perception of work group identification have positive influence on their job satisfaction level. In their meta-analysis of 40 studies on the relationship between employee's attachment to a broad range of work-related attitudes and behavior, Riketta and Van Dick (2005) report that work group identification seems to have a positive influence on an attitudinal construct like job satisfaction. However, the results are considered as preliminary as the number of analysed samples for conducting the meta-analysed comparison is small. In a study of the effects of tenure and work group on organisational commitment of employees working in a multi-line financial service organisation, Coble (2002) found that work group identification has positive influence on organisational commitment. However, the adoption of secondary data for examining the relationship between the construct restricts the ability of the researcher to test the validity of the measurement instrument. The validity of the constructs is thus questionable. Moreover, Coble (2002) report that the data are gathered for the purpose of another study of which only affective commitment is explored in terms of organisational commitment.

However, it is noticed that even though there are studies that look at the relationship between work group identification and, job satisfaction and organisational commitment, all of them are not conducted in Hong Kong or construction-related contexts. Moreover, SET has not been applied to explain for the relationships. Hence, it is postulated that the quantity surveyor's work group identification positively affects his job satisfaction and commitment level to the organisation. Generally, previous findings (e.g. Beck and Wilson, 2000; Bishop and Scott, 2000; Rafferty and Griffin, 2009; Sagie,1998) have also shown that organisations benefit from a workforce that has a high level of job satisfaction and organisational commitment as the employees tend to be absent less, make positive contributions and stay with the organisation. Thus, an understanding of the relationship between work group identification and, job satisfaction and organisational commitment is desirable. Relevant hypotheses and Sub-hypotheses are developed as follows:

- Hypothesis 1: Quantity surveyor's organisation-based work group identification positively affects his job satisfaction in an organisation.

  Hypothesis 1.1: Quantity surveyor's organisation-based work group identification positively affects his extrinsic job satisfaction level in an organisation.

  Hypothesis 1.2: Quantity surveyor's organisation-based work group identification positively affects his intrinsic job satisfaction level in an organisation.

- Hypothesis 2: Quantity surveyor's organisation-based work group identification positively affects his commitment to the organisation.

  Hypothesis 2.1: Quantity surveyor's organisation-based work group identification positively affects his affective commitment to the organisation.

  Hypothesis 2.2: Quantity surveyor's organisation-based work group identification positively affects his normative commitment to the organisation.

  Hypothesis 2.3: Quantity surveyor's organisation-based work group identification positively affects his continuance commitment to the organisation.

Research Context and Methodology

This paper reports a study on the role of work group identification, in the development of job satisfaction and organisational commitment of quantity surveyors working in the Hong Kong construction industry. While the research objectives are not limited to any specific region, this study focuses on the Hong Kong context. Hong Kong has a commercialized economy and its business environment is more similar to the West than to China (Wong, 2001). Construction industry in Hong Kong accounts for around 4% of the GDP and encompasses nearly 300,000 registered construction workers (Census and Statistics Department, 2008). The construction industry is a labour-intensive industry where different types of professionals are involved in, such as architects, engineers and surveyors. Since the working nature and practice of different construction professions is rather diverse, focusing on a single profession can eliminate research result variations that may be related to the type of profession (Swailes, 2004).

In Hong Kong, chartered quantity surveyors are one of the important players in the construction industry (Liu et al. 2007). Although the professional services provided by quantity surveyors have expanded from take off of quantities to project and contract management, which is getting more diverse, stressful and complicated in nature; the working duties, roles and job duties of chartered quantity surveyors are more standardized as compared with other construction professionals. In other professions, the variation / deviation in daily job duties will have an effect on the professional's job satisfaction and organisational commitment, which may vary the result of the hypothesized relationships. In order to minimize the variation that is caused by job-related issues in mapping out the relationships between the variables, the target population for this study is confined to chartered quantity surveyors. Whilst the studies focusing on job satisfaction and organisational commitment among quantity surveyors are scanty in Hong Kong, this is also an attempt to contribute to knowledge to fill the gap.

Data Collection Method and Survey Sample

A quantitative approach using questionnaire survey is employed in this study for data collection. The questionnaire is sent to the chartered quantity surveyors working in consultant companies in the construction industry in Hong Kong. "Chartered quantity surveyors" refers to those quantity surveyors who have obtained professional membership from the HKIS or the Royal Institution of Chartered Surveyors (RICS) (i.e. quantity surveyors who are full members in at least one of the professional institutions). Each quantity surveyor is asked to complete one self-administrated questionnaire. The contact information of the chartered quantity surveyors is sorted out from the Royal Institute of Chartered Surveyors (RICS) Member List book. Although a total of 884 chartered quantity surveyors are found in the member list book, only 509 are contacted with correct mailing address. Thus, the 509 contacted quantity surveyors become the target respondents for this study.

Questionnaire Design and Measures

The questionnaire is divided into three sections: (1) work group identification, (2) job satisfaction and (3) organisational commitment and (4) demographic information.

The work group identification measurement scale used in this study is developed by adaptation of statements in several existing scales. The reason for combining the existing measurement statements developed from various measurement scales is because different weaknesses (e.g. the scale put too much focus on affective aspects but ignore other aspects that constitute the identification, the scale is not originally used for measuring work group identification, the wordings are too ambiguous for people to understand etc.) are found in different existing measurement scales. The existing instruments involved in the new scale development are Ashforth and Mael (1989), Lipponen et al. (2003) and Van Dick et al.

(2004). The new measurement scale contains 10 statements, with three statements are modified statements and the rest of the statements are directly adopted from existing scales. The statements are used to measure the respondent's view about the cognitive connection between oneself and the work unit in which one carries out day-to-day activities. The respondents are asked to indicate their level of agreement with each statement on a seven-point Likert scale ranging from "*Strongly Disagree*" to "*Strongly Agree*".

The short form of the Weiss et al. (1967) Minnesota Satisfaction Questionnaire (MSQ) - MSQ20 is used in this study for the measurement of job satisfaction. The MSQ 20 originated from MSQ long version, which uses 100 items to assess 20 facets of job satisfaction (each facet contains 5 items). The item that loaded the highest on each facet in the long version is selected to represent the facet in the short form. Thus, the MSQ20 includes 20 facets of job satisfaction and are classified into two factors/dimensions: intrinsic satisfaction (IJS) and extrinsic job satisfaction (EJS) (Hopper, 2009). Intrinsic job satisfaction refers to the nature of job tasks and how people feel about the job task that they are assigned (characteristics associated with the task) (Weiss et al. 1967). On the other hand, extrinsic job satisfaction concerns other aspects that have little direct linkage with the job tasks (non-task characteristics of the job). In the instrument, those of intrinsic nature are – activity, independence, variety, social status, moral values, security, social service, authority, ability utilization, responsibility, creativity and achievement. Those of extrinsic nature are – supervision-human relations, supervision–technical, company policies and practices, compensation, advancement, working conditions, co-workers and recognition. The items use a seven-point Likert-type response scale ranging from "*Very Dissatisfied*" to "*Very Satisfied*" to indicate respondents' level of job satisfaction.

The Organisational Commitment Scale developed by Allen and Meyer (1990) is used in this research. The commitment scale consists of three dimensions, namely affective, continuance and normative commitment; of which each contains eight items. The affective commitment (AC) refers to the employee's psychological attachment, identification with and involvement in the organisation (Meyer and Allen, 1991). The continuance commitment (CC) refers to the investments (e.g. time, cost and effort) that an individual would be lost when leaving the organisation. The normative commitment (NC) is about the employee's feeling of obligation to remain with the organisation (Allen and Meyer, 2000). Seven items in the scale are written in negative wording to ensure respondents read the questions carefully before providing answers. The remaining ones are written in positive wording. All items employ a seven-point Likert-type response scale ranging from "*Strongly Disagree*" to "*Strongly Agree*" to indicate respondents' level of commitment to the organisation.

A demographic information section is included in the survey to collect some background information for the data analysis. The demographic information section contains ten questions. Both structured and filled-in questions are used. Seven questions (e.g. education qualification, age and gender etc.) are structured items of which multiple options are provided for respondents to select. Three filled-in questions (e.g. Years of working experience in the construction industry) ask the respondents to fill in appropriate answers.

Questionnaire Pretesting

Before the main survey is conducted, questionnaires were sent to 36 chartered quantity surveyors through personal contacts for pretesting. It was felt necessary to carrying out the pilot for two reasons. First, the measurement scales adopted were either developed from western scholars or specifically developed for the study. Thus, the pretesting was to ensure the questions set in the measurement scales were well-defined understood by the research target and, suitable to be applied in the Hong Kong construction context. Second, the reliability of the measurement scales would be established and any item in the measurement scales that needed to be modified or eliminated for the main study could also be identified.

In the 10 statements developed for measuring work group identification, one statement was eliminated after the pretesting due to its inconsistency with other items. The alpha coefficient of the revised work group identification scale is 0.851, indicating acceptable reliability. Regarding the MSQ20 used to measure job satisfaction, three items used to measure intrinsic job satisfaction were eliminated due to their low consistency with other items. The alpha coefficient of the revised scale is 0.901, indicating good reliability. On the other hand, the items used to measure extrinsic job satisfaction achieved a good reliability score with an alpha value of 0.920. Concerning the OCS scale that used to measure the three dimensions of organisational commitment, the pretesting result shows that a total of six items (two for each dimension) should be eliminated because of their low consistency with other items. The alpha coefficient of the revised scale is acceptable (0.708 for affective commitment, 0.790 for normative commitment and 0.849 for continuance commitment). Those revised scale are used in the main study questionnaire survey for data collection.

Statistical Methods for Data Analysis

To examine whether work group identification improves job satisfaction and organisational commitment, bivariate correlation and multiple regression (MR) analysis are performed. The bivariate correlation test is performed to identify and understand the strength, direction and significance of associations amongst the variables. For the regression analysis, the hierarchical multiple regression technique is adopted. MR has been extensively used as data analytical tools for indicating relationships among variables in social science research (e.g. Cohen et al. 2003; Wong, 2006 and Wu, 2009). In this study, two reasons account for the choice of MR. The first and most important reason is that the number of questionnaire responses obtained is able to fulfill the minimum sample number requirements proposed by different researchers (e.g. Hair et al. 1998; Howell, 2002 and Stevens, 1996) for a MR analysis. The second reason is that MR is a simpler technique than other statistical techniques and is able to serve the research objectives adequately. That is, the independent variable – work group identification – can be evaluated in terms of its predictive power, over and above that offered by all other variables that exist in the model for enhancing job satisfaction and organisational commitment (Ho, 2006). According to Best and Smith (2005) and Westen and Rosenthal (2003), a primary principle for the selection and use of statistics for data analysis is to use a simple and minimally sufficient statistics. The choice of MR in this study is accorded to this rule.

Main Study Results and Discussion

There are 71 returned questionnaires from a total of 509 contacted survey sample; giving a response rate of 13.95%. All returned questionnaires are acceptable for analysis as the number of missing data is small and affect a few questionnaires only. The demographic variables (tenure, age, gender and education level) are included in the correlation matrix as control variables to restrain the effects of possible respondent bias on ratings of other variables (Pallent, 2001; Yip, 2007) and to check for the problems of multicollinearity. The results of the bivariate correlation analysis are summarized in Table 1.

Table 1 shows that the bivariate correlations between the variables are less than 0.70, indicating that multicollinearity is unlikely to cause problems in formulating the multiple regression model for analysis. The correlation results show that at the 0.01 significance level, work group identification is significantly correlated with both extrinsic job satisfaction ($r = .484$), intrinsic job satisfaction ($r = .474$), affective commitment ($r = .404$) and normative commitment ($r = .458$). However, work group identification does not show significant correlation with continuance commitment. Accordingly, continuance commitment is not considered in the regression analysis.

VARIABLES	Mean	S.D.	TENURE	GENDER^2	AGE^2	EDCUATION^2	WGI^3	EJS^3	IJS^3	AC^3	NC^3	CC^3
TENURE	11.54	8.95	-									
GENDER^2	.14	.35	-.101	-								
AGE^2	.49	.50	.646**	-.237*	-							
EDCUATION^2	.51	.50	-.023	-.168	.183	-						
WGI^3	5.61	.89	.125	-.094	.058	.053	-					
EJS^3	4.95	1.02	.170*	.015	.102	-.009	.484**	-				
IJS^3	5.08	.88	.135	.034	.087	-.004	.474**	.000	-			
AC^3	4.24	1.33	.239*	-.050	.135	-.111	.404**	.333**	.170	-	.000	.000
NC^3	4.15	1.08	.038	-.128	.116	-.060	.458**	.396**	.289*	.000	-	.000
CC^3	4.46	1.11	-.050	-.056	-.040	-.060	-.197	-.050	-.290*	.000	.000	-
**. Correlation is significant at the 0.01 level (2-tailed).												
*. Correlation is significant at the 0.05 level (2-tailed).												

Table 1: Correlation1 Matrix between study variables

NOTES:

1 Correlations involving age, gender and education level are Spearman rank-ordered statistic, all others are Pearson product-moment statistics, pairwise, N = 71

2. Variable Coding is as follows: Gender is recorded as 0 = Male, 1 = Female. Age is recorded as 0 = less than or equal 40, 1 = more than 40. Education level is recorded as 0 = undergraduate or below, 1 = postgraduate. N = 71.

3. Except demographic variables, all other variables are scored on a 7-point Likert-type scale: 1 = strongly disagree / dissatisfied, 7 = strongly agree / satisfied. N = 71.

4. EJS = Extrinsic Job Satisfaction; IJS = Intrinsic Job Satisfaction; AC = Affective Commitment; NC = Normative Commitment; CC = Continuance Commitment

Table 2 shows that work group identification has significant positive influence on both extrinsic (beta = 0.488, p<0.001) and intrinsic (beta = 0.482, p<0.001) job satisfaction. This means that work group identification explained as much as 20.2% and 20.6% of the variance in extrinsic job satisfaction and intrinsic job satisfaction respectively. This also means that Hypothesis 1 has received full support (as sub-hypotheses 1.1 and 1.2 are supported) from the sample data collected in the questionnaire survey. This implies that work group identification is a predictor of job satisfaction. In essence, similar perceptions for the positive relationship between work group identification and job satisfaction are also found in some previous studies. For example, Van Dick et al. (2004, p. 353) point out that '*individuals who are strongly identified with their work group will perceive their working condition and situation positively, leading to higher job satisfaction.*' In the quantity surveying profession, cooperation and collaboration is a common practice among work group members, which imposes a positive feeling on the quantity surveyors that they are not working alone. Moreover, quantity surveyors in different position hierarchy are used to help each other within the work group for project task completion (goal attainment within the work group). Exchange relationships are built among work group members in undertaking project tasks. The content of exchange is not limited to the requirement of project task completion, but also extends to the exchange of valued resources and support that goes beyond what is required for task completion (e.g. sharing of knowledge / past experience learnt from other projects, information attainment). This is especially important when difficulties or problems occur in the project which can be solved together. When the problems are solved, this also induces feelings of success among work group members. The work group members would be happy and their job satisfaction level would also be improved. In this study, one possible reason for the positive relationship is because the work group identification induce chartered quantity surveyors' positive perception and feeling about the working condition and the job task, such as better working environment (e.g. better communication among colleagues; able to seek help from each other) or feeling involvement in the project, which contributes to high job satisfaction level.

Predictors / Variables	EJS	IJS	AC	NC
Control Variables				
Age	.100	.101	.139	.073
Gender	.078	.082	-.006	-.067
Education level	-.039	-.125	-.146	.012
Work Group Identification	.488***	.482***	.404***	.447***
Adjusted R^2	.202	.206	.151	.175
F	5.444***	5.552***	4.121**	4.721**

Table 2 Regression Results on Job Satisfaction and Organisational Commitment

**. Correlation is significant at 0.01 level (2-tailed)
***. Correlation is significant at 0.001 level (2-tailed)
EJS = Extrinsic Job Satisfaction; IJS = Intrinsic Job Satisfaction; AC = Affective Commitment; NC = Normative Commitment

Concerning commitment, the results in Table 2 show that work group identification has positive influence on both individual's affective (beta = 0.404, p<0.001) and normative commitment (beta = 0.447, p<0.001) to the organisation. Work group identification explained around 15.1% and 17.5% variance in affective commitment and normative commitment respectively. This also means that sub-hypotheses 2.1 and 2.2 have received support from the sample data collected in the questionnaire survey. One possible reason for the positive relationship is because work group identification can induce in the chartered quantity surveyor positive feeling about the working experience within the work group and working environment. As the work group is also part of the organisation, quantity surveyors who have positive perception about the work group and working environment would extend this belief to the organisation and develop a reciprocal behaviour that benefits the organisation, which become a motivation for quantity surveyors to stay with the organisation. In essence, similar perceptions for the positive relationship between work group identification and organisational commitment are also found in some previous studies. For instance, Coble (2002, p. 62) points out that

> *Employee's relationship with co-workers and membership in the work unit may contribute to an employee's perception of the social and emotional habitat in which they work... The experience that the employee has within the local work group may account for the difference in individual commitment level to the organization.*

The correlation results show that work group identification does not have significant relationship with continuance commitment. This means sub-hypothesis 2.3 is not supported. There are two possible reasons for this. First, the number of quantity surveyors involved in a project work group is small, which implies that chartered quantity surveyors working in the construction industry think that not much time and effort is needed for building working relationships again in the newly joined work group. Thus, work group identification does not constitute the cost for quantity surveyors to leave the company. Second, the projects that the quantity surveyor mostly is involved are not exceptionally big and well-known. In general, the involvement in the work group has no implication for quantity surveyors' working status, as the job duties that they handle in every work group and company are routine and repetitive (such as preparation of Bills of Quantities and Schedule of Rates, cost planning, monthly payment validation etc.). However, the situation may be different when the quantity surveyor is involved in well-known big projects. It is because the involvement can improve the professional value of the quantity surveyor and has positive implications for his future career prospects. In this situation, work group identification constitutes a cost for quantity surveyor's decision to leave the organisation.

Conclusion, Implication and Future Research Direction

Job satisfaction and organisational commitment have been widely studied in different disciplines for several decades. However, there are limited investigations in the quantity surveying context. The purpose of this paper is to report on a study examining the role of work group identification, in the development of job satisfaction and organisational commitment of quantity surveyors working in the Hong Kong construction industry. SET is applied to provide conceptual underpinnings for the study. Work group identification is about the identification with the work unit in which individuals conduct most of their day-to-day activities. Job satisfaction is used to reflect employee's feeling about job elements. Organisational commitment is about an individual's attachment to the organisation.

Quantitative research approach and survey data collection method are adopted in this study to achieve the stated objectives. The empirical results from this study positioned work group identification as a potential antecedent of job satisfaction and organisational commitment. Further analysis show that work group identification has positive influence to job satisfaction and the two dimensions (affective and normative) of organisational commitment. Thus, this study provide initial evidence supporting the premise that work group identification plays a role in raising job satisfaction and bringing about organisational commitment in the Hong Kong quantity surveying profession.

This study is the first to apply the SET to study the relationship between work group identification and, job satisfaction and organisational commitment. The adoption of SET provides a new approach to understanding the improvement of job satisfaction and organisational commitment in the company. Implications of the findings of this study are that organisations / senior managers should focus their efforts on fostering the work group identification among quantity surveyors by establishing positive relationships in the work place and implementing other non-work related activities (e.g. regular group meeting; lunch gathering; happy hour etc.) for enhancing communication between work group members and understanding between each other.

Whilst the present regression analysis provides a basis to support that work group identification has positive influences on job satisfaction and organisational commitment, other contextual factors that may also influence job satisfaction and organisational commitment are not taken into consideration. Future model development is therefore suggested for a more comprehensive understanding of other possible antecedents of job satisfaction and organisational commitment. It is suggested that the model can be tested by means of structural equation modeling based on larger scale data collection. In addition, this study is also limited by its cross-sectional design. Therefore, it fails to address the causality of the relationship between individual's work group identification, job satisfaction and organisational commitment and the stability of the relationship across time. The longitudinal research design is recommended for future research as it can provide an in-depth analysis about the causality of the relationships across variables and observe whether changes occur over time, which would complement the results obtained in this study.

References

Allen, N.J., & Meyer, J.P. 1990, 'The measurement and antecedents of affective, continuance and normative commitment to the organization', Journal of Occupational Psychology, **63**, 1-18

Allen, N.J., & Meyer, J.P. 2000, 'Construct validation in organizational behavior research: the case of organizational commitment', Goffin, R.D. & Helmes, E., (eds) Problems and Solutions in Human Assessment, Kluwer Academic Publishers, London, 285-314

Ashworth, A.A. & Mael, F. 1989, 'Social identity theory and the organization', Academy of Management Review, **14**, 20-39

Barlett, K.R. 2001, 'The relationship between training and organizational commitment: a study in the health care field', Human Resource Development Quarterly, **12** (4), 335-352

Beck, K., & Wilson, C. 2000, 'Development of affective organizational commitment: a cross-sequential examination of change with tenure', Journal of Vocational Behavior, **56**, 114-136

Becker, H.S. 1960, 'Notes on the concept of commitment', American Journal of Sociology, **66**, 32-42

Bernerth, J.B. & Walker, H.J. 2009, 'Propensity to trust and the Impact on Social Exchange', Journal of Leadership and Organizational Studies, **15** (3), 217-226

Bishop, J.W. & Scott, K.D. 2000, 'An examination of organizational and team commitment in a self-directed team environment', Journal of Applied Psychology, **85**, 439-450

Blau, P. 1964, Exchange and power in social life, Wiley, New York

Best, R. & Smith, N. 2005, 'Statistics and the peer review process in construction research', The Queensland University of Technology Research Week International Conference. 1027-1033

Chen, Z.X.. & Francesco, A.M. 2000, 'Employee demography, organizational commitment and turnover intentions in China: do cultural differences matter?', Human Relations, **3** (6), 869-887

Coble, D.F. 2004, Employee commitment and organizational influences: A study of the ways in which tenure, work group and job level influence employee commitment, unpublished PhD Thesis, Fielding Graduate Institute

Cohen, J., Cohen, P., West, S.G. & Aiken, L.S. 2003, Applied Multiple Regression / Correlation Analysis for the Behavioral Sciences (3rd Edition), Erlbaum Associates, Mahwah, NJ

Cranny, J., Smith, P. & Stone, E. 1992, Job Satisfaction: How people feel about their jobs and how it affects their performance, Lexington, New York

Farrell, D. & Rusbult, C.E. 1981, 'Exchange variables as predictors of job satisfaction, job commitment and turnover: The impact of rewards, costs, alternatives and turnover', Organizational Behavior and Human Performance, **28**, 78-95

Fellows, R., Liu, A. & Cheung, M.F. (2003), 'Leadership style and power relations in quantity surveying in Hong Kong', Construction Management and Economics, **29**, 809-818

Foote, N.N. 1951, 'Identification as the basis for a theory of motivation', American Sociological Review, **16**, 14-21

Gould, S.B. 1975, Organizational identification and commitment in two environments. unpublished PhD Thesis, Michigan State University, Lansing

Gould-Williams, J. & Davies, F. 2005, 'Using social exchange theory to predict the effects of hrm practice on employee outcomes: An analysis of public sector workers', Public Management Review, **7** (1), 1-24

Guzzo, R.A. & Shea, G.P. 1992, 'Group performance and intergroup relations in organizations,' Dunnette, M.D. & Hough, L.M. (eds.), Handbook of industrial and organizational psychology, 2nd Edn. Vol. 3., Consulting Psychologists Press, Palo Alto, CA , 269-313

Hair, J.F. Jr., Anderson, R.E. & Tatham, R.L. 1998, Multivariate data analysis, Prentice Hall, Upper Saddle River

Ho, R. 2006, Handbook of Univariate and Multivariate Data Analysis and Interpretation with SPSS, Taylor and Francis Group, USA.

Hoppock, R. 1935, Job satisfaction, Harper and Brothers, New York

Howell, D.C. 2002, Statistical Methods or Psychology (5th Edition), Duxbury, Pacific Grove, CA

Hirschfield, R.R. 2000, 'Does revising the intrinsic and extrinsic subscales of the Minnesota Satisfaction Questionnaire Short Form make a difference?', Educational and Psychological Measurement, **60**, 255-270

Kang, D.S. & Stewart, J. 2007, 'Leader-member exchange (LMX) theory of leadership and HRD: Development of units of theory ad laws of interaction', Leadership & Organizational Journal, **28** (6), 531-551

Kanter, R.M. 1968, 'Commitment and social organization: a study of commitment mechanisms in utopian communities', American Sociological Review, **33** (4), 499-517

Loosemore, M., Dainty, A. & Lingard, H. 2003, Human Resource Management in Construction Projects: Strategic and Operational Approaches, Spon Press, London

Laschinger, H. 2001, 'The impact of workplace commitment, organizational trust on staff nurses' work satisfaction and organizational commitment', Health Care Management Review, **26** (3), 7-24

Lipponen, J., Helkama, K. & Juslin, M. 2003, 'Subgroup Identification, Superordinate Identification and Intergroup Bias between the Subgroups', Group Processes and Intergroup Relations, **6** (3), 239-250

Liu, A.M.M., Chiu, W.M. & Fellows, R. 2007, 'Enhancing commitment through work empowerment', Engineering, Construction and Architectural Management, **14** (6), 568-580

Lok, P. & Crawford, J. 2001, 'Antecedents of organizational commitment and the mediating role of job satisfaction', Journal of Managerial Psychology, **16**, 594-613

Lingard, H. & Francis, V. 2004, 'The work-life experiences of office and site-based employees in the Australian construction industry', Construction Management and Economics, **22**, 991-1002

Lu, H., While, A.E. & Barriball, K.L. 2005, 'Job satisfaction among nurses: a literature review', International Journal of Nursing Studies, **42**, 211-227

Mael, F.A. & Tetrick, L.E. 1992, 'Identifying organization identification', Educational and Psychological Measurement, **52**, 813-824

Martin, A. & Roodt, G. 2008, 'Perceptions of Organizational Commitment, Job Satisfaction and Turnover Intentions in a Post-Merger South African Tertiary Institution', Journal of Industrial Psychology, **34** (1), 23-31

Meyer, J.P. & Allen, N.J. 1991, 'A three-component conceptualization of organziational commitment', Human Resource Management Review, **1**, 61-89

Meyer, J.P. & Allen, N.J. (1997) Commitment in the workplace: Theory, research and application, Sage, Thousands Oaks

Meyer, J.P., Becker, T.E. & Vandenberghe, C. 2004, 'Employee Commitment and Motivation: A Conceptual Analysis and Integrative Model', Journal of Applied Psychology, **89** (6), 991-1007

Moorman, R.H., Niehoff, B.P. & Organ, D.W. 1993, 'Treating employees fairly and organizational citizenship behavior: Sorting the effects of job satisfaction, organizational commitment and procedural justice', Employee Responsibilities and Rights Journal, **6**, 209-225

Moreland, R.L. & Levine, J.M. 2001, 'Socialization in organizations and workgroups,' Turner, M.E., (ed.), Groups at work: theory and research, Erlbaum, Hillsdale, NJ, 69-112

Mowday, R.T., Porter, L.W. & Steers, R. 1982, Organizational linkages: The psychology of commitment, absenteeism and turnover, Academic Press, Dan Diego, CA.

Olkkonen, M. E. & Lipponen, J. 2006, 'Relationships bwtween organizational justice, identification with organization and work unit, and group-related outcomes', Organizational Behaviour and Human decision Processes, **100**, 202-15

Pallent, J. 2001, SPSS Survival Manual: A step by step guide to data analysis using SPSS for Windows (Version 10), Open University Press, Buckingham

Park, R., Appelbaum, E. & Kruse, D. 2010, 'Employee involvement and group incentives in manufacturing companies: a multi-level analysis', Human Resource Management Journal, in press

Poon, J.M.L., Salleh, A.H.M. & Senik, Z.C. 2007, 'Propensity to trust as a moderator of the relationship between perceived organizational support and job satisfaction', International Journal of Organizational Theory and Behavior, **10** (3), 350-365

Porter, L.W., Steers, R.M., Mowday, R.T. & Boulian, P.V. 1974, 'Organizational commitment, job satisfaction and turnover among psychiatric technicians', Journal of Applied Psychology, **59**, 603-609

Rad, A.M.M. & Yarmohammadian, M.H. 2006, 'A study of relationship between managers leadership style and employees' job satisfaction', Leadership in Health Services, **19** (2), xi-xxviii.

Rafferty, A.E. & Griffin, M.A. 2009, 'Job Satisfaction in Organizational Research,' Buchanan , D.A. & Bryman, A., (eds), The SAGE Handbook of Organizational Research Methods, Sage Publications, London, 197-212

Riketta, M. & Van Dick, R. 2005, 'Foci of attachment in organizations: A meta-analytic comparison of the strength and correlates of workgroup versus organizational identification and commitment', Journal of Vocational Behavior, **67**, 490-510

Riordan, C.M. & Weatherly, E.W. 1999, 'Defining and Measuring Employee's Identification with their work groups', Educational and Psychological Measurement, **59** (2), 310-324

Robbins, S.P. 2005, Organizational Behavior, Pearson Prentice Hall

Sagie, A. 1998, 'Employee, Absenteeism, Organizational Commitment, and Job Satisfaction: Another Look', Journal of Vocational Behavior, **52**, 156-171

Sarantinos, V.S.T. & Co, J.S.C. 2007, 'Flexibility in the workplace what happens to commitment?', Journal of Business and Public Affairs, **1** (2), 10.

Setton, R.P., Bennett, N. & Liden, R.C. 1996, 'Social exchange in organziations: perceived organizational support, leader-member exchange, and employee reciprocity', Journal of Applied Psychology, **81**, 219-227

Sheldon, M.E. 1971, 'Investments and involvements as mechanisms producing commitment to the organization', Administrative Science Quarterly, **16**, 142-150.

Spector, P. 1997, Job Satisfaction, Sage Publications, Thousands Oaks, CA.

Stevens, J. 1996, Applied multivariate statistics for the social sciences (3rd Edition), Lawrence Erlbaum, New Jersey

Tolman, E.C. 1943, 'Identification and the post-war', Journal of Abnormal and Social Psychology, **31**, 141-148

Tse, H.H.M. & Dasborough, M.T. 2008, 'A Study of Exchange and Emotions in Team Member Relationships', Group and Organization Management, **33** (2), 194-215

Tyler, T. R. & Blader, S.L. 2001, 'Identity and cooperative behavior in groups', Group Processes and Intergroup Relations, **4**, 207-226

Van Dick, R., Christ, O, Stellmacher, J., Wagner, U., Alhswede, O., Grubba, C., Haputmeier, M., Hohfeld, C., Moltzen, K. & Tissington, P.A. 2004, 'Should I Stay or Should I Go? Explaining Turnover Intentions with Organizational Identification and Job Satisfaction', British Journal of Management, **15**, 351-360

Van Dick, R., Van Knippenberg, D., Kerschreiter, R., Hertel, G. & Wieseke, J. 2008, 'Interactive effects of work group and organizational identification on job satisfaction and extra-role behavior', Journal of Vocational Behavior, **72**, 388-399

Van Knippenberg, D. & Sleebos, E. 2006, 'Organizational identification versus organizational commitment: self-definition, social exchange and job attitudes', Journal of Organizational Behavior, **27** (5), 571-584

Van Knippenberg, D. & Van Schie, E.C.M. 2000, 'Foci and correlates of organizational identification', Journal of Occupational and Organizational Psychology, **73**, 137-147

Wayne, S.J., Shore, L.M., Bommer, W.H. & Tetrick, L.E. 2002, 'The role of fair treatment and rewards in perceptions of organizational support and leader-member exchange', Journal of Applied Psychology, **87**, 590-598

Weiss, D., Dawis, R., England, G. & Lofquist, L. 1967, Manual for the Minnesota Satisfaction Questionnaire (Minnesota Studies on Vocational Rehabilitation, Vol 22), University of Minnesota, Minneapolis

Westen, D. & Rosenthal, R. 2003, 'Quantifying construct validity: two simple measures', Journal of Personality and Social Psychology, **84** (3), 608-618

Wiener, Y. & Vardi, Y. 1980, 'Relationships between job, organization and career commitments and work outcomes: An integrative approach', Organizational Behavior and Human Performance, **26**, 81-96

Wong, M.M. 2006, Satisfying all stakeholders in evaluating the feasibility of public-private partnership projects: a structural equation model approach, unpublished M.Phil Thesis, The University of Hong Kong, Department of Civil Engineering

Wu, J. 2009, A Behavioral Analysis of Learning Processes amongst Construction Project Team Members in China, unpublished PhD Thesis, The University of Hong Kong, Department of Real Estate and Construction

Yip, B.L.P. (2007), Job burnout among construction professionals in Hong Kong: A moderator model with coping strategies, unpublished PhD Thesis, The University of Hong Kong, Department of Real Estate and Construction

Factors Influencing the Adoption of Information Technology in a Construction Business

14

Kimberley Sargent, Paul Hyland and Sukanlaya Sawang, (Queensland University of Technology, Australia)

Abstract

Construction firms are increasingly utilizing information technologies to better manage geographically dispersed projects. Often these technologies involve changes to existing working practices and processes and are viewed as disruptive by members of the organization. Understanding the factors that can influence individuals' intention to utilize technology can assist managers to implement strategies to increase and improve the uptake of technologies and improve the innovation adoption process. Using a case study organization, factors identified in the Unified Theory of Acceptance and Use of Technology (UTAUT) are examined and the UTAUT is extended by including resistance to change and top management support. The findings indicate effort expectancy, internal facilitating conditions and top management support all influence individuals' intention to use information technology. The results also show that resistance to change or fear of change does not always play a role in innovation adoption. The findings reinforce the need to support new technologies from both a managerial and technical perspective.

Keywords: Technology adoption, Resistance, Organizational change, Top management support

Introduction

Information technologies can assist project and construction managers to standardize routine tasks so that available organizational resources are utilized both effectively and efficiently (Adam et al., 2007). Past research has shown organizational benefits from IT adoption for construction management (Bjork, 2003; Kaner et.al., 2008; Oyediran & Odusami, 2005) The project management profession in recent years has grown exponentially and information technologies are used by many organizations to meet business requirements (Kendra and Taplin, 2004). Information technologies are commonly used to assist project managers with direct control over business functions, personnel and other resources (Hobday, 2000). As project managers oversee resource coordination and allocation it can be difficult to coordinate business functions across various projects. Information technologies are one of the key innovations that are frequently implemented to assist this process (Hobday, 2000). Peansupap and Walker (2005) maintain that information technologies are often implemented as they are believed to facilitate communication and improve integration (Bjork, 1999) as well as enhance productivity and service delivery.

Peansupap and Walker (2005) also maintain that the benefits of innovations such as information technologies can be limited if the adoption and use is poorly diffused, as effective diffusion requires user acceptance. Poor user acceptance can occur when transitioning from an existing system to a new system such as transitioning from a paper-based to a fully electronic environment; such a transition requires users who will readily adopt and utilize the information technology (Peansupap and Walker, 2005).When organizations implement a new technology, commonly they are not ready to adopt that technology and employees resist its introduction.

The technology adoption decision within organizations is usually authorized by a group of senior managers (Peansupap and Walker, 2005), therefore a key question of information technology adoption in construction firms should be how to ensure that users accept and utilize information technology in their work processes. However, studies indicate that the rate of unsuccessful information technology implementations is growing and further, the adoption rate is very slow (Acar et al., 2005; Mole et al., 2004; Shin, 2006). To address this poor success rate in adoption and contribute to Gambatese and Hallowell's (2011) call to identify the characteristics of innovation adopters, individual and managerial factors that impact on intention to utilize a specific information technology in a construction firm were studied.

Information Technologies

There is a positive relationship between IT adoption and organizational performance in the construction industry. For example, Construction Industry Institute (CII) and the National Institute of Standards and Technology (NIST) empirically examined the impact of IT adoption and found that cost and schedule performance improve with increased IT use (Thomas 1999; Thomas et al. 2001). Similar to these studies, Kang et.al (2008) employed CII data from 139 projects from 74 companies examining the impact of IT adoption and project performance. Their findings also supported previous studies that level of IT adoption positively associated with improved performance. It is not sufficient to simply select the appropriate technology and implement it throughout an organization. According to Hussain and Wearne (2005), the construction and defence industries have adopted project management technologies to deal with the growth in scale, complexity and financial risks of capital projects, yet the number of defence projects that fail or blow out in terms of time and budget is well publicised. While Yang (2007) maintains that the use of technologies to enhance project performance has been widely supported, simply having innovative information technologies does not guarantee a project will be delivered on-time or within budget.

Yang (2007) maintains that studies have shown that the construction industry is reluctant to apply new technologies and employs lower levels of technology than other industries. One issue related to poor uptake that has emerged is organizational inertia, which Lawrence and Scanlan (2007) found means organizations tend to become accustomed to their own tools and technologies, and they find it difficult to give up and change established procedures and familiar information technology products. Therefore, changing existing ways of doing things or resistance to change (RTC) can pose problems. While Gambatese and Hallowell (2011) found that fear of change was a significant barrier to implementation of innovations such as information technologies unless this fear is translated into active or passive resistance it will not have a major impact on implementation or intention to use a technology. There is a need therefore to measure the RTC rather than fear of change. Critically on large projects it is essential to establish common integrated information technologies, as well as common methods and processes (Lawrence and Scanlan, 2007). Even when information technologies are available and are being used they are interpreted and utilized in different ways by different people.

Guss (1998) also found that project management professionals are inclined to be at different phases of acceptance and willingness to change in relation to adopting, planning and communicating information technologies. Often existing project management materials are too technical for the novice project manager and team (Longman and Mullins, 2004). The benefits of project management tools such as information technologies do not appear to justify the investment of time and energy and in some cases project managers perceive these technologies to be unproven or ineffective (Longman and Mullins, 2004; Guss, 1998). It is critical that project managers in the construction industry continually seek innovative information technologies to overcome limitations and inefficiencies in managing projects. However better information technologies alone will not ensure more effective management of projects. It has been argued that user acceptance is a critical success factor for technology implementation and can be tested and predicted by several factors (Hu et al., 1999).

Factors Influencing Information Technology Adoption and Use

The Unified Theory of Acceptance and Use of Technology (UTAUT) model developed by Venkatesh et al. (2003) is a useful starting point to investigate technology adoption. The UTAUT can be used to identify factors that influence the intention to use information technology to be adopted by an organization (Venkatesh et al., 2003). The UTAUT model was based upon conceptual and empirical similarities across user acceptance models. The UTAUT has been used in the IT adoption literature, including IT adopting for construction management (Adriaanse, Voordijk & Dewulf, 2010; Hjelt & Björk, 2007; Samuelson, 2011). However, these studies did not include additional constructs such as top management support (TMS) (Dong et al., 2009) or resistance to change (Venkatesh et al., 2000), which have been identifies as influencing technology adoption. TMS plays a crucial role in determining technology implementation success and failure (Neufeld, Dong, & Higgins, 2007). Similarly, resistance to change is also a vital factor for technology adoption because resistance is often the main reason a technology fails (Venkatesh et.al., 2001). This study, thus, adopts the UTAUT to examine IT adoption in construction management, similar to previous studies, but we also include two important constructs TMS and resistance to change which have also been identified as important to IT adoption.

According to the UTAUT model, seven constructs were found to be significant determinants of intention or usage in one or more of the individual user acceptance models examined (Venkatesh et al., 2003). Of the seven, Venkatesh et al. (2003) theorised that four would play a significant role as direct determinants of user acceptance and usage behaviour, namely performance expectancy, effort expectancy, social influence and facilitating conditions. In their study (Venkatesh et al., 2003), the UTAUT model was shown to be a good predictor of user acceptance and adoption of information technologies.

Performance expectancy is defined by Venkatesh et al. (2003, p.447) as 'the degree to which an individual believes that using a particular technology will help him or her to attain gains in job performance'. Venkatesh et al. (2003) propose that performance expectancy captures the constructs of perceived usefulness, extrinsic motivation, job fit, relative advantage and outcome expectations. Perceived usefulness has been strongly related to usage intentions in various studies (Davis, Bagozzi & Warshaw, 1989) and Davis et al. (1992, p.1112) define it as 'a person's expectation that using the technology will result in improved job performance:

Hypothesis 1: Performance expectancy will have a significant positive influence on individuals' intention to use an information technology.

Effort expectancy is the 'degree of ease associated with the use of the system' (Venkatesh et al., 2003, p. 450). There are three existing constructs that capture the essence of effort expectancy, namely ease of use, complexity and perceived ease of use. Ease of use is defined by Moore and Benbasat (1991) and Davis et al. (1989) as the 'degree to which an individual believes that using a particular system would be free of physical and mental effort'. Complexity relates to the degree to which a technology is perceived as relatively difficult to understand and use (Thompson & Higgins, 1991):

Hypothesis 2: Effort expectancy will have a significant positive influence on individuals' intention to use an information technology.

Social influence is 'the degree to which an individual perceives that important others believe he or she should use the new system' (Venkatesh et al., 2003, p.451). Three items contribute to the social influence construct: subjective norms, social factors and image. Subjective norms are a construct that has been widely researched in technology adoption literature. Ajzen (1991) and Mathieson (1991) define the construct as the perceived social pressure to perform or not to perform a behaviour:

Hypothesis 3: Social influence will have a significant positive influence on individual's intention to use an information technology.

Facilitating conditions is the final construct identified by Venkatesh et al. (2003, p.453) and is 'the degree to which an individual believes that organizational and technical infrastructure exists to support use of the system'. Three items are used to measure the facilitating conditions construct, namely perceived behavioural control, facilitating conditions and compatibility. The facilitating conditions construct was examined by Thompson and Higgins (1991) and was described as the objective factors present that make an act easy to do. Thompson and Higgins (1991) examined information technologies and found that training users and assisting them when they encounter difficulties is an example of a facilitating condition that can influence technology utilisation:

Hypothesis 4: Facilitating conditions will have a significant positive influence on individuals' intention to use an information technology.

Influence of Top Management Support and Resistance to Change

In spite of the empirical applicability of the UTAUT, additional efforts are needed to validate existing research results, in particular those involving different industry sectors, users or organizational contexts. Gallivan (2000) argued that generic innovation adoption models may not be applicable under the several circumstances:

- adoption within organizations where expected users are mandated to adopt;
- adoption is dependent on multiple adopters; and
- adoption requires extensive training to upgrade users' skills.

As generic adoption models rely on voluntary adoption decisions by individuals, they may be less suitable in explaining complex organizational adoption decisions (Gallivan, 2000). So in the context where a technology is mandated it is important to consider non-generic factors such as the influence of senior managers. Successful technology adoption by expected users in construction firms requires implementation support and encouragement from senior managers if individuals are to adopt and utilize the technology (Peansupap and Walker, 2005).

As senior managers can influence the implementation and use of new technologies, which according to Young and Jordan (2008) involves managers devoting time to the technology in proportion to its costs and potential, as well as reviewing plans, monitoring results and facilitating the management problems involved with integrating the technology with the management process of the business. Dong et al. (2009) also maintain that TMS encourages technology usage and better user performance, influences positive user perceptions and improves the overall technology adoption uptake. Similarly Gambatese and Hallowell (2011) found that effective upper management support was one of the strongest enablers on innovation implementation in construction firms. TMS is viewed as a clear commitment and allocation of sufficient resources to the innovation and if needed active involvement in managing change and innovation adoption (Gambatese and Hallowell 2011). According to Neufeld et al. (2007), TMS plays a crucial role in determining technology implementation success and failure. Furthermore, Ifinedo (2008) argues that a substantial body of knowledge has linked TMS to large scale technology implementation success. In terms of the UTAUT model, the facilitating conditions construct does not specifically measure TMS. According to Neufeld et al. (2007), the TMS construct has not been sufficiently integrated into existing user adoption theories, nor has literature defined specific top management behaviours that are associated with technology implementation success:

Hypothesis 5: Top management support will have a significant positive influence on individuals' intention to use an information technology.

Individual RTC is defined as an 'individual's dispositional inclination to resist change and predict reactions to specify change' (Oreg, 2003, p.680). It is a natural part of the organizational change process and individual resistance occurs because change involves going from the known to the unknown (Bovey and Hede, 2001). Individual RTC is an important factor for technology implementation because resistance is often one of the main reasons a technology fails (Venkatesh et al., 2000). When technologies are implemented, people differ from each other in their inclinations to resist or implement changes, therefore the theory surrounding RTC and the RTC scale can help explain why some people are more inclined than others to try to implement and use new technologies (Oreg et al., 2005). The RTC scale, developed by Oreg (2003, p.680), was designed to assess 'an individual's tendency to resist or avoid making changes'. Oreg (2003) states that using the RTC scale is far more economical as it uses a broad range of measures that tap into different aspects of RTC, rather than broader measures. Four personality-based aspects are assessed by the RTC scale: routine seeking, emotional reaction to change, short-term focus and cognitive rigidity (Oreg et al., 2009). Routine seekers are individuals who enjoy and seek out stable and lasting routines in their lives. Emotional reactors to change are how individuals respond to imposed changes. An individual who has short term focus focuses on short term hassles that are part of most changes; otherwise they focus on the potential long term benefits of change. Cognitive rigidity refers to an individual's tendency to maintain one's view. The combination of the four dimensions captures a variety of elements that predispose individuals to resist and avoid changes:

Hypothesis 6: Resistance to change will have a significant negative influence on individuals' intention to use an information technology.

Drawing together the UTUAT model, RTC and TMS an extended model emerges as can be seen in Figure 1. Figure 1 also includes age, gender, education and computer experience which are control variables, but may influence a users intention to adopt an information technology and are worth investigating in this context.

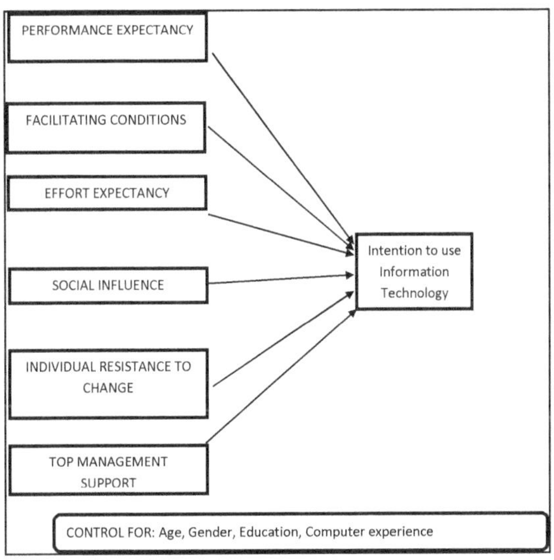

Figure1 Modified unified theory of acceptance and use of technology model

Method

Participants

The case organization is a privately owned and operated Australian civil engineering and building construction company specializing in project management, infrastructure, engineering, building, mining services and precast. It is a medium-sized (650 employees) organization that has successfully delivered projects such as civil engineering construction, complex bridgeworks, large scale concrete works and retaining wall construction across geographically dispersed locations. The project management unit was introduced in early 2009 and it provides internal and external clients with excellence in project advisory services. The project management unit has since been continually seeking to improve training, project management qualifications and delivery. The business had recently identified issues in storing and accessing their working files, often from remote locations. The organization introduced an information management system which offers a comprehensive document control system and the software enables employees to save time and money by leveraging and reusing existing data, business processes and applications. Management believes the introduction of the document control system will enable end users to better manage files and file access.

Throughout the data collection process, the researchers received emails from respondents who were unable to complete the survey due to being offsite and not able to sustain an internet connection to complete the survey. The information obtained indicated that employees had experienced internet connection problems whilst working at rural locations. Further, these rural locations only had limited internet access at the site during peak hour. This situation restricted employees from retrieving information that they had stored on the document control system. Employees indicated that they were frustrated with the way technology restricted and hindered the efficient operation of the project.

Ruigrok and Gibbert (2010) argue that case study data can be collected using both qualitative and quantitative methods and argue that the use of questionnaires with validated measures increases the rigour of case study research. In this case study both qualitative and quantitative techniques were used to collect data, however, here results focus on quantitative analysis as two new measures are put forward and tested. The quantitative data was collected from 147 respondents who completed the survey (22.61% response rate). The respondents were 104 males (70%) and 43 females (30%). Average respondents' age was 35 years (SD = 9) ranging from 19 to 60 years. Average computer technology usage experience was 15 years (SD = 7). The majority of respondents held undergraduate (39%) and postgraduate (27%) degrees. Respondents had an average organizational tenure of three years (SD = 3.25), ranging from one to 20 years.

Measures

Measures were adapted from well established and validated scales of UTAUT (Venkatesh et al., 2003), and the RTC scale, which was adapted from Oreg's (2003) study. This scale consists of 4 constructs; routine seeking, emotional reaction, short term focus and cognitive rigidity. TMS was measured using 5 items relating to management commitment to using the technology, management's commitment to support staff efforts and their commitment to encouraging the use of technology, as well as their general support for technology and there emphasis on the importance of the technology to the business. A five-point Likert scale was employed to measure the attitudes designed to allow respondents to signify how strongly they agree or disagree with established measures that range from very positive to very negative toward an attitudinal object.

Data Analysis

Data collected using multidimensional scales (e.g. UTUAT, TMS and resistance to change) were initially analysed using a principal components analysis with varimax rotation. The

Exploratory Factor Analysis was conducted using SPSS/PASW (18.0) to determine how a range of change items loaded onto factors derived from a combined data set (N=147). For this analysis, a Principal Axis Factoring (PAF) extraction method was used with a varimax rotation. Ten series of the EFA were conducted and as a result, eight factors loaded cleanly and clearly. Two constructs (facilitating conditions and social influence) loaded onto one factor. It appeared that respondents perceived the social influence (1-item: senior management at our company has been helpful in the use of the document control system) and facilitating conditions (3-item: resources, knowledge and technical assistance) as their internal facilitator assisting them or encouraging them to adopt the document control system. Therefore, the two constructs were combined and renamed as 'internal facilitating conditions'. Hypothesis 3 and Hypothesis 4 were amended:

> *Hypothesis 3/4: Internal facilitating conditions will have a significant positive influence on individuals' intention to use the information technology.*

Table 1 exhibits the correlations between the variables studied. Performance expectancy is significantly and positively associated with behavioural intention (r=.36, p<.001); such that an increase in perceived performance expectancy is associated with an increase in behavioural intention to adopt the PM tool.

A significant positive relationship exists between effort expectancy and behavioural intention (r=.47, p<.001) such that when perceived effort expectancy increases, perceived behavioural intention increases. A significant positive relationship exists between internal facilitating conditions and behavioural intention (r=.48, p<.001) such that when perceived internal facilitating conditions increase, behavioural intention increases. A significant positive relationship exists between TMS and behavioural intention (r=.40, p<.001) such that when perceived TMS increases, behavioural intentions increase. A significant negative relationship exists between emotional reaction and behavioural intention (r=-.19, p<.05), such that when perceived emotional reaction increases, behavioural intention increases.

To examine the internal reliabilities for the measures, the Cronbach alpha was examined. Table 1 indicates that seven out of the nine internal reliabilities are above the threshold (α >0.70), demonstrating that performance expectancy, effort expectancy, internal facilitating conditions, TMS, emotional reaction, short term focus and behavioural intention have demonstrated high consistency among items. Routine seeking and cognitive rigidity were amongst the few that displayed low internal reliabilities (below .70).

Modelling procedures together with multiple regression analyses were undertaken to determine the extent to which internal facilitating conditions, performance expectancy, effort expectancy, TMS and individual RTC predicted intention to use the document control system. Hierarchical regression analyses was utilized rather than structural equation modelling (SEM), as according to Buhi et al. (2007) having an adequate sample size is a major concern in SEM utilisation. Small samples are most likely to result in unreliable and untrustworthy parameter estimates and fit statistics yielding models that are nonreplicable. They suggested at least 200 cases are needed for adequate model specification.

On Block 1 of the series of multiple hierarchical regressions, all control variables (gender, education, computer experience and age) were entered as the independent variables (IVs). The behavioural intention dependent variable (DV) was also entered on Block 1. Having specified the first Block in the hierarchy, all UTAUT IV's (performance expectancy, effort expectancy and internal facilitating conditions) were entered on Block 2 of the regression. As noted in the literature review, theory indicates that the UTAUT constructs are significant predictors of the behavioural intention DV. On Block 3 of the regression, individual RTC and TMS were entered as IVs. Individual RTC and TMS were entered on the last Block to examine the significant effect of this additional regression equation. The results of the multiple hierarchical regressions follow.

Table 2 displays the results of the multiple hierarchical regressions conducted to examine the effect that the eight IVs had on the behavioural intention DV while controlling for gender, age, education and computer experience.

The entry of the covariates (computer experience, education, gender and age) in Step 1 did not further explain variance on behavioural intention (*Adj.* $R^2 = -.02$, $F(4,117) = .53$, *ns*). A positive insignificant relationship existed between gender and behavioural intention ($\beta = .13$, *ns*), and computer experience and behavioural intention ($\beta = .09$, *ns*). A negative insignificant relationship exists between age and behavioural intention ($\beta = -.01$, *ns*), and education and behavioural intention ($\beta = -.00$, *ns*).

On Block 2 of the regression, the UTAUT constructs as IVs (effort expectancy, internal facilitating conditions and performance expectancy) were entered to examine the direct effect they have on behavioural intention. The entry of these three IVs explained an additional 16.9% variance to the model ($\Delta R^2 = .17$, $F(3,114) = 7.93$, $p < .001$). A positive significant relationship exists between effort expectancy and behavioural intention ($\beta = .19$, $p < .05$) such as when perceived effort expectancy increases, perceived behavioural intention increases. Furthermore, a significant positive relationship exists between internal facilitating conditions and behavioural intentions ($\beta = .37$, $p < .001$) whereby when perceived internal facilitating conditions increase, perceived behavioural intentions increase. Finally, a positive non-significant relationship was observed between performance expectancy and behavioural intention ($\beta = .16$, *ns).*

On Block 3 of the regression, the TMS and RTC 1 to RTC 4 as IVs (TMS, RTC 1: Emotional reaction, RTC 2: short term focus, RTC 3: routine seeking, and RTC 4: cognitive rigidity) were entered to examine the additional effect they have on behavioural intention. The entry of these IVs explained an additional 7% variance to the model ($\Delta R^2 = .07$, $F(5,109) = 2.03$, *ns*). The TMS construct significantly and positively predicted behavioural intention ($\beta = .18$, $p < .05$) such that when perceived TMS increases, perceived behavioural intention increases. The emotional reaction component of the RTC scale negatively but non-significantly predicted behavioural intention ($\beta = -.08$, *ns*), as well as the short term focus construct negatively but non-significantly predicted behavioural intention ($\beta = -.12$, *ns*). The routine seeking construct positively but non-significantly predicted the behavioural intention DV ($\beta = .15$, *ns*). The cognitive rigidity construct also negatively but non-significantly predicted behavioural intention ($\beta = -.08$, *ns*).

In conclusion, after undertaking the multiple hierarchical regressions to test the proposed hypotheses, the following hypotheses were supported:

> *H2: Effort expectancy positively influenced individuals' intention to use the information technology.*
>
> *H3/4: Internal facilitating conditions positively influence individuals' intention to use the information technology.*
>
> *H5: Top management support positively influences individuals' intention to use the information technology.*

	Mean (SD)	1	2	3	4	5	6	7	8	9	10	11
1. Performance expectancy	1.61 (.77)	(.89)										
2. Effort expectancy	2.93 (.94)	.55***	(.90)									
3. Internal facilitating conditions	3.23 (.87)	.44***	.55***	(.80)								
4. Top management support	4.09 (.79)	.34***	.33***	.63***	(.88)							
5. Resistance to change – Routine seeking	1.34 (.52)	-.14*	-.24**	-.11	-.06	(.57)						
6. Resistance to change – Emotional reaction	1.73 (.58)	-.04	-.21**	-.09	-.00	.33***	(.76)					
7. Resistance to change – Short term focus	1.35 (.52)	-.06	-.11	-.11	-.06	.21**	.37***	(.74)				
8. Resistance to change – Cognitive rigidity	2.82 (.82)	.00	.08	.09	-.02	.05	-.03	.09	(.61)			
9. Behavioural intention to adopt the document control system	3.97 (.92)	.36***	.47***	.48***	.40***	-.14	-.19**	-.06	.09	(.70)		
10. Computer experience	14.79 (6.58)	.11	.17**	-.10	.00	-.20***	-.22***	-.09	.19**	.14*		
11. Education	3.52 (1.15)	.11	.06	.07	.07	-.01	-.04	-.23***	-.12	.02	.20**	
12. Age	35.32 (8.91)	.05	-.09	-.17**	-.11	-.12	-.12	-.04	.15**	.02	.53***	.07

Table 1 Descriptive data for focal variables for the study

Note. Cronbach alpha reliability coefficients appear in the diagonals. $*p < .05$; $** p < .01$; $***p < .001$

Predictors	Intention to adopt the document control system *β*
Regression 1	
Step 1 – Covariates	
Age	-.01
Gender	.13
Computer experience	.09
Education	-.00
Adj. R^2	-.02
Step 2 – Main effect - UTAUT	
Effort expectancy	$.19^{*}$
Internal facilitating conditions	$.37^{***}$
Performance expectancy	.16
R^2 Ch.	$.17^{***}$
Step 3 – Main effect - Top management support and Resistance to change	
Top management support	$.18^{*}$
Resistance to change 1: Emotional reaction	-.08
Resistance to change 2: Short term focus	-.12
Resistance to change 3: Routine seeking	.15
Resistance to change 4: Cognitive rigidity	-.08
R^2 Ch.	.07

Table 2 Hierarchical multiple regression analysis showing main effects of UTAUT, resistance to change and TMS on intention to adopt the document control system

$^{*}p < .05$; $^{**}p < .01$; $^{***}p < .001$

Discussion

This study examined individual and managerial factors which impact on intention to utilize a specific information technology in a project-based organization, extending the UTAUT model. Effort expectancy and internal facilitating conditions positively predicted an intention to use the document control system. The literature supports that users' perceptions of ease of use and usefulness are amongst some of the most important indicators for behavioural intention to adopt and implement a new technology (Lu et al., 2005). What is interesting about this finding pertains to the demographics of the respondents. It is very likely that the high levels of computer experience were contributory to this hypothesis being supported. The demographics of this sample had an average of 15 years computer experience and most respondents (39%) undergraduate (27%) postgraduate had obtained an undergraduate degree. It can be assumed that the respondents were a highly skilled and educated cohort therefore adapting to the new technology and learning how to use it did not pose a significant problem.

Performance expectancy did not influence an intention to use the document control system. This can be explained by the fact that information technology is intended to minimize effort and time required for planning and monitoring projects (Ali, Anbari, & Money, 2008);

however, as the case organization is a project based organization (PBO) and project managers frequently find themselves away from the office, it is understandable that employees did not agree that the document control system enabled them to accomplished tasks more efficiently as access to the Internet in many remote locations was very limited.

In summary, in this case study the UTAUT model was partially supported (that is the effort expectancy and internal facilitating conditions constructs). It is noted here that this is the first time that the document control system has been tested as the technology to be adopted and it is not surprising that the majority of the UTAUT constructs fit well and predict intention to adopt; however, the UTAUT model was extended and this provides interesting insights and contributions.

Top Management Support and Resistance to Change

TMS influenced individuals' intention to adopt the document control system. This finding is supported by Dong et al. (2009), who argue that TMS encourages technology usage and better user performance, influences positive user perceptions and improves the overall technology adoption uptake. Similarly Gambatese and Hallowell (2011) found that effective upper management support was one of the strongest enablers of innovation implementation in construction firms. As the internal facilitating conditions also predicted a positive intention to adopt the document control system, it is logical that TMS is one of the factors that employees value when adopting a new technology.

TMS appears in two forms in the case study organization. Firstly, the document control system was implemented gradually to allow employees to transition from the old project management technology to the new technology. Eventually, the old system was to be discarded and employees had few options but to use the new system. This was a clear example of how management directly ensured technology uptake. Secondly, indirectly the institutional structures were manipulated as employees' current systems and the way of doing things were changed so that employees had little choice but to accept the change. Consistent with the literature presented by Murphy and Ledwith (2007), this case organization had clear goals and objectives for the technology implementation as well as perceived TMS. As the technology implementation was evolutionary in nature and these supporting factors were in place, it was more likely that the employees would intend to use the document control system if management were supportive (Murphy & Ledwith, 2007).

While it has been suggested that with any organizational change, people are likely to develop some resistance to that change, in this case, a significant result was not found between RTC and intention to use the document control system. The overall RTC score was relatively low (1.94 out of 5). This is most likely due to the nature and structure of the organization, as it is project-based and continually evolving, therefore stable routines are far from apparent. The development of change resistance could be at an early stage and it may not be recognizable. Furthermore, the average length of service for employees at this case organization is three years with the most common length of service being two years. It could be assumed that because employees have not been employed for a long duration, they have not developed inertia to change. Contributory to this, 54% of the respondents are project-based indicating that their roles are continually changing due to projects in different locations. The literature reveals that RTC in PBOs refers directly to the lack of acceptance of change by the people affected and this causes projects to fail. It is evident that in the case organization, employees are well educated, experience continual change and operate in a dynamic environment. This may explain why there is not a significant relationship between RTC and intention to utilize the document control system.

Future research that should be considered includes firstly the testing of the UTAUT with other project management technologies. Secondly, because TMS was seen as a significant predictor of behavioural intention, this construct should be tested in other construction firms to verify its importance to be included in the UTAUT model. How crucial TMS is for technology adoption and implementation is already understood, but it would be interesting to test this construct alongside the UTAUT model in a different organizational context or perhaps other PBOs. Thirdly, the RTC should not be disregarded as an unimportant predictor of behavioural intention to adopt a technology. Due to the project-based context of the case organization, the organization is continually changing and is therefore used to change turbulence. The RTC should be tested along the UTAUT and TMS to further confirm or to disregard as a predictor of behavioural intention.

The research was limited in a number of important respects. The possibility of common methods bias always exists with self-report survey data of this kind. Notwithstanding this, future research should utilize varied data collection methods to overcome this potential source of bias. The research was undertaken as a baseline survey in a single PBO in which information technology was being adopted; further research needs to be conducted in a number of case study construction organizations. Also, further research needs to examine other factors identified by Gambatese and Hallowell (2011) such as organizational culture, the presence of an innovation champion and communication.

Conclusions

The theoretical implications that have developed from this study include the importance of TMS in a project-based environment and the RTC scale being tested in difference contexts. The two components of the UTAUT model that are still relevant to technology adoption in construction firms are effort expectancy and internal facilitating conditions. Although performance expectancy was not seen as a predictor of intention to use the project management technology, it is still seen as useful to test in construction organizations.

From a management perspective, RTC does not appear to be a major cause for concern. Management needs to support any technology adoption initiatives, for example encouraging employees to use the technology and demonstrating its benefits. The practical implications resulting from the RTC is that although it does not directly affect intention to adopt a new technology, it is a construct that definitely could prevent the general adoption of a technology. In particular, this could be the case for those employees who have longer service lengths and are set in their own ways. The emotional reaction construct did, however, influence the behavioural intention to use the information technology. Therefore, practical implications from this finding imply that management must be aware that if they change the processes that employees use regularly, their employees will feel stressed.

References

Acar, E., Kocak, I., Sey, Y. and Arditi, D. (2005) 'Use of information and communication technologies by small and medium-sized enterprises (SMEs) in building construction', *Construction Management and Economics,* **23**, 713-722

Adam, F., Carton, F. and Sammon, D. (2007) 'Project management: a case study of a successful ERP implementation', *International Journal of Managing Projects in Businesses*, **1**, 106-124

Adriaanse, A., Voordijk, H. and Dewulf, G. (2010) 'The use of interorganisational ICT in United States construction projects', *Automation in Construction,* **19** (1), 73-83

Ajzen, I. (1991) 'The theory of planned behavior', *Organizational Behavior and Human Decision Processes*, **50** (2), 179-211

Ali, A.S.B, Anbari, F.T and Money, W.H. (2008) 'Impacts of organisational and project factors on acceptance and usage of project management software and perceived project success', *Project Management Journal,* **39** (2), 5-33

Buhi, E. R, Goodson, P. and Neilands, T. B. (2007) 'Structural equation modelling: a primer for health behavior researchers', *American Journal of Health Behavior*, **31** (1), 74 – 85

Bovey, W. H. and Hede, A. (2001) 'Resistance to organisational change: the role of defence mechanisms', *Journal of Managerial Psychology,* **16** (7/8), 534-548

Bjork, B.-C. (1999) 'Information technology in construction: domain definition and research issues', *International Journal of Computer Integrated Design and Construction,* SETO, **1** (1), *3-16*

Bjork, B.-C. (2003) 'Electronic document management in construction: research issues and results'. *ITcon*, **8**, 105-117

Davis, F.D. (1989) 'Perceived usefulness, perceived ease of use, and user acceptance of information technology', *MIS Quarterly*, **13** (3), 319-340

Davis, F.D., Bagozzi, R.P. and Warshaw, P. R. (1989) 'User acceptance of computer technology: a comparison of two theoretical models', *Management Science,* **35** (8), 982-1003

Davis, F. D., Bagozzi, R.P. and Warshaw, P. R. (1992) 'Extrinsic and intrinsic motivation to use computers in the workplace', *Journal of Applied Social Psychology,* **22** (14), 1111-1132

Dong, L., Neufeld, D. and Higgins, C. (2009) 'Top management support of enterprise systems implementations', *Journal of Information Technology,* **24** (1), 55-80

Gallivan, M. J. (2000) 'Examining workgroup influence on technology usage: a community of practice perspective', *Proceedings 2000 ACM SIGCPR,* The conference on Computer personnel research, ACM Press Chicago, IL, 54–66

Gambatese, J. A. and Hallowell, M. (2011) 'Enabling and measuring innovation in the construction industry', *Construction Management and Economics*, **29** (6), 553-567

Guss, C. L. (1998) 'Virtual project management: tools and the trade', *Project Management Journal,* **29** (1), 22-31

Hjelt, M. and Bjork, B-C. (2007) 'End user attitudes towards EDM use in construction project work: a case study', *American Society for Civil Engineering Journal of Computing in Civil Engineering*, **21** (4), 289-300

Hobday, M. (2000) The project-based organisation: an ideal form for managing complex products and systems?', *Research Policy,* **29** (7-8), 871-893

Hu, P. J. Chau, P. Y. K, Liu Sheng, O.R and Kar Yan, T. (1999) 'Examining the technology acceptance model using physician acceptance of telemedicine technology', *Journal of Management Information Systems,* **16** (2), 91-112

Hussain, R. and Wearne, S. (2005) 'Problems and needs of project management in the process and other industries', *Chemical Engineering Research and Design,* **83** (4), 372-378

Ifinedo, P. (2008) 'Impacts of business vision, top management support, and external expertise on ERP success', *Business Process Management Journal,* **14** (4), 551-568

Kang, Y., O'Brien, W. J, Thomas, S. and Chapman, R. E. (2008) 'Impact of information technologies on performance: cross study comparison', *Journal of Construction Engineering and Management*, **134** (11), 852-863

Kaner, I., Sacks, R., Kassian, W. and Quitt, T. (2008) 'Case studies of BIM adoption for precast concrete design by mid-sized structural engineering firms', *ITcon*, **13** (Special Issue), 303-323

Kendra, K. A. and Taplin, L. J. (2004) 'Change agent competencies for information technology project managers', *Consulting Psychology Journal: Practice and Research,* **56** (1), 20-34

Lawrence, P. and Scanlan, J. (2007) 'Planning in the dark: why major engineering projects fail to achieve key goals', *Technology Analysis & Strategic Management,* **19** (4), 509-525

Longman, A. and Mullins, J. (2004) 'Project management: key tool for implementing strategy', *Journal of Business Strategy*, **25** (5), 54 – 60

Lu, J, Yu, C-S and Liu, C. (2005) 'Facilitating conditions, wireless trust and adoption intention', *The Journal of Computer Information Systems,* **46** (1), 17-24

Mathieson, K. (1991) 'Predicting user intentions: comparing the technology acceptance model with the theory of planned behavior', *Information Systems Research,* **2** (3), 173-191

Mole, K. F., Ghobadian, A., O'Regan, N and Liu, J. (2004) 'The use and deployment of soft process technologies within UK manufacturing SMEs: an empirical assessment using logit models', *Journal of Small Business Management,* **42** (3), 303-324

Moore, G. C. and Benbasat, I. (1991) 'Development of an instrument to measure the perceptions of adopting an information technology innovation', *Information Systems Research,* **2** (3), 192-222

Murphy, A. and Ledwith, A. (2007) 'Project management tools and techniques in high-technology SMEs', *Management Research News,* **30** (2), 153-166

Neufeld, D., Dong, L. and Higgins, C. (2007) 'Charismatic leadership and user acceptance of information technology', *European Journal of Information Systems,* **16** (4), 494-510

Oreg, S. (2003) 'Resistance to change: developing an individual differences measure', *Journal of Applied Psychology,* **88** (4), 680-693

Oreg, S., Goldenberg, J. and Frankel, R. (2005) 'Dispositional resistance to the adoption of innovations', *Proceedings of the Annual Meeting of the European Association of Work and Organizational Psychology*, May 2005, Istanbul, Turkey

Oreg, S., Nevo, O., Metzer, H., Leder, N. and Castro, D. (2009) 'Dispositional resistance to change and occupational interests and choices', *Journal of Career Assessment,* **17** (3), 312-323

Oyediran, O. S. and Odusami, K. T. (2005) 'A study of computer usage by Nigerian quantity surveyors', *ITcon*, **10**, 291-303

Peansupap, V. and Walker, D. H. T. (2005) 'Factors affecting ICT diffusion: a case study of three large Australian construction contractors', *Engineering Construction and Architectural Management*, **12** (1), 21-37

Ruigrok, W. and Gibbert, M. (2010) 'The "what" and "how" of case study rigor: three strategies based on published work', *Organizational Research Methods*, **13** (4), 710 – 737

Samuelson, O. (2011) 'Adoption processes for EDM, EDI and BIM technologies in the construction industry', *Proceedings CIB W78-W102 2011,* International Conference, 26-28 October, Sophia Antipolis, France

Shin, I. (2006) 'Adoption of enterprise application software and firm performance', *Small Business Economics*, **26**, 241-256

Thomas, S. (1999) 'Impacts of design/information technology on project outcome'. *Proceedings NIST GCR 99-786*, 46 p, The Conference of the National Institute of Standards and Technology, Washington, D.C.

Thomas, S., Macken, C. and Lee, S. (2001) 'Impacts of design/information technology on building and industrial projects', *Proceedings NIST GCR 01-828*, The Conference of the National Institute of Standards and Technology, Washington, D.C.

Thompson, R. L. and Higgins, C.A. (1991) 'Personal computing: toward a conceptual model of utilization', *MIS Quarterly,* **15** (1), 125-143

Venkatesh, V., Morris, M. G. and Ackerman, P. L. (2000) 'A longitudinal field investigation of gender differences in individual technology adoption decision-making processes', *Organizational Behavior and Human Decision Processes,* **83** (1), 33-60

Venkatesh, V., Morris, M. G., Davis, G. B. and Davis, F. D. (2003) 'User acceptance of information technology: toward a unified view', *MIS Quarterly,* **27** (3), 425-478

Yang, L-R. (2007) 'Exploring the links between technology usage and project outcomes', *Construction Management and Economics*, ***25*** (10), 1041-51

Young, R. and Jordan, E. (2008) 'Top management support: mantra or necessity?', *International Journal of Project Management,* **26** (7), 713-725

An IPD Approach to Construction Education

15

Jennifer MacDonald, (University of Technology Sydney, Australia)

Julie Mills, (University of South Australia, Australia)

Abstract

The construction industry worldwide is moving towards more collaborative working practices, such as integrated project delivery (IPD). The era of the Master Builder has passed; modern construction projects are too complex for one person to understand all aspects. New specialist disciplines are emerging, with experts engaged at earlier stages of the design process. However, this collaborative approach is not reflected in the current education of architecture, engineering, and construction (AEC) professionals. Students of the separate disciplines are generally educated in isolation from the others. On graduation, they are then expected to be able to work in integrated teams without prior training in multidisciplinary teamwork. This contributes to the poor levels of trust and inadequate information sharing that plague the industry and prevents it from fully embracing the productivity savings that BIM (building information modelling) and IPD promise. Academics of the AEC disciplines also work in isolation, and the same courses are developed from scratch every time a new academic arrives at an institution. This paper proposes an integrated approach to the teaching of AEC subjects, including a framework to assist academics in adapting their existing curricula, and reports on some initial trials carried out in three Australian Universities.

Keywords: Building Information Modelling (BIM), PLIM, IPD, Collaborative education framework

Introduction

Academics in the architecture, engineering and construction (AEC) disciplines are currently facing enormous changes due to disruptive technologies and processes emerging in both the higher education sector and the wider construction industry. Governments are demanding that universities prove their worth as education providers and research generators in order to secure further public funding. At the same time, universities in general are also facing pressure from the "Google Generation": the rise of Massive Open Online Courses (MOOCs) and other free or low-cost courses provided over the internet means that students have more choice in their courses and may be less willing to pay high fees for traditional chalk-and-talk transmission-style classes in Universities. The construction industry is also facing disruptive changes, due to the on-going global financial crisis and the introduction of new technologies and processes, such as building information modelling (BIM). However, in general, AEC curricula have not been adapted to reflect these industry changes.

Construction is a litigious, inefficient industry, with low levels of trust and poor information sharing. Various sources estimate that inadequate documentation contributes an average of 10-15% to typical project delivery costs (e.g. Woolstenholme (2009); Allen Consulting Group (2010)). Innovative construction projects employing collaborative, Integrated Project Delivery (IPD) methods and BIM technologies have been shown to deliver better, energy efficient products for their clients, while saving costs on delivery through reduced requests-for-information (RFIs) and re-work (e.g. Bernstein, 2012). However, the current shortage of building design professionals, trained in collaborative design and construction practices, remains a barrier to universal adoption of collaborative working practices in the industry.

Just as industry must undergo a paradigm shift from its old combative culture to one of integration and information sharing, so must academia.

The need for a framework to support adoption of collaborative design education by Architecture, Engineering and Construction (AEC) schools has been stated previously (Macdonald & Mills, 2011). A framework (called the "IMAC Framework"), to support academics in adapting their curricula to include collaborative learning and working techniques, with the assistance of BIM tools and processes, has been developed with the aid of an OLT (Australian Office of Learning and Teaching) grant. This paper discusses current trends in multidisciplinary AEC education and the progress of the IMAC trial in three Australian partner institutions.

Innovate to Survive: Reality for Both Industry and Academia

The global financial crisis has driven many businesses to the wall and many of the survivors have realised they need to innovate to win further work, and to be more efficient in the way they work in order to survive. Interestingly, this trend seems most apparent in countries whose economies, and construction industries, have been hardest hit (in the US and UK in particular) and less so in better-off economies such as Australia's (Allen Consulting Group, 2010). This effect is illustrated by Teicholz' graph (Teicholz, 2013) (Fig. 1), which shows a moderate increase in real value added per US construction employee in the years after the 2007 downturn. However, the graph also illustrates the poor performance of construction compared to the manufacturing industry over the same period (the upper line indicates manufacturing). Even without factoring in the effects of the downturn, construction productivity has declined over the past 30 years (Teicholz, 2004).

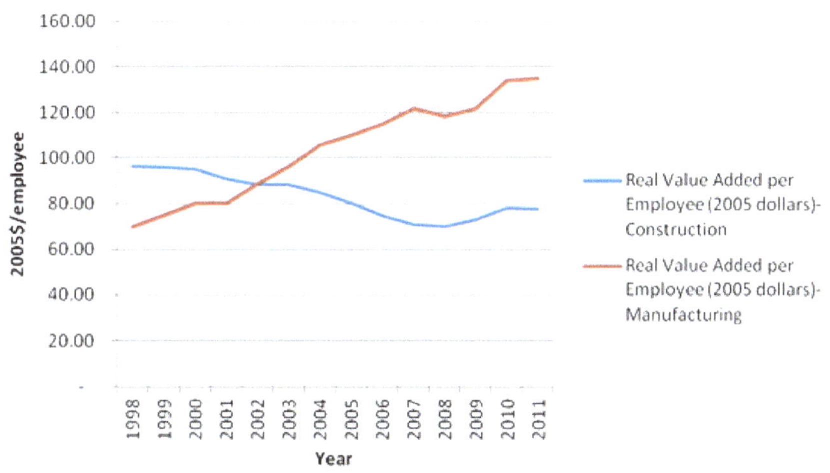

Figure 1 Value Added per US Employee in 2005 dollars Source: Teicholz (2013)

The construction industry is extremely fragmented and lacking integration (NIST, 2004; Woolstenholme, 2009). Additionally, reports show that the quality of project documentation has declined over the past 20 years and that poor documentation is contributing an additional 10 to 15% to project costs. 60-90% of all variations are due to poor design and documentation (QCIF, 2005). It is worth noting, here, that the introduction of computer aided drafting (CAD) tools, far from improving the quality of documentation, has actually coincided with its decline.

In an attempt to achieve greater efficiencies, and echoing earlier changes adopted by the car and manufacturing industries, the construction industry worldwide is moving towards more collaborative working practices, aided by building information modelling (BIM) tools and processes. Collaboration between multidisciplinary, trans-global AEC teams is becoming commonplace (O'Brien et al., 2003). Globalisation means that consultants can work from almost anywhere in the world and building elements can be prefabricated offshore and simply assembled on site. In the future, construction workers are likely to become assembly workers as prefabrication takes over. The increasing complexity of our buildings also means that our construction professionals must collaborate at earlier stages on jobs to produce the best result for clients. However, for the industry to move wholesale into integrated, collaborative practices, a seismic change is required in its culture. The culture of the construction industry has traditionally been very pugilistic, with minimal trust between parties on projects, aided by an atmosphere of litigation and punitive contracts. This lack of trust does not encourage information sharing and collaboration. Industry needs to move from the current culture of fragmentation, litigation, mistrust, and withholding of information to one of open-ness, collaboration, teamwork, and trust if it really wants to maximise the potential of BIM and improve overall productivity.

The current models for teaching students of the AEC disciplines do not support the new demands from both academia and industry. However, there is also a great opportunity to rethink the way AEC courses are developed and to become more efficient in delivering them. Holland et al. (2010) describe the need for "T-shaped people" in the construction teams of the future; that is to say, people who have deep knowledge of their own professional discipline, but who also have broad knowledge of the other disciplines in the team.

Awash in Acronyms: Defining BIM or PLIM or..?

The uptake of BIM has snowballed in recent years but at the same time "BIM" has become a fashionable marketing term, promoted by software vendors. Unfortunately, these vendors have led some professionals to think that to "do BIM" they simply have to purchase a 3D-rendering software package. Some industry figures have coined the term "BIMwash" to describe the phenomenon whereby many firms are claiming to be "doing BIM" but very few are using it to its full potential and integrating with all the other construction disciplines (Miller, 2009).

Despite the glossy marketing materials produced by various software manufacturers, there is still no universally accepted definition for BIM. BIM is not a software product, but an approach to creating and managing information about projects. The term "Building Information Modelling" appeared in print (in English) in 1992 (van Nederveen & Tolman) though, as stated previously, the tools and processes and ideas have been evolving since at least the 1970s (e.g. Eastman, 1975) and many of the processes mimic those adopted by the car industry decades ago, such as just-in-time manufacturing (JIT). BIM can be a noun (the building information *model*) or a verb (the modelling *process*). Additionally, the individual words making up the acronym can mean different things to different professionals:

(B) **Building** can apply to a single building envelope or the wider built environment;

(I) **Information** is really the most important part of the acronym, as the concept is all about creating and sharing quality information about a project and not losing or recreating information when moving between phases;

(M) **M** originally stood for **Modelling** or **Model** but current literature tends towards the more appropriate definition of **Management** (e.g. Race, 2012).

The major problem with the term "building information modelling" is that it is inadequate for describing the complexity involved in creating and managing information over the lifecycle of construction projects, whatever their scale. Race (2012) proposes borrowing from the acronym PLM (product lifecycle management) and replacing BIM with **PLIM** (Project Lifecycle Information Management), as this is a more accurate indication of the intent to create a comprehensive package of information representing the life of a facility.

Overcoming Institutional Inertia

> *Many academic programs still produce students who expect they will spend their careers working as heroic, solitary designers. But integrated practice is sure to stimulate a rethinking of that notion. Pedagogy must focus on teaching not only how to design and detail, but also how to engage with and lead others, and how to collaborate with the professionals they are likely to work with later.* Pressman (2007), (p3)

Since engineering and architecture emerged as separate professional disciplines, AEC students have been educated in isolation from each other. Starzyk and McDonald (2010) note that the focus of architectural education in the past was on developing individual skills such as being able to draw. Now, they state, "the importance of personal skill is yielding to the primacy of collective knowledge".

The complexity of modern building projects and technologies means that nobody can be a master of all anymore. Often the professions do not have a deep understanding of the information that each requires at different stages of a project. Time is thus wasted stripping out and even rebuilding models, when the models could have been set up more efficiently from the start of the process and unnecessary detail excluded prior to model exchange. If students are educated to work collaboratively and to learn the requirements of the other disciplines before they graduate, this level of misunderstanding is likely to be removed in future and trust improved. However, In order to bridge the disciplinary silos in industry, we need to start by breaking down the silos that exist in academia.

A Thorny Question: Are We Educating or Training?

The concept of creating job-ready graduates brings to the fore the "training vs. educating" debate. Traditionally, Universities have focused on teaching theory, with many faculty members believing that it is industry's role to train incoming graduates in job-related skills. The authors have frequently heard the refrain "we're not teaching students to press buttons" being used among educators who believe that BIM is just another CAD tool. However, the uptake of BIM is facilitating process, technological, and cultural changes, and its benefits extend far beyond mere visualization: students cannot be expected to "teach themselves BIM" any more than they could be expected to "teach themselves structural engineering". BIM actually provides a great opportunity to engage students more effectively and to aid understanding of how buildings are constructed. Hardy, quoted in Deutsch (2011, p 202) states:

> *When I look at the logic of construction means and methods that BIM inherently teaches, I see the potential to educate…*

Any major change process is likely to encounter resistance. Some of the difficulties for academia in introducing BIM may include:

1. Questions about how to fit new topics into a crowded curriculum.
2. Reluctance to change teaching habits established over many years.
3. For those who may have developed their own niche or expertise, there may be resistance to take on a new subject, about which they are not an expert, or to retrain in an area they are not familiar with.
4. As the technologies supporting BIM evolve at a rapid pace, academics who have been out of industry for some time may feel overwhelmed trying to keep abreast of them.
5. The traditional silos of architecture, engineering and construction schools can be difficult to bridge. As in industry, mistrust of the other professions also exists in academia, and questions can arise as to who is responsible for (and who will pay for) cross-disciplinary courses. *"The biggest hurdle is with true interdisciplinary efforts due*

to conflicting student/faculty schedules and lack of compensation for more than one faculty member involved in a course" (Architectural Engineering Lecturer quoted in Vogt (2010, p 26)

6. Size of classes. Particularly in Australia, many academics face class minimum class sizes of 80 students, and the resources and time required to convert large cohort standard lecture-based courses into smaller multidisciplinary teamwork-based courses may seem an insurmountable challenge.

Motivation may also play a factor in the success of developing integrated curricula. The main motivation for industry to move towards collaborative working and the use of BIM has been pressure from major Clients and various governments, and the opportunity for improved profits and competitiveness. AEC educators are not generally subject to these same pressures. However, the construction industry has expressed a need for graduates skilled in collaborative building design and BIM. For example, the Built Environment Industry Innovation Council (BEIIC) wrote to all the Deans of Australian Built Environment Faculties in June 2010 to enquire as *"to what extent the universities are embracing new technologies such as BIM and equipping our future professionals with cutting edge experience"* (BEIIC, 2010). The Australian National BIM Initiative defines multi-disciplinary AEC education as one of six key areas needing support from both industry and government (buildingSMART, 2012).

The AEC professional bodies are also beginning to apply pressure to academic institutions to develop integrated courses, through proposed changes to accreditation criteria. For example, the Australian Institute of Architects (AIA) and Consult Australia set up a joint BIM and IPD task force in 2012 (AIA, 2012), including a dedicated BIM in Education Working Group.

CODE BIM: Collaborative Design Education using BIM

The Australian Office of Learning and Teaching (OLT) has funded a project to develop a framework to assist AEC academics in introducing collaborative courses utilising BIM tools and processes. The Australian partner institutions in this project are the University of Technology, Sydney, the University of South Australia and the University of Newcastle.

The initial stages of the project involved benchmarking current practice in academic institutions worldwide. According to Barison and Santos (2010), the majority of US institutions (excepting pioneers such as Georgia Tech) began to introduce BIM in their courses from 2002, and it is a rapidly developing field. For example, Morse (2009) lists eight US Academic Institutions and records survey responses from them indicating that 82% were providing formal teaching in BIM. A current web search reveals that 100% of the institutions listed by Morse now teach BIM in some of their courses, at least 17 of the 18 (94%) ABET-accredited Architectural Engineering programs in the US now offer some level of BIM instruction, and at least 9 out of the 12 ABET accredited Construction Engineering institutions offer BIM courses (75%) (Macdonald, 2011a).

To explore further the current practice and understanding of BIM and collaborative AEC education amongst AEC faculty, interviews were conducted with senior AEC academics and researchers from four leading Universities in Australia, three in the UK, one in Sweden, two in the Netherlands, one in Canada and five in the USA, over a period of two years. Some results of this study were reported previously (Macdonald & Mills 2011). In 2010, the author was also involved in conducting a large on-line industry survey of AEC professionals and educators across Australia and New Zealand. The responses from the academic sector to this survey indicated that around 70% were using BIM at some level within their schools.

The most successful implementations of collaborative courses have been between Schools housed within the same financial unit (whether that is a faculty or department). For example,

Penn State has implemented extremely successful capstone courses involving collaborative groups of students from its four architectural engineering disciplines, architecture and landscape architecture, as described by Solnosky *et al.* (2013). It appears that academics around the world face the same issues in relation to overcoming bureaucratic hurdles, such as which school pays for teaching of another school's students. Some schools have found that it is easier to develop collaborative courses between institutions than across faculties within their own institution. For example, the University of Southern California (USC) has run graduate-level courses with students from Virginia Tech and the University of Texas at Austin. Stanford University has collaborated with Twente University, Netherlands, to develop graduate-level courses in construction management with BIM support (Peterson, *et al.*, 2011) Other institutions have found that collaboration is easier when creating Masters-level courses, such as the Master of Design-Build program run jointly between Auburn University's Schools of Architecture and Building Science, and Salford University's Master of BIM and Integrated Design. So far, no university appears to have implemented fully collaborative courses across the three separate AEC disciplines at undergraduate level.

A striking difference observed between the US schools visited and those in Australia was typical class size. Class sizes continue to grow in Australia, boosted by the removal of government caps on entry numbers. However, staff numbers have not increased proportionally and it appears that many institutions have embraced larger and larger lecture-format classes to cope. BIM requires multidisciplinary collaboration, but it is very difficult to foster collaborative, technology-intensive teamwork in a classroom format consisting of a 3-hour weekly lecture to over 90 students. The largest multi-disciplinary class observed in the US contained 25 students, and some had as few as six.

Lectures, both physical and online, involve a transmission style of teaching that tends to focus on the lower levels of the taxonomy of learning proposed by Bloom *et al* (1956). Where small group, collaborative, problem-based learning approaches excel is in promoting the higher levels of learning in the taxonomies. Saljo (1979) also defines deeper skills that students should develop, including making meaning and reinterpreting. The development of these types of skills is what Universities and other training providers should be focusing their efforts on. Transmission teaching just involves the broadcasting of data to a largely passive audience: knowledge cannot be transmitted into someone else's head.

IMAC: Illustration, Manipulation, Application, Collaboration

The development of a framework to support academics to incorporate collaborative design principles in their curricula has been described in detail previously (Macdonald, 2012). The framework comprises four key stages: Illustration, Manipulation, Application and Collaboration (IMAC). These stages do not necessarily always follow the years of a degree course, but can be viewed as building blocks of learning. The framework comprises two components: a benchmarking tool and a separate guide to implementation.

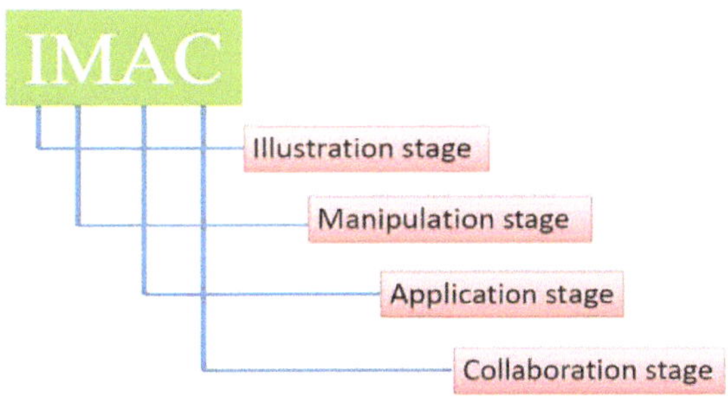

Figure 2 The four stages making up the IMAC framework

The stages have been mapped to various levels on the taxonomy of learning proposed by Bloom *et al* (1956) and expanded by Krathwohl *et al* (1964). As the IMAC framework aims to assist development of both technical (I.T and discipline-specific) and interpersonal (collaborative and teamwork) skills, it straddles the cognitive and affective domains.

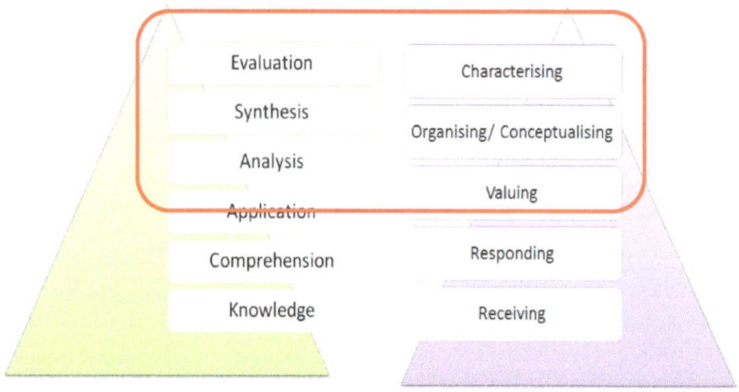

Figure 3 Bloom's Taxonomy of Learning: Cognitive (L) and Affective (R) Domains
(Bloom et al., 1956; Krathwohl et al, 1964)

The framework does not dictate in which academic year each stage should be introduced. Students from the different AEC disciplines study courses of varying lengths and some skills are introduced earlier in some courses than others. For example, students of architecture tend to be introduced to modelling tools from first year whereas students of structural engineering might only be introduced to them in third year. It may also be possible to progress between stages within one academic year. The framework also considers suitable delivery methods at each stage, aiming to achieve deeper levels of learning as students progress through their education.

The benchmarking component of the IMAC framework has been used to benchmark existing courses at the three institutions and to plot targets for future curriculum developments. The framework recognises that the different disciplines will not be aiming to achieve full collaboration in all courses or areas – for example, architecture graduates will be expected to be able to create full BIM models from scratch whereas engineers and construction managers would usually only be expected to be able to manipulate existing models for their own analysis purposes. The framework tool thus suggests different targets (shaded, refer Fig 4) for the different discipline areas, and it is expected that these will be mapped to professional accreditation criteria as these are developed.

	BUILDING TECHNOLOGY	ENVIRONMENT		MANAGEMENT			IT		SPECIALISM		
	Framing	Materials	Sustainability	People	Communication	Teamwork	General IT	BIM tools	A	E	C
Collaboration											
Application											
Manipulation											
Illustration											
0 Not Used											

Figure 4 IMAC chart for an Engineering Course (suggested Target Levels shaded)

The project considers two separate aspects of collaboration: collaborative working and collaborative learning. Collaborative working involves student teams tackling project-based tasks that simulate the new collaborative industry working practices. In other words, they are collaborating to achieve a goal (e.g. the design of a building). By contrast, for collaborative learning it is the learning itself that is the goal. There is a large body of pedagogical research showing that students engage and learn much better through teamwork and team learning. The Stoic philosopher Seneca stated that, "by teaching we learn", and the theory that students learn more from teaching has been proven through research (Annis, 1983; McKeachie *et al*, 1986). The teacher acts more like a peer in the collaborative environment.

Assessing the IMAC Curriculum Framework

The IMAC framework is being trialled at three Australian partner institutions, across a variety of year levels and disciplines. Data will be collected from these trials during the Spring and Autumn 2013-2014 semesters. The curriculum assessment methodology that will be followed is similar to that described by Mills (2002), based on earlier research by Rosier and Keeves (1991) and Treagust (1987). This methodology is based on dividing the curriculum into four stages, namely: (1) intended; (2) implemented, (3) achieved and (4) perceived. Mills (2002) summarises the details of the stages as shown in **Table 1**, below. Jackson (1992) also describes a similar four-step approach to evaluating the curriculum, for which he suggests the terms (1) official; (2) enacted; (3) delivered and (4) experienced. Jackson's terms are shown in brackets in Table 1.

Stage	Description
Intended (Official) Curriculum	The original vision underlying a curriculum, possibly in the form of a national curriculum document or accreditation criteria. Within a particular course, this may be presented in the form of a course or subject document
Implemented (Enacted) Curriculum	The actual instructional process as implemented
Perceived (Delivered) Curriculum	The actual learning experiences as perceived or experienced by the students
Achieved (Experienced) Curriculum	The resulting learning outcomes of the students

Table 1 The Four Stages of Curriculum for Evaluation (Mills, 2002)

In terms of assessing the effectiveness of the IMAC curriculum framework, it will not be possible (due to the two-semester time constraint) to assess students going through the entire four-stages of the framework. Instead, selected existing courses at different stages have been selected to trial best-practice collaborative modifications. These will be used as case studies for the research. Course co-ordinators will be asked to supply their course objectives (learning objectives) in order to determine the intended curriculum. The other stages of the curriculum will be assessed using pre- and post- course surveys, classroom observation, journals, reviews of artefacts produced and focus groups.

Conclusions and Some Proposals

In response to some of the questions raised earlier in this paper, integrating principles of collaboration and BIM technologies into existing classes throughout the curriculum should reduce the need to develop completely new courses. Bolt-on capstone courses, though useful as a first foray into collaborative courses, do not constitute curriculum renewal. In order to encourage true curriculum renewal, the professional bodies should update their accreditation criteria to reflect the industry need for graduates skilled in BIM and collaborative working. Accreditation criteria provide the greatest incentive for academic institutions to instigate changes to their curricula.

Ironically, the industry problem that is creating huge demand for BIM-related courses is also one that affects employers of academics: how to recruit (or train) and then retain teaching staff skilled in the areas of collaborative working and BIM technologies? This is a particularly difficult question given the high demand, and consequently high salaries, on offer in industry. Many universities are reliant on assistance from guest tutors from industry when introducing technical courses in BIM, but it would be wise for these institutions to develop strategies to capture this external knowledge and train faculty members in addition to their students. AEC education should perhaps take its cue from Integrated Practice in the construction industry by looking at ways to become more efficient and collaborative. Why, for example, should institutions not pool their resources for the teaching of lower level courses and use their valuable faculty to deliver more intensive courses at higher levels? For example, engineering mathematics and statistics classes tend to cover the same content all over the world. An excellent lecturer could be nominated from one institution to provide a definitive online course on one topic that students in all engineering courses are required to pass in their own time before progressing to more advanced in-class courses. This is not really too dissimilar to many institutions setting textbooks that have been written by academics in other universities.

The author has attended many BIM workshops and conferences over the past few years and it seems that general questions from industry have moved on from *"what is BIM and why should we adopt it?"* to *"we accept that we need to adopt BIM, now how do we go about doing so?"* Although AEC academics (with notable exceptions) generally appear to be at the earlier stage of questioning, it is likely that they will also move towards the question of implementation, and the framework described should provide assistance in this. The future belongs to the integrators!

> *I am looking for a lot of men who have an infinite capacity to not know what can't be done...* [Henry Ford]

The IMAC framework will be trialled in the Australian partner institutions over the Spring and Autumn Semesters 2013-2014 and the results published in late 2014. More information can be viewed at http://codebim.com

References

AIA (2012) *BIM and IPD Steering Committee*, Documents produced by the four working groups are available online: http://www.bim.architecture.com.au/groups/

Allen Consulting Group (2010) *Productivity in the buildings network: assessing the impacts of Building Information Models*, report to the Built Environment Innovation and Industry Council, Sydney, October 2010

Annis, L. F. (1983) 'The process and effects of peer tutoring', *Human Learning,* **2**, 39-47

Barison, M.B., Santos, E.T. (2010) 'BIM teaching strategies: an overview of the current approaches' in *Proc. of the International Conference on Computing in Civil and Building Engineering*, Nottingham, 2010

BEIIC (2010) *Letter to all Deans of Australian Built Environment Faculties*, Unpublished

Bernstein, H.M. (e.d.) (2012) *SmartMarket Report: The Business Value of BIM in North America*, McGraw-Hill Construction, Bedford, MA., USA

Bloom, B.S., Englehart, M.D., Furst, E.J., Hill, W.H., Krathwohl, D.R. (1956) *The Taxonomy of Educational Objectives, The Classification of Educational Goals, Handbook I: Cognitive Domain,* David McKay Company, New York, NY

buildingSMART (2012) *The National BIM Initiative*, Report to DIISRTe, June 2012, Available http://buildingsmart.org.au/nbi-folder/NationalBIMIniativeReport_6June2012.pdf

Deutsch, R. (2011) *BIM and Integrated Design: Strategies for Architectural Practice*, John Wiley & Sons, Inc., New Jersey

Eastman, C. (1975) 'The use of computers instead of drawings in building design', *AIA Journal*, March, **63** (3), 46-5

Holland, R., Messner, J., Parfitt, K., Poerschke, U., Pihlak, M., Solnosky, R., (2010) 'Integrated Design Courses Using BIM as the Technology Platform', National Institute of Building Sciences, *Annual Meeting of EcoBuild America Conference*, December 7, 2010, Washington, DC

Jackson, P. (1992) Conceptions of curriculum and curriculum specialists, In P. Jackson (Ed.), *Handbook of Curriculum Research*, 3-40, New York: MacMillan

Krathwohl, D.R., Bloom, B.S., Masia, B.B. (1964) *The Taxonomy of Educational Objectives: The Classification of Educational Goals, Handbook II: Affective Domain*, David McKay Company, New York, NY

Macdonald, J.A. & Mills, J.E. (2011) 'The Potential of BIM to Facilitate Collaborative AEC Education', American Society for Engineering Education Annual Conference, Vancouver, Canada, June 2011, in *Proceedings of the 118th ASEE Annual Conference*, American Society of Engineering Education, Vancouver, Canada

Macdonald, J.A. (2011a) 'BIM – Adding Value by Assisting Collaboration', LSAA 2011 Conference, Novotel, Sydney Olympic Park, October 2011 in *Proceedings of LSAA 2011 Conference, ed. Kneen, P.,* Lightweight Structures Association Australia, Sydney. Paper available at http://codebim.com/bim-library/codebim-project-publications/

Macdonald, J.A. (2012) 'A framework for collaborative BIM education across the AEC disciplines. *37th Annual Conference of Australasian University Building Educators Association (AUBEA)*, 4-6 July 2012, Sydney Australia

McKeachie, W. J., Pintrich, P. R. Lin, Y.-G, & Smith, D. A. F. (1986) *Teaching and Learning in the College Classroom: A Review of the Research Literature*. Ann Arbor: University of Michigan, National Center for Research to Improve Postsecondary Teaching and Learning

Miller, R. (2009) 'The BIM Mirage or BIMwashing', online blog entry in *The Commercial Real Estate Revolution*, October 2009, available http://thecrerevolution.com/2009/10/the-bim-mirage-or-bimwashing/

Mills, J.E. (2002) *The Effectiveness of Project-Based Learning in Structural Engineering*, PhD Thesis, Curtin University, available at http://espace.library.curtin.edu.au/R?func=dbin-jump-full&local_base=gen01-era02&object_id=12603

Morse, E. J. (2009) *An Online Case Study Resource For Building Information Modeling In College Education.* (Worcester Polytechnic Institute). Available at http://www.wpi.edu/Pubs/E-project/Available/E-project-090109-010641/unrestricted/EJM-BIMinEducation.pdf

NIST (National Institute of Standards and Technology) (2004) *Cost Analysis of Inadequate Interoperability in the U.S. Capital Facilities Industry*. NIST GCR 04-867. Gaithersburg, Md.: NIST

Pressman, A. (2007) 'Integrated Practice in Perspective: A New Model for the Architectural Profession', *Architectural Record*, May 2007, http://archrecord.construction.com/practice/projDelivery/0705proj-3.asp

QCIF (Queensland Construction Industry Forum) (2005) *Getting it Right First Time*

O'Brien, W., Soibelman, L., Elvin, G. (2003) 'Collaborative Design Proceses: An Active- and Reflective- Learning Course in Multidisciplinary Collaboration', *Journal of Construction Education*, Summer 2003, **8** (2), 78-93

Peterson F., Hartmann T., et al. (2011) 'Teaching construction project management with BIM support: Experience and lessons learned.' *Automation in Construction*, **20**, 115-125

Race, S. (2012) *BIM Demystified: An Architect's guide to Building Information Modelling/Management (BIM)*, RIBA Publishing, London

Rosier, M.J. & Keeves, J.P. (Eds.) (1991) *The IEA study of science I: Science education and curricula in twenty-three countries*, Pergamon Press, London

Säljö, R. (1979) 'Learning in the learner's perspective: 1. Some common-sense conceptions', *Reports from the Institute of Education, University of Gothenburg*, 76

Solnosky, R., Parfitt, M., and Holland, R. (2013) 'An IPD and BIM Focused Capstone Course Based on AEC Industry Needs and Involvement.' *J. Prof. Issues Eng. Educ. Pract.*, 10.1061/(ASCE)EI.1943-5541.0000157 (Feb. 28, 2013)

Starzyk, G.F. and McDonald, M. (2010) 'The Collaborative Dance: Only Three Steps' in *Proc of the BIM-Related Academic Workshop*, December 7-9 2010, Washington DC, eds. Salazar, G. and Issa, R.

Teicholz, P. (2004) 'Labor productivity declines in the construction industry: causes and remedies'. *AECbytes Viewpoint*, Issue 4, April 14

Teicholz, P. (2013) 'Labor-Productivity Declines in the Construction Industry: Causes and Remedies (Another Look)'. *AECbytes Viewpoint*, Issue 67, March 14

Treagust, D.F. (1987) 'Exemplary Practice in High School Biology Classes' in Tobin, K.G. & Fraser, B.J. (eds.) *Exemplary practice in science and mathematics education,* 29-44, Curtin University of Technology, Perth

van Nederveen, G.A. & Tolman, F. (1992) 'Modelling Multiple Views on Buildings', *Automation in Construction*, December, Vol 1, Number 3, pp215-224

Vogt, B.A. (2010) *Relating Building Information Modeling & Architectural Engineering Curricula*, Department of Architectural Engineering, Kansas State University, MSc Thesis

Wolstenholme (2009) *Never Waste a Good Crisis: A Review of Progress since Rethinking Construction and Thoughts for Our Future*, Constructing Excellence, October 2009, available online: www.constructingexcellence.org.uk/pdf/Wolstenhome_Report_Oct_2009

Textual Complexity of Standard Conditions used in the Construction Industry

16

Raufdeen Rameezdeen, (University of South Australia, Australia)

Anushi Rodrigo, (University of Northumbria, UK)

Abstract

Clearly written communication aids the understanding of construction contracts, resulting in less disputation. Past research, using opinion surveys rather than objective criteria, shows that construction contracts lack clarity and standard forms have become complex over time. The study outlined in this paper uses three objective measures of clarity developed by linguists to establish the readability of construction contracts. In addition, thirty industry professionals participated in a Cloze Test which measured the level of comprehension of clauses concerning disputes. The study verifies that contract conditions are very difficult to read, with college level reading skills needed to comprehend half of the clauses. However, the hypothesis that standard forms have become complex over time was not supported by the study. The study establishes a linear relationship between readability and comprehension, proving the hypothesis that improved readability increases the comprehension of a contract clause.

Keywords: Readability, Clarity, Comprehension, Construction, Contract

Introduction

Successful delivery of a construction project depends on proper communication between the parties (Emmitt and Gorse 2003). A contract stipulates privileges and commitments as well as procedures to be followed by the contracting parties (Ndekugri et al. 2007). A good contract develops an acceptable level of communication ensuring smooth functioning of a project (Kennedy et al. 1997). Lawyers and arbitrators are not the primary users of a contract but the parties to a project and the contract administrators. Therefore, these non-legal professionals' comprehension is fundamental for contract performance (Haapio 2011). According to Smith and Taffler (1992), the usefulness of a narrative is decided by the intricacy of the text (readability) and the ability of users to capture its meaning (comprehension). Furthermore, if the message intended by the writer is to be successfully conveyed, the receiver must be able to both read and understand it (Russell 2007). For construction-related technical communication, such as conditions of contract, it is obvious that both readability and comprehension are vital for determining the contract's effectiveness in practice (Broome & Hayes 1997; Chong et al. 2011). According to Rameezdeen and Rajapakse (2007), when the readability of a contract clause is high, its comprehension by different readers is also high.

As readability and comprehension of contract conditions can improve communication between the parties, an examination of whether contract conditions can be easily read and understood is warranted. While the theme of readability is a mainstream research area in linguistics, research dealing with readability in the context of contract interpretation is minimal. Prior research, using opinion surveys, shows that contract conditions used in construction lack clarity (Broome & Hayes 1997; Bunni 2003; Chong & Zin 2010). Studies also show that standard forms have become complex over time (Bunni 2003). However, these past studies have not used objective criteria to support their claim. In addition, they do not make it clear whether clarity can improve comprehension. Given that the cost of disputes to the construction industry is enormous (Jergeas & Hartman 1994) a focus on clarity of contract conditions may help partly bridge this gap in understanding. In that context, this paper reports the findings of a research study carried out on the clarity of contract conditions and its relationship to comprehension.

Clarity of Contract Conditions

Broom and Hayes (1997) note that standard forms of contract used in construction are plagued with many problems; lack of clarity is the most significant. By clarity, the authors mean the design and layout of the whole contract document, as well as the use and order of words within a sentence (Broom & Hayes 1997, p. 255). Bubshait and Almohawis (1994, p. 134) define clarity as 'the ease with which the language of the general conditions can be understood, and the absence of ambiguities'. According to Broome and Hayes (1997, p. 256) lack of clarity in traditional contract conditions is mainly attributable to long sentence length, poor layout and the presence of many redundant legal expressions. This contention is supported by Ali and Wilkinson (2010), Chong and Zin (2010), and Wright and Fergusson (2009). In their experiment, Masson and Waldron (1994) found that comprehension of a contract document among non-legal experts increased when redundant terms were removed or replaced, sentence structures were simplified and legal jargon were replaced by simplified terms. In order to achieve legal precision, drafters have ignored laypersons' ability to understand the language contained in these documents (Greene et al. 2012). Additionally, most of the legal jargons had derived historically from Latin or French languages and this tradition continues to prevail today (Greene et al. 2012).

Bunni (2003) is of the view that original standard forms were error-free and drafted using precise language. With later editions, the original documents were tinkered with to fill gaps in an *ad hoc* manner making the document more complex and ambiguous. Thus, Bunni (2003) asserts that contract forms have become more complex and difficult to read with each successive edition. According to Wright and Ferguson (2009) this process of constant revision has led to contracts causing disagreements and disputes. According to Broom and Hayes (1997) these revisions are normally carried out by committees and they are to be blamed for lack of clarity in subsequent editions.

Researchers believe that lack of clarity in contract documents can adversely affect the relationship between client and contractor by making it more adversarial (Bresnen & Marshall 2000; Cheung & Yiu 2006). Kennedy et al. (1997) finds this to be true between contractors and sub-contractors as well. Lack of clarity in contract conditions can even lead to disputes (Broome & Hayes 1997; Harmon 2003; Chong & Zin 2010). Most often an ambiguous clause may be the focal point of a dispute (Thomas et al. 1994). Thus, better interpretation and understanding of contract terms by parties would help avoid a dispute from festering (Cheung & Yiu, 2007; Chong & Zin, 2010). Broome and Hayes (1997) attribute interpretation errors mainly to contract clarity and legalese. Legalese may obscure the overall meaning due to unnecessary formality (Wang & Yang 2005; Ali & Wilkinson 2010). However, in some instances legalese helps convey a message in certain terms (Thomas et al. 1994; Candlin et al. 2002).

Legal scholars generally agree that proper drafting of standard forms could improve readability and comprehension (Valentine 1992; Ali & Wilkinson 2010). Researchers widely believe that New Engineering Contracts (NEC) and Engineering and Construction Contracts (ECC) are clearer than other standard forms and constitute a way forward in contract drafting (Broome & Perry 1995; Rameezdeen & Rajapakse 2007; Wright & Fergusson 2009). Chong and Zin (2010) propose many recommendations to improve the clarity of contract conditions; the most profound being improvement of the language structure. This reinforces the observations made in other studies which call for 'clear' and 'plain' language for contract conditions (Broome & Hayes 1997; Rameezdeen & Rajapakse 2007; Ali & Wilkinson 2010).

Measures of Clarity and Comprehension

Linguists have tried to develop measures to determine the clarity of documents. In the beginning different readability formulae were used and later alternative measures such as the texture index, transitivity index and diction scores were developed (Sydserff & Weetman 2002). Though many new measures have evolved, readability formulae are considered to be the most suitable in measuring clarity (Glick et al. 2010). The literature provides many definitions for readability; Klare's definition - 'the ease of understanding or comprehension due to the style of writing' - is the most popular (DuBay 2004, p.3). Other definitions include McLaughlin's (1969, p.640) 'a measure of the linguistic characteristics of a text' and Calderȏn et al.'s (2006, p.49) 'the semantic and syntactic attributes of the written word'. Stamboltzis and Pumfrey (2000, p.58) consider readability to be a cognitive constructive process through which individuals make meaning. DuBay (2004, p.3) explains readability simply as what makes some texts easier to read than others.

Numerous methods are used to obtain the readability of a text. All methods rely on the complexity of words in sentences (Koo et al. 2003). Readability formulae can predict the level of difficulty of a text (Hall 2006). Some of the popular readability formulae are:

- Flesch Reading Ease Score,
- Flesch–Kincaid Grade Level,
- Simple Measure of Gobbledygook,
- Automated Readability Index,
- Spache Readability Formula,
- Coleman Liau Index,
- Gunning Fog Index,
- Raygor Estimate Graph, and
- Fry Readability Graph

Flesch Reading Ease Score (FRES) is a popular, tested and reliable formula for readability testing (Rameezdeen & Rajapakse, 2007), being consistent and highly associated with other indices (Paz et al. 2009). FRES ranges from 0–100, where a score of 0 indicates a text that is very difficult to read, while a score of 100 is very easy to read. Table 1 provides a guide to FRES using an example of the difficulty level based on different reading grades.

FRES	Difficulty Level	Estimated Reading Grade
0–30	Very difficult	Postgraduate
30–50	Difficult	College
50–60	Fairly difficult	High school
60–70	Standard	8th to 9th Grade
70–80	Fairly easy	7th Grade
80–90	Easy	5th to 6th Grade
90–100	Very easy	3rd to 4th Grade

Table 1 FRES guide to comparisons of readability (Source: Rameezdeen & Rajapakse 2007)

The drawbacks of readability formulae are well documented in the literature. Readability tests do not factor in meaning and they should not be considered perfect measures of readability (Harrison & Bakker, 1998). A readability formula evaluates only the text that could be quantified (Orlow et al. 2003). It does not take word order or grammar into consideration (Rutherford, 2003) and neither are reader characteristics (Curtis 1998). Though there are limitations, Velez and Ashworth (2007) argue that the formulae still offer valuable predictions on readability. The popularity is mainly because they are inexpensive, objective and reliable, and can be helpful in detecting certain obvious errors such as excessive sentence length (Sydserff & Weetman 2002).

Comprehension is the understanding of a text and is quite different from readability (Jones 1997). Readability measures the textual complexity of a document while comprehension the reader's understanding (Jones 1997). Thus, comprehension is influenced by the reader's characteristics in addition to the textual complexity (Fanguy et al. 2004). According to the Construction-Integration model developed by Kintsch (1988), text comprehension is achieved by two levels of mental representation: firstly, extraction of information from the text: and secondly, linking it to prior knowledge on the subject. The first level involves a micro level text-driven process of establishing semantic connections between various text elements (van den Broek et al. 2005). When a text lacks clarity, the reader has to make more effort to deal with the micro level textual analysis.

Different methodologies have been used to measure comprehension with the Cloze test, developed by Wilson L. Taylor in 1953, being the most popular (Smith & Taffler 1992). A Cloze test employs several paragraphs with words deleted and asks the respondent to fill in the gaps with appropriate words. The percentage of words correctly filled is known as the Cloze score. When the score is low, the text is considered to be difficult (Smith & Taffler 1992). Table 2 provides an interpretation of the results of a Cloze test.

Cloze	Comprehension level	Estimated reading level
60%–100%	Easy	Unassisted reading
35%–60%	Standard	Instructional, assisted reading
0%–35%	Difficult	Frustration level

Table 2 Cloze guide to comparisons of comprehension (Source: Jones 1997)

The literature identifies several limitations of Cloze tests. McKamey (2006) finds that Cloze tests may function differently for different language groups. A reader's prior knowledge is the key variable in comprehension and a Cloze test provides a means to measure how an 'average' reader actually understands key sections of written material (Fanguy et al. 2004). Therefore, a Cloze test is only a proxy measure to capture the average reader's comprehension (Fanguy et al. 2004). Despite these limitations, linguists prefer the test due to its simplicity and ease of interpretation (McKamey 2006).

Aim and Objectives

The literature review shows that traditional construction contracts lack clarity and this could lead to issues of interpretation and comprehension. However, most studies used opinion surveys and other qualitative methods to arrive at this conclusion. The literature review also shows that standard forms have become complex and difficult to read in their successive editions. This claim is not verified by an objective criterion and there is no clear evidence regarding clarity improving comprehension. Therefore, clarity and its relationship to comprehension of contract conditions remains an understudied area. To overcome this shortfall, the present study analyses the clarity of contract conditions and their comprehension using objective measures that have been developed in the field of linguistics. The study focuses on the second to the most recent editions of Fédération Internationale des Ingénieurs-Conseils (FIDIC) Conditions of Contract for Works of Civil Engineering (FIDIC 1969; FIDIC 1977; FIDIC 1987; FIDIC 1999). FIDIC has a suite of contracts and the Conditions of Contract for Works of Civil Engineering is popularly known as the *Red Book*. To achieve the above aim, this study devised the following three objectives:

1. To measure the clarity of the FIDIC *Red Book* from the second to the most recent editions in order to test the hypothesis that standard forms have become complex over time.
2. To measure the comprehension level of clauses in the FIDIC *Red Book* latest edition (FIDIC 1999) that were related to disputes.

3. To establish the relationship between readability and comprehension to test the hypothesis that improved readability increases comprehension of a contract clause.

Research Methods

The basis of the study was the FIDIC Conditions of Contract *Red Book*. The first edition of the FIDIC Conditions of Contract (International) for Works of Civil Engineering was published in 1957. The second (FIDIC 1969), third (FIDIC 1977) and fourth (FIDIC 1987) editions followed. FIDIC 1987 dropped 'international' from its title (FIDIC 1987; Rameezdeen & Rajapakse 2007). Most recently FIDIC 1999 saw a major change in the layout, text and title (FIDIC 1999; Bunni 2005). The title was changed to Conditions of Contract for Building and Engineering Work – Designed by Employer and is called *New Red Book*. The study analysed the last four editions of FIDIC in order to obtain the following measures of clarity on a clause-by-clause basis (or sub-clause as appropriate). The methodology used for this stage was a specific type of quantitative content analysis similar to the analysis by Raj et al. (2009). All clauses of the four FIDIC versions were analyzed using these quantitative content analysis tools.

1. Flesch Reading Ease Score (FRES)

   FRES = 206.835 – (1.015 x ASL) – (84.6 x ASW) ....(1)

   Where,

   ASL = Average sentence length (the number of words divided by the number of sentences)

   ASW = Aaverage number of syllables per word (the number of syllables divided by the number of words)

2. Average Sentence length (ASL)
   Sentence length measures the number of words in a sentence as in formula 2. The running texts of each clause were counted for words from period to period. This gives the average sentence length for each clause of the standard conditions.

   $$\mathbf{ASL} = \frac{NW}{NS} \qquad \text{.......... (2)}$$

   Where,

   NW = Number of words in a clause

   NS = Number of sentences in a clause

3. Average Packet Length (APL)

   Packet length refers to the number of words between any syntactical punctuation marks in a clause. Zero counts caused by adjacent punctuation marks have been excluded. Hyphenated words, abbreviations, acronyms, and quantities were counted as one word. The punctuation marks used were full-stops, commas, colons, semi-colons, exclamation marks, question marks, long dashes and parentheses. The measure gives the average packet length of each clause.

   $$\mathbf{APL} = \frac{\mathbf{NW}}{\mathbf{NP}} \qquad \text{.......... (3)}$$

   Where,

   NW = Number of words in a clause

   NP = Number of packets in a clause

An interview survey was conducted with 11 experts who have more than five years' experience in construction dispute resolution and dealt with FIDIC 1999 at least twice. The objective was to find the FIDIC 1999 clauses that related to contractual disputes. The sampling was based on peer selection of professionals from the industry (snowball sampling) who routinely worked in dispute resolution and are familiar with contracts based on FIDIC conditions (Robson 2011). The interviewees were asked to identify frequently referred contract clauses in disputes and rank them in the descending order of frequency. The purpose of the interview survey was to reduce the number of clauses used in the subsequent Cloze test. A Cloze test is time-consuming and all clauses of FIDIC 1999 cannot be subjected to it. Therefore, only a sub-set of clauses, the most relevant in terms of disputes, were selected.

Cloze test has a structured procedure that must be followed so that comprehension level of a text can be measured in an objective manner. In total, 30 professionals from the industry who had experience working with FIDIC conditions participated in the Cloze test (refer Table 3 for the participant profile). These participants were selected using Purposive Sampling methodology. They were asked to fill the missing words of each clause as per the standard Cloze test guidelines. The test was administered in a place that was convenient for the respondents (most were done in their offices) but in front of the Research Assistant so that it was similar to a closed-book examination environment. The average Cloze scores were obtained for these clauses as a measure of comprehension.

Category	Number of participants
Quantity Surveyor	24
Project Manager	3
Engineer	2
Architect	1
Total	30

Table 3 Profile of professionals who took part in the Cloze test

Results

As discussed above, the current and all previous editions of the FIDIC *Red Book*, except the very first edition were subjected to different measures of clarity. These measures were obtained for all clauses of the Standard Conditions and the results are summarized in Tables 5–7. The tables clearly indicate that the most recent edition, FIDIC 1999, is the easiest to read having the highest average FRES, lowest average sentence length and lowest average packet length.

To understand how difficult it is to read a Standard Condition, the FRES values obtained from the study have to be viewed in the light of comparable readability levels. Table 8 gives the percentages of clauses belonging to each readability level for the four documents under consideration. It shows that even the easiest document among them, FIDIC 1999, contains 95% of clauses which have college level readability or above. More than 50% of its clauses are 'very difficult' to read which is equivalent to postgraduate level.

Descriptive statistic	FIDIC 1969 (n=119)	FIDIC 1977 (n=142)	FIDIC 1987 (n=175)	FIDIC 1999 (n=165)
Median	21.1	22.9	22.3	28.9
1st Quartile	12.1	13.0	11.4	22.5
3rd Quartile	31.8	33.5	35.15	37.2

Table 5 Distribution of FRES values of FIDIC clauses

Descriptive statistic	FIDIC 1969 (n=119)	FIDIC 1977 (n=142)	FIDIC 1987 (n=175)	FIDIC 1999 (n=165)
Median	56	50	40	24
1st Quartile	39	33	24	16
3rd Quartile	83	78	63	36

Table 6 Distribution of Average Sentence Length of FIDIC clauses

Descriptive statistic	FIDIC 1969 (n=119)	FIDIC 1977 (n=142)	FIDIC 1987 (n=175)	FIDIC 1999 (n=165)
Median	28	12	10	8
1st Quartile	13	6	6	5
3rd Quartile	49	21	18	14

Table 7 Distribution of Average Package Length of FIDIC clauses

FRES	Difficulty level	Percentage of clauses			
		FIDIC 1969	FIDIC 1977	FDIC 1987	FIDIC 1999
0–30	Very difficult	71.7	66.9	72.2	53.6
30–50	Difficult	25.8	29.8	21.8	41.0
50–60	Fairly difficult	2.5	3.3	3.6	3.6
60–70	Standard	-	-	1.8	1.2
70–80	Fairly easy	-	-	0.6	0.6
80–90	Easy	-	-	-	-
90–100	Very easy	-	-	-	-

Table 8 Readability levels of FIDIC clauses

The second objective of this study was to measure comprehension levels of clauses of FIDIC 1999 identified as frequently referred in disputes by an expert panel. The clauses identified (see Table 9) generally matched with the results of similar studies elsewhere (Iyer et al. 2002; Seifert 2005). This enabled the Cloze test to be performed on only 20 clauses as opposed to all the clauses of FIDIC 1999. The average Cloze values of the 20 clauses were between 49.2-65.2, as given in Table 9, with the clause 'Employer's Entitlement to Termination' being the easiest to comprehend while the clause dealing with 'Programme' proved to be the most difficult. There were 6 clauses belonging to the 'easy' level of comprehension while the rest found to be at the 'standard' level. It is interesting to note that none of the identified clauses were at the 'difficult' level of comprehension.

Clause number	Title	Cloze score
Easy		
15.5	Employer's Entitlement to Termination	65.2
8.8	Suspension of Work	64.4
8.10	Payment for Plant and Materials in Suspension	64.1
14.8	Delayed Payment	63.6
14.6	Issue of Interim Payment Certificates	62.2
8.1	Commencement of Works	61.5
Standard		
12.4	Omissions	59.6
13.3	Variation Procedure	59.6
20.1	Contractor's Claims	59.4
8.6	Rate of Progress	58.1
13.1	Right to Vary	57.5
12.3	Evaluation	57.3
8.9	Consequences of Suspension	56.7
14.3	Application for Interim Payment Certificates	56.3
14.7	Payment	54.6
8.4	Extension of Time for Completion	52.9
8.5	Delays Caused by Authorities	52.8
8.2	Time for Completion	51.1
16.2	Termination by Contractor	50.4
8.3	Programme	49.2

Table 9 Mean Cloze scores of FIDIC 1999 clauses that are related to construction disputes (n=30)

The third objective was to establish the relationship between readability and comprehension in order to test the hypothesis that improved readability increases comprehension of a contract clause. To operationalize this research question, Pearson's Correlation Coefficient was used as the measure of association to express the extent to which the two variables, FRES and Cloze scores, were related. The value of Pearson's Correlation Coefficient (r) vary between +1 and -1. Both extremes represent a perfect relationship and 0 absence of a relationship (Tan 2002).

$$r = \frac{N\sum XY - (\sum X)(\sum Y)}{\sqrt{[N\sum X^2 - (\sum X)^2]\,[N\sum Y^2 - (\sum Y)^2]}} \quad (5)$$

Where;

N = Number of subjects

X = FRES score

Y = Cloze score

The direction of the relationship is indicated by the sign of the Correlation Coefficient. A positive sign indicates that high scores of one variable tend to be associated with high scores of the other. A 'r' value of +0.165 was obtained between Cloze and the relevant FRES scores of the clauses under consideration. It implies that when the readability level of a clause is high it is easier to comprehend. However, the strength of the relationship is weak, indicating that there may be other factors influencing comprehension such as prior knowledge, experience, etc.

Discussion

The study used quantitative measures to demonstrate that FIDIC 1999 is the most successful document compared to its predecessors with regard to ease of reading. Nevertheless, it requires a minimum of college level reading skills to comprehend 46% of its clauses. More than 50% of its clauses are found to be 'very difficult', which is equivalent to postgraduate level. The above findings are consistent with the findings of Bunni (2003, p.9), who reports that only 4% of the population could comprehend about 86% of the FIDIC 1977 clauses. Bunni (2003, p.9) places the readability level of FIDIC equivalent to an IQ level of 130 or above. This IQ level equates to a 'difficult' level of standard FRES (Mason and Morris, 2000). The results of the present study show that about 96% of FIDIC 1977 clauses are within this level of readability.

According to Ali and Wilkinson (2010, p. 326), experts suggest 20 words or less per sentence is ideal for clarity. However, the present study finds that the four FIDIC editions having an average of 56, 50, 40 and 24 words per sentence respectively. Bunni (2003, p.9) observes that 86% of FIDIC 1977 sentences are of more than 28 words. A study by Raj et al. (2009, p.217) discovered that the average sentence length of the FIDIC *White Book* was 32 for the first three editions published between 1990 and 1998, increasing to 34 in its latest edition published in 2006. Ali and Wilkinson (2010, pp. 332–333) show a GC/Works/1 and JCT Design and Build Contract clauses with 156 and 134 words, respectively, in a single sentence with a FRES score of 0.

Is this status quo conducive for the wellbeing of the construction industry? Does the contract document inhibit communication between parties to a contract? These questions are particularly pertinent to the construction industry where parties to the contract, client and contractor, are laymen when it comes to legal scholarship (Rhys Jones 1994; Ali & Wilkinson 2010). The parties are supposed to sign and agree on matters they can read and comprehend. While both parties might have support from professionals familiar with these documents, the parties themselves nevertheless need to be fully aware of their rights and obligations prior to the signing of a contract. Unfortunately, FIDIC has failed to provide a readable and understandable document to its users. In terms of the first hypothesis of the study, the results clearly demonstrate that the readability levels of FIDIC have increased over its successive editions. It provides a clue that the drafters are beginning to understand the necessity of clarity in their document and that future editions could be expected to be clearer and simpler than the current edition. The results, therefore, do not support the hypothesis that successive editions are more complex and difficult to read.

In defence of the present status quo with regard to clarity, one could argue that plain language does not suit a narrative that requires a very specific meaning to be conveyed to the reader (Cornes 1996). Cornes (1996) contends that use of plain language could blur legal certainty. Perry (1995) observes that it can cause lack of precision and legal scrutiny required for a legal document. Plain language can bore people and make them lose the significance of a clause (Broome & Hayes, 1997). The other reason for maintaining the status quo stems from familiarity and precedence (Broome & Hayes, 1997). With the use of Standard Conditions, practitioners become familiar with its strengths and weaknesses. In addition, a long trail of precedence can guide a practitioner to better comprehension of the document.

Nevertheless, there seems to be enough room to reduce the difficulty of reading by enhancing the readability of FIDIC documents. A good move in this regard could be observed in the increased readability levels of the NEC/ECC contract clauses (Perry 1995; Broome & Hayes 1997; Rameezdeen & Rajapakse 2007). NEC is a standard contract specifically developed with a focus on reducing legalese and the drafting of plain easy to read language structure. A comparison of summary readability statistics of FIDIC 1999 and NEC 1993 obtained from one of our previous studies is given in Table 10. It clearly shows that NEC is a smaller document with better language structure and readability.

If there is potential to improve communication as a result of increased readability, it is the responsibility of the drafting committee of these documents to look into this aspect seriously in future editions (Raj et al. 2009; Ali & Wilkinson 2010). However, the literature review shows there is a lack of understanding about the benefits of improved clarity on the comprehension of a contract clause. In order to shed some light on this issue, the present study tested the hypothesis that improved readability increases comprehension of contract clauses. The result supports the hypothesis but the relationship between readability and comprehension was found to be weak.

Description	FIDIC	NEC
Counts		
Number of characters	158,024	41,870
Number of words	29,891	8,248
Number of sentences	860	309
Averages		
Sentences per paragraph	1.50	1.40
Words per sentence	28.90	22.10
Characters per word	5.10	4.90
Readability		
FRES	29.70	40.70

Table 10 Summary readability statistics of FIDIC and NEC (Source: Rameezdeen & Rajapakse 2007)

Conclusions

FIDIC 1999 emerged as having the highest average readability level with the lowest Average Sentence Length and Average Packet Length among all FIDIC editions. The average FRES value of FIDIC 1999 is around 29, whereas the other three documents have a value of 21–23. This shows that FIDIC 1999 is the easiest to read of the four editions. Nevertheless, 54%

and 41% of its clauses are of postgraduate and college level readability, respectively. Taken together, about 95% of clauses need at least college level reading skills to comprehend the document. The hypothesis that standard forms have become complex over the years was not supported by this study. In fact successive standard forms have become easier to read. It verifies the self-stated claim of drafters of the latest edition that they wanted to make FIDIC 1999 easier to read than the previous editions. The association between FRES and Cloze scores was found to be directly proportional, with a weak relationship proving the hypothesis that improved readability increases the comprehension of a contract clause.

References

Ali, N.A.N.A. and Wilkinson, S. (2010) 'Modernising construction contract drafting – a plea for good sense'. *Proceedings of the 18th CIB World Building Congress*, 10-13 May, Salford, United Kingdom, 323–345

Bresnen, M. and Marshall, N. (2000) 'Partnering in construction: a critical review of issues, problems, and dilemmas', *Construction Management and Economics*, **18**, 229–237

Broome, J. C. and Hayes, R.W. (1997) 'A comparison of the clarity of traditional construction contracts and of the New Engineering Contract', *International Journal of Project Management*, **15**, 255–261

Broome, J.C. and Perry, J.G. (1995) 'Experiences of the use of the New Engineering Contract', *Engineering, Construction and Architectural Management*, **2**, 271–285

Bubshait A.A. and Almohawis, S.A. (1994) 'Evaluating the general conditions of a construction contract', *International Journal of Project Management*, **12**,133–136

Bunni, N. G. (2003) *Risk and insurance in construction*, Spon Press, London

Bunni, N. G. (2005) *The FIDIC form of contract*, Third ed., Blackwell, Oxford

Calderȍn, J.L., Morales, L.S., Liu, H., and Hays, R.D. (2006) 'Variation in the readability of items within surveys', *American Journal of Medical Quality,* **21**, 49-56

Candlin, C.N., Bhatia, V.K., and Jensen, C.H. (2002) 'Developing legal writing material for English second language learners: problems and perspectives', *Journal of English for Specific Purposes*, **2**, 299-320

Cheung, S.O. and Yiu, T.W. (2006) 'A study of construction mediator tactics – part I: Taxonomies of dispute sources', *IEEE Transactions on Engineering Management*, **53**, 456–470

Cheung, S.O. and Yiu, T.W. (2007) 'Are construction disputes inevitable?', *Building and Environment*, **42**, 752-761

Chong, H. Y. and Zin, R.M. (2010) 'A case study into the language structure of construction standard form in Malaysia', *International Journal of Project Management*, **28**, 255–261

Chong, H. Y., Balamuralitharan, B. and Chong, S.C. (2011) 'Construction contract administration in Malaysia using DFD: A conceptual model', *Industrial Management & Data Systems*, **111**, 1449-1464

Cornes, D. (1996) 'The second edition of the New Engineering Contract', *International Construction Law Review*, **13**, 97-119

Courtis, J.K. (1998) 'Annual report readability variability: tests of the obfuscation hypothesis', *Accounting, Auditing and Accountability Journal*, **11**, 459–471

DuBay, W.H. (2004) *The principles of readability*, Impact Information, California

Emmitt, S. And Gorse, C. (2003) *Construction communication*, Blackwell, London

Fanguy, R., Kleen, B.A. and Soule, L.C. (2004) 'Privacy policies: cloze test reveals readability concerns', *Issues in Information Systems*, **5**, 117–123

FIDIC (1969) *Conditions of Contract for Works of Civil Engineering Construction*, Second ed., Fédération Internationale des Ingénieurs-Conseils, Geneva

FIDIC (1977) *Conditions of Contract for Works of Civil Engineering Construction*, Third ed., Fédération Internationale des Ingénieurs-Conseils, Geneva

FIDIC (1987) *Conditions of Contract for Works of Civil Engineering Construction*, Fourth ed., Fédération Internationale des Ingénieurs-Conseils, Geneva

FIDIC (1999) *Conditions of Contract for Building and Engineering Works- Designed by the Employer*, First ed., Fédération Internationale des Ingénieurs-Conseils, Geneva

Glick, A., Taylor, E.D.D., Valenza, J.A. and Muhammad, F.W. (2010) 'Assessing the content, presentation, and readability of dental informed consent', Journal *of Dental Education*, **74**, 849–861

Greene, E., Fogler, K. and Gibson, S.C. (2012) 'Do people comprehend legal language in wills?', *Applied Cognitive Psychology*, **26**, 500-507

Haapio, H. (2011) 'Contract clarity through visualization: preliminary observations and experiments', *Proceedings of the 15th International Conference on Visualization*, IEEE Computer Society, 13-15 July, London

Hall, J.C. (2006) 'The readability of original articles in surgical journals', *ANZ Journal of Surgery*, **76**, 68–70

Harmon, K.M.J. (2003) 'Resolution of construction disputes: a review of current methodologies', *Leadership and Management in Engineering*, **3**, 187–201

Harrison, S. and Bakker, P. (1998) 'Two new readability predictions for the professional writer: pilot trials', *Journal of Research in Reading*, **21**, 121–138

Iyer, C. K., Satyanarayan, K.N. and Ganesh, L.S. (2002) 'Dispute prone contract clauses – a basis for operational flexibility in contract administration', *Global Journal of flexibility systems Management*, **4**, 55–63

Jergeas, G.F. and Hartman, F.T. (1994) 'Contractors' construction–claims avoidance', *Journal of Construction Engineering and Management*, **120**, 785–795

Jones, M.J. (1997) 'Methodological themes – critical appraisal of the cloze procedure's use in the accounting domain', *Accounting, Auditing and Accountability Journal,* **10**, 105–128

Kennedy, P., Morrison, A. and Milne O. (1997) 'Resolution of disputes arising from set-off clauses between main contractors and subcontractors', *Construction Management and Economics*, **15**, 527–537

Kintsch, W. (1988) 'The role of knowledge in discourse comprehension: a construction-integration model', *Psychological Review*, **95**, 163-182

Koo, M.M., Krass, I. and Aslani, P. (2003) 'Factors influencing consumer use of written drug information', *The Annals of Pharmacotherapy*, **37**, 259–267

Mason, J. and Morris, L. (2000) 'Improving understanding and recall of the probation service contract', *Journal of Community and Applied Social Psychology*, **10**, 199–210

Masson, M.E.J. and Waldron, M.A. (1994) 'Comprehension of legal contracts by non-experts: Effectiveness of plain language redrafting', *Applied Cognitive Psychology*, **8**, 67-85

McKamey, T. (2006) 'Getting closure on Cloze: a validation study of the rational deletion', *Second Language Studies*, **24**, 114–164

McLaughlin, G.H. (1969) 'SMOG grading – a new readability formula', *Journal of Reading*, **22**, 639–646

Ndekugri, I., Smith, N. and Hughes, W. (2007) 'The engineer under FIDIC's conditions of contract for construction', *Construction Management and Economics*, **25**, 791–799

Orlow, M.K.P., Taylor, H.A. and Brancati, F.L. (2003) 'Readability standards for informed-consent forms as compared with the actual readability', *The New England Journal of Medicine*, **348**, 721–726

Paz, S.H., Liu, H., Fongwa, M.N., Morales, L.S. and Hays, R.D. (2009) 'Readability estimates for commonly used health-related quality of life surveys', *Quality of Life Research*, **18**, 889–900

Perry, J.G. (1995) 'The new engineering contract: principles and design of risk allocation', *Engineering, Construction and Architectural Management*, **2**, 197–208

Raj, S., Hilling, J. and Hughes, W. (2009) 'Responsiveness to change by standard-form contract drafters in the construction industry: A case study of the FIDIC White Book', *International Journal of Law in the Built Environment*, **1**, 205–220

Rameezdeen, R. and Rajapakse, C. (2007) 'Contract interpretation: the impact of readability', *Construction Management and Economics*, **25**, 729–737

Rhys Jones S. (1994) 'How constructive is construction law?', *Construction Law Journal*, **10**, 28–38

Robson, C. (2011) *Real world research: A resource for users of social research in applied settings*, Third edition, John Wiley, London

Russell, D.R. (2007) 'Rethinking the articulation between business and technical communication and writing in the disciplines: useful avenues for teaching and research', *Journal of Business and Technical ommunication*, **21**, 248–277

Rutherford, B.A. (2003) 'Obfuscation, textual complexity and the role of regulated narrative accounting disclosure in corporate governance', *Journal of Management and Governance*, **7**, 187–210

Seifert, B.M. (2005) 'International construction dispute adjudication under international federation of consulting engineers conditions of contract and the dispute adjudication board', *Journal of Professional Issues in Engineering Education and Practice*, **131**, 149–157

Smith, M. and Taffler, R. (1992) 'Readability and understandability: different measures of the textual complexity of accounting narrative', *Accounting, Auditing and Accountability Journal* , **5**, 84–98

Stamboltzis, A. and Pumfrey, P. (2000) 'Reading across genres: a review of literature, *Support for Learning*, **15**(2), 58–61

Sydserff, R. and Weetman, P. (2002) 'Developments in content analysis: a transitivity index and DICTION scores', *Accounting, Auditing and Accountability Journal*, **15**, 523–545

Tan W. (2002) *Practical Research Methods*, Prentice Hall, Singapore

Thomas, H.R., Smith, G.R. and Mellot, R.E. (1994) 'Interpretation of construction contracts', *Construction Engineering and Management*, **120**, 321–336

Valentine, D.G. (1992) 'How not to draft a contract', *The International Construction Law Review*, **9**, 526–529

van den Broek P., Rapp, D.N. and Kendeou, P. (2005) 'Integrating memory-based and constructionist processes in accounts of reading comprehension', *Discourse Processes*, **39** (2&3), 299-316

Velez, P. and Ashworth, S.D. (2007) 'The impact of item readability on the endorsement of the midpoint response in surveys', *Survey Research Methods*, **1** (2), 69–74

Wang, W.C. and Yang, J.B. (2005) 'Applications of electronically facilitated bidding model to preventing construction disputes', *Automation in Construction*, **14**, 599–610

Wright, J.N. and Fergusson, W. (2009) 'Benefits of the NEC ECC form of contract: A New Zealand case study', *International Journal of Project Management*, **27**, 243–249

The Role of Business Models in Finnish Construction Companies

Aki Pekuri, Laura Pekuri and Harri Haapasalo, (University of Oulu, Finland)

17

Abstract

Business models are seen as an essential part of successful businesses as they define the way companies create value for their customers and subsequently make profit from their operations. A good business model has a potential to separate a company from its competitors by creating a competitive advantage. The purpose of this research is to explore how managers understand and deploy the business model concept in construction. For this, eight experienced managers have been interviewed and their answers have been analysed using the theoretical aspects of business models as a reference. In the interviews, managers deployed business model concept very differently than what is accustomed in practice and in academia; they relate these models to different project delivery and contract structures or to companies' business segments rather than to analysing how their companies provide value for customers in the selected markets. The results of the interviews indicate that the managers in construction do neither understand the concept properly nor exploit any similar value creation analysis in their business. The interviewees had significant problems describing their companies' business models and value creation logic, pointing out the lack of analysis and understanding of customer values and needs in the project delivery process. This may be one of the overtones of the persistent client dissatisfaction in the construction industry.

Keywords: Business models, Value creation, Business development, Competitive advantage, Strategy

Introduction

Business models - in other sectors a widely used managerial practice for designing, comparing and analysing an organisation's value creation logic - are a less discussed and researched concept in the field of construction. Current literature sees a business model as an essential part of successful business, as its main purpose is to separate a company from others and to give it an advantage over its competitors (Johnson et al. 2008; Teece 2010). A good business model defines the way a company operates, how it creates value for its customers and how it captures value from its operations to make a profit (Magretta 2002; Morris et al. 2005; Suikki et al. 2006; Casadesus-Masanell & Ricart 2011). Thus, it forces managers to think about their business and how it works as a whole. In other words, how different elements of a business model create synergies together. Recent developments in this field of research have also led the creation of visual representations of business models, resulting in a better understanding of different business logics and a position where business models are used as an effective management tool to communicate and execute strategy (Osterwalder & Pigneur 2009; Pekuri et al. forthcoming).

The construction industry has begun to see exceptional innovation in the offerings, delivery systems and overall operating philosophies of some construction companies. For example, IKEA Houses and Toyota Homes may well revolutionise the family house business with their cost-efficient and fast-delivered modular houses, or the hotel chain citizenM with their compact and prefabricated accommodation facilities; these represent a combination of product and process innovation. In California, companies like Herrero Builders Inc. have an organisational philosophy by which they create value for customers through intense collaboration, integrated project deliveries and adopted lean practices. This development puts pressure on traditional construction companies, urging them to sharpen their competitive edge beyond price-related issues. But whether it is this or some other issue that triggers the change, finding a new competitive edge should begin with the critical scrutiny of one's own way of doing things - namely, the business model.

The aim of this research is to explore the way the business model concept is understood and utilised in the Finnish construction industry. This will hopefully shed some light on why the vast majority of construction companies do not seem to differentiate from each other. Proper exploitation of business model thinking in the management of a company necessitates managers to think about the way they create value for their customers. Could it be that this kind of deep analysis of value creation systems has resulted in seemingly similar outcomes, or is it that this kind of comprehensive analysis is lacking altogether? The specific research question this paper aims to address is:

- How do managers in construction understand and deploy the concept of the business model?

The research is focused on Finnish construction companies, who typically serve as a main contractor in construction projects; thus, a vast number of construction actors, e.g. design firms and consultants, are not within the scope of this study. The research seeks to develop an understanding between researchers and construction practitioners. Even though the current business model literature is still somewhat scattered in terms of what specific elements constitute a business model or how it is should be defined and presented, it is unanimous in that all of its aspects are integrated into value creation. The research question is addressed by interviews that explore how managers in construction understand the concept of the business model and whether they associate it with the analysis of value creation.

The Business Model Concept

Focus in Value Creation

The business model concept builds upon central ideas in the field of business strategy and its associated theoretical traditions (Headman & Kalling 2003; Morris et al. 2005). Literature on business models attempts to draw insight from different schools of thought, such as Porter's (1980, 1996) market-based view on competition and strategic positioning and the resource-based view, which conceptualises a firm as a bundle of resources (Prahalad & Hamel 1990; Barney 1991). The difference is that whereas strategy emphasises competition, business models build more on the creation of value for customers (Morris et al. 2005) while still being aligned with the specific strategy, structures and culture of the focal firm (Casadesus-Masanell & Ricart 2010; Wikström et al. 2010). Hence, a business model can be described as the way a company operationalises its strategy to concrete business activities or initiatives (Chesbrough & Rosenbloom 2002; Casadesus-Masanell & Ricart 2010). In addition to its ties with strategy, business models relate to the value chain concept (Porter 1985) and the extended notions of value systems such as the choice of value configuration presented by Stabell and Fjeldstad (1998).

One of the first authors to discuss the concept of business models was Timmers (1998). He identified multiple emerging Internet-based business models and mapped them along two dimensions - their degree of innovation and their extent of functional integration. He concluded that these business models were only feasible because of the openness and connectivity of the Internet. Indeed, business models have often been used in generating and articulating plans to make money through innovations or certain technologies. Chesbrough and Rosenbloom (2002), for example, discussed the role of business models in capturing value from innovations at Xerox. They highlighted the point that it is not enough to merely create value for customers; it is important to also find the right revenue models to capture value, i.e. to make profit. This indicates the systematic nature of the business model - it is not about details of isolated elements, but how the elements are interconnected and how they reinforce each other to form a well-functioning whole.

Key Elements as a Common Language

The term 'business model' has steadily gained widespread popularity since the 1990s, when business models and the changing firm boundaries were discussed in the context of the Internet (Afuah & Tucci 2000; Afuah 2003; Osterwalder 2004; Burkhart et al. 2011). In recent years, the business model concept has been used as a general construct explaining how a firm interacts with suppliers, customers and partners (Zott & Amit 2007). Currently, there is a lack of consensus in regard to what constitutes a business model and how it should be defined, classified or represented among business model researchers (Osterwalder 2004; Morris et al. 2005; Shafer et al. 2005; Nenonen & Storbacka 2010). This lack of consensus has been attributed to the fact that the concept draws from and integrates a wide range of academic and practical disciplines (Chesbrough & Rosenbloom 2002; Pateli & Giaglis 2004; Shafer et al. 2005).

As said, there are multiple studies comparing, summarising and integrating the elements that a business model should contain. Morris et al. (2005), for example, identified 24 different elements from 19 sources, while Shafer et al. (2005) studied 12 different definitions to find 42 different business model elements, categorising them into four groups: strategic choices, value network, create value and capture value. In addition, several others have presented their own frameworks for analysing and conceptualising business models (see Hedman & Kalling 2003; Johnson et al. 2008; Kujala et al. 2010; Wikström et al. 2010; Mutka & Aaltonen 2013). Osterwalder's (2004) comprehensive study proposes a business model framework containing the following elements: value proposition, target customer, distribution channel, customer relationship, value configuration, capability, partnership, cost structure and revenue model. These elements were slightly adjusted by Osterwalder and Pigneur (2009), who created the visual management tool - the business model canvas - for the design and description of business models.

Although most of the studies propose different definitions for the business model, it is possible to identify certain similarities that appear in most of them (Suikki et al. 2006; Nenonen & Storbacka 2010; Pekuri et al. forthcoming). First, the majority of business model definitions include customer value creation as one of their core elements. Second, earning logic is also mentioned in various business model definitions. Third, many business model definitions discuss the value network of a firm, illuminating the relationships that a firm has with various actors in its value network. Finally, the resources and capabilities that a firm has and the types of strategic decisions, choices or principles that it makes are also often included in business model definitions (Nenonen & Storbacka 2010). The most important thing regarding the elements of business models is that they provide a common language for those who discuss business models (for which everyone usually has a different definition and understanding). Osterwalder (2004) and Osterwalder and Pigneur (2009) provide a detailed framework for business model analysis and description, but in this research, the following three elements are adequate to describe the essence of business models:

- 'Value creation system' consists of the most important assets, competences, suppliers and key partnerships that are required to develop and produce offerings that will deliver the created value to customers (Osterwalder & Pigneur 2009).
- 'Offering' refers to the output of the value creation system. With the offering, companies aim to solve customers' problems and satisfy their needs; thus, the goal of any offering is to provide value to a specific customer segment (Osterwalder & Pigneur 2009).

- 'Revenue model' is a blueprint that defines how the company creates value for itself while providing value to the customers (Johnson et al. 2008). Revenue generation measures a company's ability to translate its value proposition into revenue streams (Osterwalder 2004) and to cover its costs.

The Business Model as a Source of Competitive Advantage

Successful firms create substantial value by doing things in ways that differentiate them from the competition, e.g. by developing core competencies, capabilities and positional advantages that are different from those of competitors. The viability of for-profit companies is also tied to the way they capture value and resultantly generate profit. In this regard, the terms 'creating value' and 'capturing value' reflect the two fundamental functions that all organisations must perform to remain viable over an extended period of time (Shafer et al. 2005).

The literature on business models is filled with great stories of how the right business model may change the fate of virtually anything. Magretta (2002) even defined business models as stories that explain how enterprises work. She went on to present a century-old success story of American Express's introduction of travel cheque - which turned the normal cycle of debt and risk on its head while having all the elements of a good business model from value creation to its capture - pointing out that every viable organisation is built on a sound business model regardless of whether this model is articulated or not. Hamel (2000) as well as Osterwalder and Pigneur (2009) described multiple business models using well-known companies to emphasise the importance of business model innovation. Hamel (2000) argued that business model innovation is the only way to avoid competition even temporarily, while Johnson et al. (2008) reported that business model innovation in general is becoming even more important for success than product or service innovation. If a business model is sufficiently differentiated to meet particular customer needs and is hard to replicate, it can result in a competitive advantage (Morris et al. 2005; Teece 2010).

The problem with innovating business models is that only a few managers understand their companies' current business models well enough to develop them further or to change them at the right time (Linder & Cantrell 2000; Johnson et al. 2008). This is where a proper business model analysis using a set of defined business model elements is useful. The defined elements give managers a common language to discuss, design and compare business elements, and most importantly, to analyse the interdependencies between the elements. According to Morris et al. (2005), a business model should be both internally and externally consistent. Internal consistency means fitness between different elements (as they affect and are affected by one another), while external consistency is how the choices in business model elements fit into the external environment.

The variety of business model descriptions found from various kinds of sources demonstrate the fact that business models are no longer discussed and used exclusively in the context of e-business; instead, the concept has been transformed into a management tool that is currently used in all kinds of organisations to create a competitive advantage. In general, business models can concretely be used to capture, visualise, understand, communicate and share the business logic of a firm. They can also be applied in measuring, observing and comparing a company to others and to improve management by ameliorating the design, planning, changing and implementation of business models (Osterwalder 2004).

Research Approach

This is a qualitative research with data consisting of eight face-to-face, semi-structured interviews conducted in Northern Finland between February and March 2013. The research follows the interpretive research tradition. The interviewees were asked about their understanding and use of the concept of the business model in their companies and their answers were analysed via the aforementioned theoretical aspects. The interviews provide a basis for understanding the current role of business models in construction companies. At the same time, they provide the basis for analysing how construction companies could benefit from introducing business model thinking into their management. The interviewees were not given a definition or structure of a business model or its suggested elements; instead, they were asked to answer the questions according to their own conception of the term and to describe the operating logic of their company.

All of the interviewees had more than 15 years of experience in the construction industry, working as chief executives or regional managers in their companies. Interviewees came from three large, two medium and three small construction companies. They were selected from the top of the organisational hierarchy as this position requires the most comprehensive view of a company's business activities. The companies were selected from the same geographical region so that their competitive environment would be similar. The interviews lasted for 1-2 hours each and all of them were recorded and transcribed. Additionally, other sources of material such as company websites were also used to formulate an understanding of the logic by which these construction companies operate.

The Role of Business Models in the Construction Companies

Table 1 depicts the business models of the companies as perceived by the interviewees. Along with the formation of business model descriptions, some interpretation by the authors was needed because the interviewees did not have any systematic way of structuring or describing the answers about their business models.

	Description of the Business Models of Studied Companies
Company 1: A regional business unit of a large international construction group	Operates regionally to understand the local circumstances and the special characteristics of economic and population structures as well as to take into account customers' needs in planning and realising construction. Focuses on project development where it can deploy its business knowledge for customers' benefit. Works in challenging circumstances.
Company 2: A regional business unit of a large international construction and project development company	A pioneer and specialist in project management contracts. The competitive edge is based on an approach that combines strong project development expertise and construction into a consistent and flexible process that allows client participation. Key competences include understanding future trends, systematic plot acquisition and project management in general. Operates in big cities in Russia, Finnish metropolitan areas and other domestic growth centres.
Company 3: A large, domestically-operating construction company	Focus in own development or turnkey projects where the company can assume overall responsibility of the projects. Contracting services offered for stable and professional customers, such as companies, investors, municipalities and cities and in the field of expertise (i.e. nurseries).
Company 4: A medium-sized and domestically-operating construction contracting company	Experienced industrial builder with innate will to concentrate on complex and unique projects. A pure contractor that builds and maintains close relationships with its customers. The current focus in industrial building is on mining projects; its other specialty is hospitals. Serves its long-time industrial customers in the paper and metal industries by executing all kinds of projects as well as offering cost-based maintenance services.
Company 5: A medium-sized and regional construction company	Aims to take on the overall responsibility for projects by operating close to its customers. Acts in multiple fields of operations to grow while in the future the focus will be on development projects. All kinds of sources are exploited to find new profitable projects.

Table 1 con't

	Description of the Business Models of Studied Companies
Company 6: A small, locally-operating construction company	Aim is to master all kinds of construction bar infrastructure within the limited geographical sphere of operations. Aims at turnkey construction where it can use local knowledge in land acquisition. Industrial building is a base for business and housing is used to level market fluctuations. Invests in training its long-term workers and modern equipment to participate in industrial projects.
Company 7: A small, locally-operating construction company	Builds low-rise houses on rented land. Targets projects that last over a year to utilise its capacity efficiently. Uses well-known designers to increase its chance of getting desired pieces of land. Increases its efficiency by using standardised flat types and by targeting certain areas for synergies in logistics and marketing.
Company 8: A small, locally-operating construction contracting company	Aims to operate in complex, unique and slightly inconvenient projects in order to learn from them to deepen its expertise. Specialises in school and care facility renovations. Competitive edge is in specialist expertise and in exploiting its leaner cost structure against construction groups. Targets projects over a certain size to leave out the smallest competing firms that have no financial resources.

Table 1 Descriptions of the business models

As displayed in Table 1, the interviewed companies are varied in terms of their size and business models. Despite the apparent variety, we encountered significant challenges in an attempt to identify any truly unique and distinctive characteristics in these models that would give these companies a competitive advantage, which Barney (1991, p. 102) defined as 'a value creating strategy not simultaneously being implemented by any current or potential competitors'. Instead, the differences stem from their diversification; the bigger the company, the more diversified its portfolio of businesses (Table 2).

	Development for a client (e.g. design-build)			Own development	Contracting of construction (e.g. design-bid-build)		
Field of operation	Business premises	Commercial buildings	Nursery schools and nursing homes	Residential units	Industrial buildings	Renovation construction	Infra-structure
Company 1:	x			x	x	x	
Company 2:	x	x		x			x
Company 3:			x	x			
Company 4:					x	x	
Company 5:		x		x			
Company 6:		x		x	x	x	
Company 7:				x			
Company 8:						x	

Table 2 The construction companies' fields of operations

Common Language

Based on the interviews, it is clear that 'business model' as a term is rarely used in these construction companies, while as a concept or a management tool it is used quite differently than what is typical in other sectors or within the literature. Questions like, 'Do you know the concept of business models?', 'Do you use the word?', 'How do you define a business model?', 'What elements compose it?' and 'How and why are such models described and used?' produced a wide variety of answers. Interviewees also had opposing views as to whether there should be one or several business models at different levels of an organisation (e.g. at the level of a firm, business unit, for each field of operation or even for each independent project). Every manager seemed to think that he or she knew what business models were, but the reality was that there were as many definitions and explanations as there were interviewed managers.

Although there was a partial conception of a business model's function as a means for creating a competitive advantage via differentiation, more emphasis was put on using it to set objectives and acquire work. The minimum requirement should be to identify the value a company is creating for its customers in the chosen markets or to explain the basis for its long-term existence and profit generation. Instead, the discussions turned to fields of operations, business segments, modus operandi or certain project delivery and contract types (i.e. what kinds of projects these companies engage in). For example, in the largest companies, business models were related to practise, where processes were defined and described in terms of quality systems and/or along certain standards such as ISO. In addition, several interviewees thought it possible to express the business model with a quantitative measure (e.g. acceptable level of risk and return for certain kinds of projects).

Value Creation and Capture

The results of the interviews indicate that the managers in construction do not understand the concept of the business model in the way that it is depicted in the literature. Quite frankly, business models do not seem to play any significant role in developing and managing these companies, as the descriptions about how the concept is deployed differ quite a lot from a rigorous analysis of how a company creates value for selected markets as a system and how its parts support each other. In the managers' discourse, 'business model' is something that is used to define what kinds of initiatives they want to engage in. Referring to the most common business model elements, the managers' conception of the business model is limited to the revenue model. Even there, the focus is mainly on how the company creates value for itself - which is expressed in terms of return on investment or some similar measure - excluding the customer from the equation.

The business model descriptions in Table 1 clearly show the managers' difficulty in explaining the way their companies create value for their customers. Only a few of the interviewees could define who their customers were, who they were targeting and whom they wished to work with. In this regard, business model descriptions are limited to a few loosely-defined strategic choices, such as where companies want to operate geographically and what kind of projects (size, delivery type and field of operation) they want to engage in. Most of the companies, for example, aim for turnkey projects (i.e. through design-build delivery), but nobody was able to explain how their company's turnkey delivery differed from another. Instead, projects are executed as they are specified by the customer, meaning that executing projects 'according to customers' wishes' means creating value for them.

By using at least the most common elements of a business model - offering, value creation system and revenue model - managers in construction would have a language and framework to analyse and describe their businesses from a systems perspective. These key elements can be further specified if needed, but the key is to have a defined framework for designing and communicating business models. Currently, the lack of such a framework produces only partial descriptions, and subsequently, a partial understanding of business models with subjective and changing definitions. Current thinking does not consider alternative ways of value creation or specify exactly what kind of value a company is bringing to its clients' projects in order to create value in a way that others do not or cannot. In the worst case, all of the companies interviewed focused on acquiring their next job from highly volatile local markets; in the best case, they attempted to gain complete responsibility for the project. Only a few companies referred to the unique value they were producing for some specific field of operation, which usually feature customers with specific characteristics.

Competitive Advantage

As already stated, it was hard to identify any unique sources of competitive advantage through the interviews. The few that we managed to identify related directly to resources or to doing something specific over a longer span of time, i.e. organisational learning. For example, one of the medium-sized companies owns a crane, which provides access to a variety of projects in the company's operating area and demonstrates the surprising fact that certain equipment may still create at least a temporary competitive advantage. Then, some companies have long been specialised in project management contracting or renovation construction, consequently building their competence in these specific areas. In addition, a proportion of back-office costs seems to be a distinguishing factor between small and large companies. Smaller companies have the cost advantage in smaller projects while larger companies have the financial resources to engage in more valuable projects. The importance of human resources was largely considered to be limited to supervisors, their suitability for certain types of projects and their ability to successfully steer projects. Construction workers were considered interchangeable.

In light of Barney's (1991) definition of sustainable competitive advantage, most of the aforementioned factors only create a temporary advantage because other firms can duplicate their benefits easily. It is also hard to consider the construction companies' activities as advantage-creating core competences which, according to Prahalad and Hamel (1990), represent the collective learning in the organisation, are characterised by the ability to provide access to a wide variety of markets and make a significant contribution to the perceived customer benefits while being difficult for competitors to imitate. Small companies' operating logics come closest to meeting all three requirements, but even these companies have to compete fiercely with other very similar companies, thus missing the requirement of being difficult to imitate. In fact, it is hard to know what extra value these companies could provide over others within the project delivery. Thus, it is understandable that customers often see price as the only distinctive factor between construction companies; hence the bidding wars

Analysing the behaviour of these companies through Porter's (1980) generic strategies unveils companies' identity crises. It is hard to find any strategy - cost leadership, differentiation, focus - that is being pursued in its purest form. For example, having awareness of lean construction and practising it actively would indicate that a company is pursuing cost leadership and trying to push the productivity frontier (see Porter 1996) ahead through advanced methods, but this is not the case for these companies. Instead, the focus is mostly on short-term cost-cutting during the projects. In addition, companies do not seem to be very differentiated from others as they mostly follow the standard procedures of the construction industry and, as can be seen from Table 2, all medium and large companies are too diversified to justify focus strategy. Interviewees thought that diversification is needed to protect companies from high market volatility, a belief observed by Junnonen (1998). This diversification means that local markets set limitations for growth and that companies expand their operating spheres to find both diversity and volume for their businesses. Yet, by utilising the business model concept and developing their value creation capabilities, construction companies could better define the exact value they are creating for their customers and how they differ from their competitors.

Research Implications

Compared to many other industries where the business model is often considered and implemented at the level of a firm or a business unit, such as manufacturing and e-business

(Hedman & Kalling 2003; Morris et al. 2005; Tikkanen et al. 2005), the construction business is far too varied (meaning that the same company can build roads, mines, business premises and residential units) in nature for a similar approach to be exploited. Similarly, we thought that the use of project- or solution-level business models, as suggested by Kujala et al. (2010), would not be practical if the objective is to understand companies as systems that can be designed, managed and further developed using business models. Thus, we suggest that business models in the construction industry should be studied and developed at the level of field of operation. This would aid understanding value creation in different types of construction projects and developing business models that would better meet the needs of specific customers or market segments while also providing a competitive edge for construction companies. This level of analysis seems appropriate as in every field there are different customers and ways in which value is created and captured, hence justifying the claim that they should have their own business models. Currently, construction projects seem to be executed using a universal recipe that leaves both customers and businessman without satisfaction.

Figure 1 provides a conceptual model of our suggestion. The objective of the research and development function is to create business models that provide maximum value for certain customer segments, i.e. clients and users in specific fields of operation. At the core of these business models are offerings - the bundle of products and services that are valuable for customers. The project-delivery process incorporates many stages, from briefing to design and from purchasing to construction, realizing the offering and providing value to the customers. This process, which consists of the specific activities, resources and methods needed and used in project execution, should also be under constant development in order to fine tune the project delivery process for future projects. The last component, revenue model, defines the way a company captures value for itself from the project, i.e. makes money. In construction, this is usually dependent on the used contract where the fees and additional incentives are defined, but managers should still understand the differences between projects (e.g. the role of sales in B2C projects or fixed fees in B2B projects) and their implications for other parts of the system.

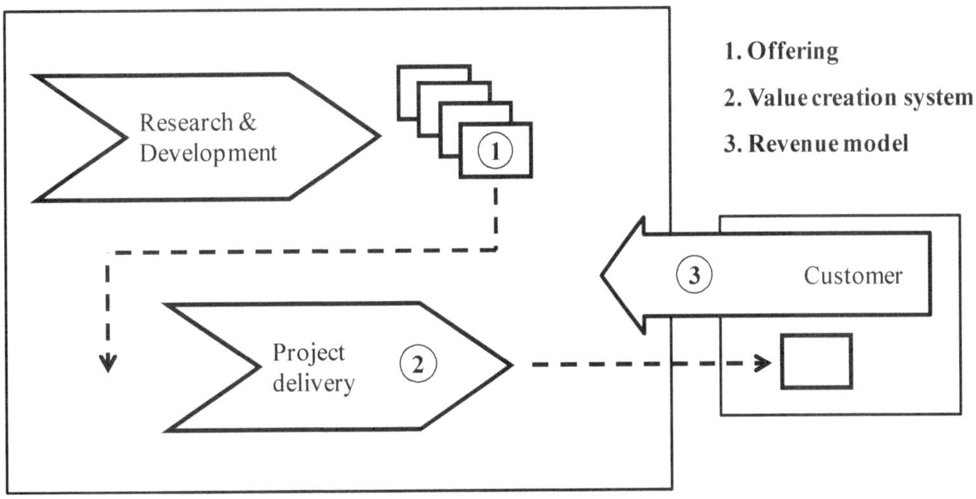

Figure 1 Interdependent business model elements in a process context

Conclusions

This paper focused on analysing managers' understanding of the business model concept and explored the business models currently used in Finnish construction companies. According to the results, managers in construction understand business models very differently than those in other sectors or within the academic literature. The main difference is that they relate these models to their companies' fields of operation, business segments, modus operandi, certain project delivery and contract types, financial objectives and securing more work. There seem to be as many definitions and purposes for business models as there are managers in construction. The results of the interviews indicate that the managers in construction do neither understand the concept properly nor exploit any similar value creation analysis in their business. The interviewees had significant problems describing their companies' business models and value creation logic, pointing out the lack of analysis and understanding of customer values and needs in the project delivery process. This may be one of the overtones of the persistent client dissatisfaction in the construction industry.

The limitations of the study are mostly concerned with the depth of interviews, sample size and the choice of companies. There were varying interpretations for the business model concept that may have affected how interviewers understood the operating logic of companies. One reason for this, perhaps, is that the interviewed managers were, due to their education and experience, more technically rather than business-oriented. This is, however, quite common in the construction industry and, thus, does not prove as a limitation of the study in itself/as such. Further, the local nature of construction markets may have unidentified consequences that would prevent further generalisation of the findings as the studied companies operate in rather small geographical area - Finland. However, the authors' general observation while studying the construction industry is that construction companies have nowhere near exploited the potential of business model innovation when making their businesses more profitable and sound. Thus both qualitative and quantitative research on business models is highly recommend in order to produce new information about the nature of business models in construction and their effectiveness from customers' and firm's perspective.

References

Afuah, A. (2003) 'Redefining firm boundaries in the face of the internet: are firms really shrinking?', *The Academy of Management Review*, **28** (1), 34-53.

Afuah, A., and Tucci, C. (2000) *Internet Business Models and Strategies: Text and Cases*, McGraw-Hill Higher Education, Boston, MA.

Barney, J.B. (1991) 'Firm resources and sustained competitive advantage', *Journal of Management*, **17** (1), 99-121.

Burkhart, T., Krumeich, J., Werth, D., and Loos, P. (2011) 'Analyzing the business model concept—a comprehensive classification of literature', in *International Conference on Information Systems 2011 Proceedings,* Shanghai, China.

Casadesus-Masanell, R., and Ricart, J. (2010) 'From strategy to business models and onto tactics', *Long Range Planning*, **43** (2/3), 195-215.

Casadesus-Masanell, R., and Ricart, J. (2011) 'How to design a winning business model', *Harward Business Review*, **89** (1/2), 100-107.

Chesbrough, H., and Rosenbloom, R.S. (2002) 'The role of the business model in capturing value from innovation: evidence from Xerox Corporation's technology spin-off companies', *Industrial and Corporate Change*, **11** (3), 529-555.

Hamel, G. (2000) *Leading the revolution*, Harvard Business School Press, Boston, MA.

Hedman, J., and Kalling, T. (2003) 'The business model concept: theoretical underpinnings and empirical illustrations', *European Journal of Information Systems*, **12** (1), 49-59.

Johnson, M., Christensen, C., and Kagermann, H. (2008) 'Reinventing your business model', *Harvard Business Review*, **86** (12), 51-59.

Junnonen, J.-M. (1998) 'Strategy formation in construction firms', *Engineering, Construction and Architectural Management*, **5** (2), 107-114.

Kujala, S., Artto, K., Aaltonen, P., and Turkulainen, V. (2010) 'Business models in project-based firms – towards a typology of a solution-specific business models', *International Journal of Project Management*, **28** (2), 96-106.

Linder, J., and Cantrell, S. (2000) 'Changing business models: surveying the landscape', working paper, The Accenture Institute for Strategic Change, 24 May, available at: http://course.shufe.edu.cn/jpkc/zhanlue/upfiles/edit/201002/20100224120954.pdf (accessed 23 April 2013).

Magretta, J. (2002) 'Why business model matter', *Harvard Business Review*, **80** (5), 86-92.

Morris, M., and Schindehutte, M., and Allen, J. (2005) 'The entrepreneur's business model: toward a unified perspective', *Journal of Business Research*, **58** (6), 726-735.

Mutka, S., and Aaltonen, P. (2013) 'The impact of a delivery project's business model in a project-based firm', *International Journal of Project Management*, **31** (2), 166-176.

Nenonen, S., and Storbacka, K. (2010) 'Business model design: conceptualizing networked value co-creation', *International Journal of Quality and Service Sciences,* **2** (1), 43-59.

Osterwalder, A. (2004) *The business model ontology - a proposition in design science approach*, doctoral thesis, Department of Business and Economics, University of Lausanne, Switzerland.

Osterwalder, A., and Pigneur, Y. (2009) *Business Model Generation*, John Wiley & Sons, New Jersey, NJ.

Pateli, A.G., and Giaglis, G.M. (2004) 'A research framework for ananysing eBusiness models', *European Journal of Information Systems,* **13** (4), 302-314.

Pekuri, A., Herrala, M., Haapasalo, H., and Pekuri, L. (forthcoming) 'Managing value creation: the business model approach in construction', *International Journal of Business Innovation and Research.*

Porter, M.E. (1980) *Competitive Strategy*. Free Press, New York.

Porter, M.E. (1985) *Competitive Advantage: Creating and Sustaining Superior Performance*. Free Press, New York.

Porter, M.E. (1996) 'What is strategy', *Harvard Business Review,* **74** (6), 61-78.

Prahalad, C.K., and Hamel, G. (1990) 'The core competences of the corporation', *Harvard Business Review,* **68** (3), 79-91.

Shafer, S., Smith, J., and Linder, J. (2005) 'The power of business models', *Business Horizons*, **48** (3), 199-207.

Stabell, C., and Fjeldstad, Ø. (1998) 'Configuring value for competitive advantage: on chains, shops, and networks', *Strategic Management Journal*, **19** (5), 413-437.

Suikki, R., Goman, A., and Haapasalo, H. (2006) 'A framework for creating business models – a challenge in convergence of high clock speed industry', *International Journal of Business Environment*, **1** (2), 211-233.

Teece, D. (2010) 'Business models, business strategy and innovation', *Long Range Planning*, **43** (2-3), 172-194.

Tikkanen, H., Lamberg, J.-A., Parvinen, P., and Kallunki, J.-P. (2005) 'Managerial cognition, action and the business model of the firm', *Management Decision*, **43** (5/6), 789-809.

Timmers, P. (1998) 'Business models for electronic markets', *Electronic Markets,* **8** (2), 3-8.

Zott, C., and Amit, R. (2007) 'Business model designs and the performance of entrepreneurial firms', *Organization Science*, **18** (2), 181-99.

Wikström, K., Artto, K., Kujala, J., and Söderlund, J. (2010) 'Business models in project business', *International Journal of Project Management*, **28** (8), 832-841.

The Effects of the late 2000s Global Financial Crisis on Australia's Construction Demand

18

Heng Jiang, (Deakin University, Geelong, Australia)

Xiao-Hua Jin, (University of Western Sydney, Australia)

Chunlu Liu, (Deakin University, Australia)

Abstract

An accurate measurement of the impacts of external shocks on construction demand will enable construction industry policymakers and developers to make allowances for future occurrences and advance the construction industry in a sustainable manner. This paper aims to measure the dynamic effects of the late 2000s global financial crisis on the level of demand in the Australian construction industry. The vector error correction (VEC) model with intervention indicators is employed to estimate the external impact from the crisis on a macro-level construction economic indicator, namely construction demand. The methodology comprises six main stages to produce appropriate VEC models that describe the characteristics of the underlying process. Research findings suggest that overall residential and non-residential construction demand were affected significantly by the recent crisis and seasonality. Non-residential construction demand was disrupted more than residential construction demand at the crisis onset. The residential construction industry is more reactive and is able to recover faster following the crisis in comparison with the non-residential industry. The VEC model with intervention indicators developed in this study can be used as an experiment for an advanced econometric method. This can be used to analyse the effects of special events and factors not only on construction but also on other industries.

Keywords: Vector error correction, Intervention analysis, Construction demand, The global financial crisis

Introduction

The construction industry is an important sector of a nation's economy. It makes a significant contribution to the economic output and provides employment and business opportunities (Li & Liu 2012). Ofori (1990) highlighted that construction is the engine of economic growth. However, the construction industry is greatly affected by the performance of the economy as the output of construction is a response to the demand for buildings, a derived demand for other sectors (Song & Liu 2006). The relationship between economic and construction cycles is widely discussed by many construction economists. A close relationship was found between the construction cycle and the economic cycle in Singapore, where the short-term business cycle has a direct influence on the demand for construction (Hua 1998). This close relationship implies that shocks from specific external events, such as global financial crises, affect not only the general economy but also the construction industry. During periods of economic austerity, the level of demand in the construction industry will be affected dramatically due to reduction of construction spending by public and private sectors (Hua 2005). With this as the backdrop, construction economists apply an advanced and accurate method to assess the impacts of such external shocks on the construction industry and its lagged effects.

Limited research has been undertaken to assess the effect of external shocks on the construction industry. For example, Hua (2005) estimated the dynamic effects of the 1997 Asian financial crisis on the demand for construction in Singapore using the auto-regressive-integrated moving average (ARIMA) model. Hua (2005) reported different net effects of the Asian financial crisis on the demand from both the public and private sectors, with a positive response in the public sector and a negative response in the private sector. A study by Fan et al. (2010), based on the ARIMA technique, discussed the effects of the SARS outbreak and the Asian financial crisis on the construction industry in Hong Kong. The study concluded that the total demand for construction was only slightly impacted by the SARS outbreak and financial crisis, but the financial crisis deeply influenced demand in the residential sector.

The global financial crisis of 2007-2009 is considered the worst financial crisis since the Great Depression of the 1930s (Jiang & Liu 2011, Murphy 2011). In July 2007, US investors lost confidence in the value of sub-prime mortgages, resulting in a liquidity shortfall in the United States banking system. This was followed by the collapse of large financial institutions and downturns in stock markets around the world. The crisis began in 2007, but fully impacted the global economy in 2008 and 2009. The annual GDP growth rates from 2007 to 2009 respectively were 3.8%, 1.6% and -2.2% for the world economy (Nayyar 2011). The International Labour Organisation claimed that worldwide unemployment rose by at least 30 million people, and reached as high as 50 million people during 2007 and 2009 (Blankenburg & Palma 2009). The collapse of Lehman Brothers on September 14, 2008 marked the beginning of a new phase in the global financial crisis (Berkmen et al. 2009). In the following month, the financial crisis spread from the US to Europe and worldwide. The Australian economy was injured deeply in the crisis. In the period between 2007 and mid-2008, the Australian economy registered a high growth rate at around 4% (ABS 2012). However, in December 2008, the Australian stock market experienced the largest fall on record, with AUD 3.4 billion, while the Australian unemployment rate increased by 2% during the 2008 and 2009 global economic downturn. At the same time, the global financial crisis also triggered unexpected shock waves in the Australian construction industry. In Australia the value of construction approvals shrunk by almost a quarter in March 2009, compared to its peak in March 2008, while house prices declined 6% from March 2008 to March 2009.

The effects of the recent financial crisis on the construction industry can be readily observed. Hence, an analysis of the effects of the crisis on growth in the construction industry can help to facilitate resource planning. This, in turn, ensures all valuable public resources can be allocated for the construction sectors, to retain labour and skills after this global recession. In previous studies, the ARIMA model was the main technique used to assess external shocks of specific events on the construction industry. This technique assesses the impacts of external shocks on the construction industry, based solely on past data of the time series. A significant issue for this measuring method is that it ignores the impacts of changes in economic conditions on the variation in construction economic indicators. Furthermore, previous research on event analysis focuses on how events affect the market or concerned indicators but doesn't assess the market or indicators in the absence of an event. This study addresses this issue by estimating construction demand based on two scenarios: A - the change of construction demand is assumed to follow the trend of economic growth in the pre-crisis period, with no effects of the global financial crisis, and B - the change in construction demand is assumed to follow the trend of economic growth in the post-crisis era. An advanced econometrics method, vector error correction (VEC) model with intervention indicators, would serve as a more appropriate modelling technique for assessing the effects of external shocks and impacts of changes of local economic conditions on construction demand. Indeed, an accurate measurement of the impacts of external shocks on construction demand will enable construction industry policymakers and developers to make allowances for future similar occurrences and develop the construction industry in a sustainable manner.

In this study the main stages for assessing the effects of global financial crises on the construction demand are: firstly, create two intervention indicators to represent the duration of the event and seasonal effect; secondly, develop VEC models with intervention indicators for the construction demand; then, estimate endogenous variables and intervention indicators in the robust VEC model; next, identify whether the global financial crisis and seasonality affect the construction demand or not with Australian Bureau of Statistics (ABS) data; subsequently solve the robust VEC model during the intervention period and estimate the intervention of special events (i.e. the global financial crisis and seasonality); and finally interpret the estimated results and the intervention effects of the event on the construction demand.

VEC Model Development

Intervention Indicators

Intervention analysis has been applied in many studies for example in assessing the intervention of institution controls on the rate of change in the US consumer price index (Box & Tiao 1975), analysing the dynamic effects of the 1997 Asian financial crisis on construction demand and tender price in Singapore (Hua 2005), and measuring the impact of the Sydney Olympic Games in 2000 and the Bali Bombing in 2005 on the tourism industry in Australia (Yap & Allen 2011). The characteristics of intervention analysis technique have been explored in other studies, such as Box and Tiao (1975), and Hua (2005). In brief, the intervention indicators can be represented in two forms; namely a step and a pulse. As exogenous variables the indicator variables take the value 0 and 1 to denote non-occurrence and occurrence of the intervention. If the intervention function is a step, then the value of the indicator is 0 until the event begins at time T. At the occurrence of the event, the intervention equals 1 and the indicator can be denoted by S_t^T where:

$$S_t^T = \begin{cases} 0, & t < T \\ 1, & t \geq T \end{cases} \tag{1}$$

If the intervention function is a pulse function, then the value of the indicator is 0 at non-occurrence of the event. At the occurrence of the event, the indicator is 1 and it remains 1 for the duration of the occurrence of the event. In some cases, a conventional pulse is only one time period, or the duration of the event. A indicator for pulse function can be denoted by P_t^T where:

$$P_t^T = \begin{cases} 0, & t \neq T \\ 1, & t = T \end{cases} \tag{2}$$

The step and conventional pulse functions are represented by input variables and these two functions are interrelated. In this study, the intervention indicator of the effects of the late 2000s global financial crisis is represented as an extended pulse function, and seasonal impact is represented as a conventional pulse function. The first sign of deterioration in Australia's construction industry was in the second half of 2008, when the demand for construction declined more than 29%. At the same time, the first significant policy response to the global financial crisis came from the Australian Commonwealth Government. The government announced it would guarantee all bank deposits, and an economic stimulus package worth AUD 10.4 billion was announced. In this package, AUD 1.5 billion was allocated to support housing construction. This announcement could be considered an ideal indicator in denoting when the financial crisis began to affect the Australian economy. Through a series of effective boost strategies, the approvals of Australian construction industry reached the same level in the September quarter of 2009 as at the beginning of the global financial crisis. However the approved value of construction work registered a dramatic decline (26%) from the December quarter of 2009 to the June quarter of 2010, and was mainly due to the sharp decline in the value of non-residential construction approvals. This can be signalled as the lagged effects of the late 2000s global financial crisis. Following this, the demand in the Australian construction industry remained at a low level but grew slightly after the June quarter of 2010 to the March quarter, 2012. This signalled an end to the lagged effects of the crisis. Indeed, the intervention period of the late 2000s global financial crisis on the Australian construction industry can be defined as starting in the September quarter, 2008 and ending in the June quarter, 2010.

The Vector Error Correction Model with Intervention Indicators

The vector error correction model is a combination of the vector autoregressive model and co-integration restrictions. Co-integration, an econometric property of time series variables, is generally used to estimate the long-run relationships between non-stationary variables. If the level of time series data is not stationary, but a linear combination of variables is stationary after the first difference, then the series are co-integrated to an order one or $I(1)$. They will tend to return to this trend in the long term, even though they deviate from each other in the short term. A prior condition for the co-integration test is that all the variables should be integrated to the same order or contain a deterministic trend (Engle & Granger 1991; Liu et al. 2009). A unit root test is conducted for each variable using the Augmented Dickey-Fuller (ADF) unit root test and the Phillips-Perron (PP) unit root test, which were introduced by Dickey and Fuller (1987), and Phillips and Perron (1988) respectively.

The general VEC model employed by Jiang and Liu (2011) is represented in Equation (3).

$$\Delta Y_t = C + \Pi Y_{t-1} + \sum_{i=1}^{k-1} \Gamma_i \Delta Y_{t-i} + \varepsilon_t \tag{3}$$

where Y_t is an independent $I(1)$ variable being integrated to an $I(0)$ vector, C is the intercept, Γ is the matrix reflecting the short-run dynamic relationship among elements of Y_t, and ε_t is residual. $\Delta = (I - L)$, L is the lag operator, k is the number of lags, where Π is the matrix containing long-run equilibrium information. If the elements of Y_t are $I(1)$ variables and co-integrated with rank $(\Pi) = r < p$, then the rank of Π can be rewritten as $\Pi = \alpha\beta' = \alpha ecm_{t-1}$, ecm_{t-1} is the error correction term and $\beta' Y_t$ is stationary. This implies that there exist $r < p$ stationary linear combinations of Y_t. β is a vector of co-integration relationships and α is a loading matrix defining the adjustment speed of the variables in Y to the long-run equilibrium defined by the co-integrating relationships.

Johansen co-integration test was introduced by Johansen and Juselius (1990) to conduct the multivariate maximum likelihood approach and revealed the number or co-integration equations without using arbitrary normalisation rules. There are five models in the Johansen co-integration test. Model one represents all series as having zero mean. Model two represents deterministic data with an intercept but no trend in the co-integration equations (CE). Model three suggests that data has a linear trend with an intercept but no trend in the CE. Model four has a linear trend with both an intercept and a trend in the CE, while model five suggests a quadratic data trend with an intercept and a trend in the CE. This paper analyses three different specifications in the Johansen co-integration estimation, excluding models 1 and 5, as they are not applicable in real life situations (Hui & Yue 2006). Lag length of the VEC model is selected for the time series in VAR modelling on the basis of a sequential modified likelihood ratio test statistic, final prediction error, Akaike information criterion, Schwarz information criterion and Hannan-Quinn information criterion. The test results of lag length selection are then input into the Johansen co-integration test for constructing VEC models with different combinations of construction demand and each economic indicator. Once all variables proved to be stationary and co-integrated, a VEC model was formulated.

The VEC model, containing intervention indicators, was earlier used by Ramey (1993) to analyse the effects of seasonality and monetary-policy disturbance on the money market. As an exogenous variable, the intervention indicator has been included in the VEC model to estimate the impacts of the 1985 United Airlines strike and the 1991 Persian Gulf War on the tourism demand and supply for Hawaii (Bonham et al. 2009). Based on the Equation (1), the VEC model with intervention indicators can be represented in Equation (4).

$$\Delta Y_t = C + \Pi Y_{t-1} + \sum_{i=1}^{k-1} \Gamma_i \Delta Y_{t-i} + \sum_{j=1}^{n} \Theta_N D_{j,t} + \varepsilon_t \quad (4)$$

Where $D_{j,t}$ is the intervention indicator j at time t, N is the number of endogenous variables, n is the number of exogenous variables, Θ_k are $n \times N$ vectors.

Estimating the VEC Model for Australian Construction Demand
Economic Indicators affecting Construction Demand

The value of construction work approved, or the value of construction approvals, has been used to represent the demand in the construction industry, as it is an indicator of changes in the level of construction demand (Ofori 1990). In this study, the value of construction work approved was selected to represent the demand in construction. This can be defined as the total monetary cost of the construction work that clients are able, and will be able, to purchase in a given period. Furthermore, the total residential and non-residential values of construction work approved were abstracted from the Australian Bureau of Statistics for assessing the effects of the recent global financial crisis on the different types of construction industries (ABS 2012). The time series of different types of construction demand are shown in Figure 1. It indicates that all types of construction demand in the Australian construction industry had a significant decline in 2008 Q3, followed by a large fluctuation to 2010 Q2. It then remained constant between 2010 Q2 and 2012 Q3.

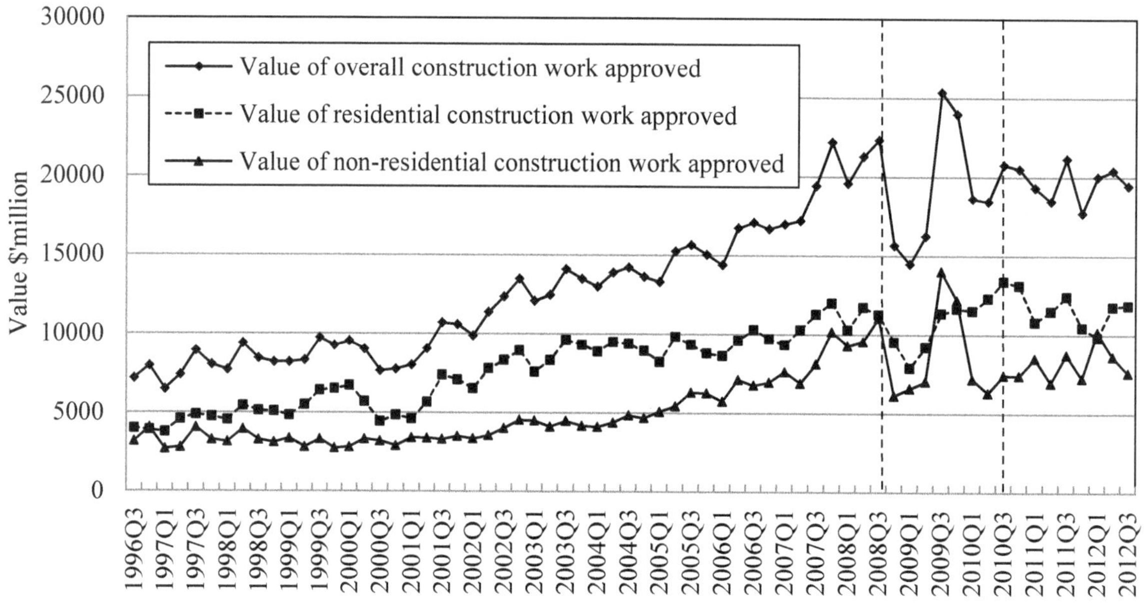

Figure 1 A plot of the time series for three types of the value of construction work approved (two vertical dot lines highlight the duration of the effect of the late 2000s financial crisis on the construction industry)

Economic indicators including construction price, national income, householder expenditure, size of population, unemployment rate, interest rate and the value of export have been widely used to model construction demand. A construction producer price reflects the movement of prices in the construction industry for each period of time (Hua 1998). Ball et al (2000) suggested that for each market sector, construction prices should be determined by the total demand. Construction prices differ by region, partly as a result of local resources and demand, because a fluctuating demand will lead to fluctuating prices and vice versa (Meikle 2001). Akintoye and Skitmore (1994) adopted construction price as one of the significant factors for modelling housing construction demand in the private sector. Any variation in the national income will affect the level of demand for construction and the construction price in both the residential and non-residential industries. Consumer demand in goods and services will soar in a period of economic prosperity, which will also trigger an increase in the level of demand for construction space (Akintoye & Skitmore 1994). The change in household expenditure demonstrates the expectation of householders with regard to the future national economy. Increasing the expenditure of householders will lessen the resources available to the construction industry, which will indirectly affect the level of demand for construction. Demographic influences have been widely cited for modelling construction economic indicators, such as demand and prices. The growth of a population raises the basic need for new dwellings and has been identified as a key determinant of the demand for residential construction (Tang et al. 1990; Hua 1998; Fan et al. 2010). The unemployment rate is measured as the total number of people not in employment who are ready and able to work (Hua 1998). An increase in unemployment may discourage investment in the construction industry, because employment is the main source of income for residents and a rise in unemployment rate represents a lowering of the purchasing power of the population. Furthermore, Ma and Liu (2013) applied demographical structure and housing market scales to develop a demographical distance for investigating the house price convergences in Australian capital cities. Fan et al. (2010) indicated that a change in interest rates can affect the lending costs of clients, contractors, developers and company profits. A lower interest rate will encourage investments in the construction industry and thus raise the level of demand for construction. In contrast, an increase in interest rates will raise the cost of bank lending for construction projects and lead to a decline in purchasing power. It can be seen that the average export of goods and services in Australia is more than 20% of the total GDP (ABS 2012). This means that the export industry is a vital sector, not only for the national economy, but also for other sectors.

The quarterly data series of selected economic indicators and the demand for construction were abstracted from the ABS in the period of September 1996 to September 2012. The data series from September 2008 to June 2010 were used to undertake an intervention analysis to analyse the impact of the recent global financial crisis on the construction demand in Australia. All the time series data used in this study has been expressed as natural logarithm variables. In collecting the variables for estimation, the following important issues were taken into account (Akintoye et al. 1998); economic plausibility of their leading character; availability of the time series with as few interruptions as possible; and availability of the data with minimum delay.

The Estimates of VEC Models with Intervention Indicators

The VEC model with the late 2000s financial crisis intervention ($D_{1,t}$) and seasonal intervention ($D_{2,t}$) for different types of construction demand (CD_t^q) can be constructed in Equation (5).

$$\Delta CD_t^q = C + \alpha(ecm_{t-1}Y_{t-1} + \rho_0) + \sum_{i=1}^{k} \theta_{1,i}\Delta CD_{t-1}^q + \sum_{i=1}^{k} \theta_{2,i}\Delta CPI_{t-i} + \sum_{i=1}^{k} \theta_{3,i}\Delta NI_{t-i} + \sum_{i=1}^{k} \theta_{4,i}\Delta HHE_{t-i} + \sum_{i=1}^{k} \theta_{5,i}\Delta UR_{t-i} + \sum_{i=1}^{k} \theta_{6,i}\Delta POP_{t-i} + \sum_{i=1}^{k} \theta_{7,i}\Delta IR_{t-i} + \sum_{i=1}^{k} \theta_{8,i}\Delta VOE_{t-i} + \delta D_{1,t} + \varphi D_{2,t} + \varepsilon_t \quad (5)$$

Where α is the adjustment coefficient, ρ_0 is the intercept of co-integrating equations. $\theta_{j,i}$ reflects the short-run aspects of the relationships between the independent variables and the target variable. At time t, CPI_t is the construction producer price, NI_t is the national income,

HHE_t is the household expenditure, UR_t is the unemployment rate, POP_t is the size of population, IR_t is the interest rate and VOE_t is the value of export. δ is the coefficient of the late 2000s financial crisis dummy variable, φ is the coefficient of the seasonal dummy variable, $D_{1,t}$ and $D_{2,t}$ are the one-off event dummy variables for the financial crisis and seasonal effect respectively.

Before constructing the VEC models, the time series data are first tested for unit root or the order of integration. The ADF and PP unit root tests were carried out to test stationarity for all the variables. The results are summarised in Table 1 and suggest that all the variables were stationary after the first difference at the 0.01 and 0.05 significance levels. Based on the VAR lag length selection system, the smallest values of the LR FPE, AIC, SC and HQ tests indicate that the lag length for the VEC models for overall construction demand, residential and non-residential construction demand all equalled three. Following that, co-integration tests were carried out. The results of the trace statistics indicate that each variable Y_t has a linear trend with an intercept, and however three types of construction demand have no trend in the co-integrating relations. The deterministic trend in model three and one co-integration relationship was identified and implemented into these three VEC models. These results suggest that the three types of construction demand and selected economic indicators do not move independently of each other in the long-term and also share some common trends.

Indicators	**ADF unit root test**				**PP unit root test**			
	Level		First difference		Level		First difference	
	T-Stat	P-value	T-Stat	P-value	T-Stat	P-value	T-Stat	P-value
CD	2.02	0.99	-8.62	0.00**	-1.25	0.65	-11.95	0.00**
CPPI	2.43	1.00	-3.18	0.03*	-0.26	0.92	-3.18	0.03*
NI	0.34	0.98	-4.92	0.00**	0.41	0.98	-6.01	0.00**
HHE	-1.73	0.41	-6.35	0.00**	-1.74	0.41	-6.27	0.00**
UR	-1.12	0.24	-4.11	0.01*	-1.26	0.19	-4.11	0.01*
POP	1.97	0.99	-3.68	0.03*	1.18	1.00	-7.54	0.00**
IR	-0.38	0.54	-4.63	0.00**	-0.75	0.39	-4.49	0.00**
VOE	2.47	1.00	-8.12	0.00**	-1.87	0.34	-8.79	0.00**
RCD	-1.52	0.51	-7.02	0.00**	-1.25	0.65	-11.51	0.00**
NRCD	-1.79	0.38	-6.12	0.00**	-1.48	0.54	-15.70	0.00**

Table 1 Summarised results of ADF and PP unit root tests

* denotes rejection of null hypothesis of unit root based on their P-value at the 0.05 significance level.

** denotes rejection of null hypothesis of unit root based on their P-value at the 0.01 significance level.

Having found the long-term relationships between three types of construction demand and selected economic indicators, three VEC models with global event and seasonal intervention indicators were constructed based on Equation (5). The estimates of these three VEC models, with intervention indicators for overall, residential and non-residential construction demand, are reported in Table 2. The estimates of VEC models for overall and non-residential construction demand show that overall and non-residential construction demand is affected by construction price, national income, householder expenditure, unemployment rate, population, interest rates and the value of export. In contrast, the residential construction demand is not affected by the construction price. In addition, the lagged construction demand, the growth of population and the interest rate have significant roles in explaining the level of demand in the construction industry.

Variables	ΔCD_t	ΔRCD_t	$\Delta NRCD_t$
CD_{t-1}	1	1	1
$CPPI_{t-1}$	-1.394 (-2.744)***	-0.572 (-1.029)	2.843 (4.905)***
NI_{t-1}	-2.283 (-5.117)***	-4.107 (-11.291)***	-5.638 (-4.707)***
HHE_{t-1}	7.307 (10.990)***	8.323 (6.431)***	4.785 (3.322)***
UR_{t-1}	0.734 (1.966)**	1.077 (12.523)***	0.995 (3.001)***
POP_{t-1}	-8.421 (-5.163)**	7.363 (5.790)**	-7.620 (-3.603)**
IR_{t-1}	0.674 (8.665)***	0.556 (1.974)**	-0.837 (-5.608)***
VOE_{t-1}	-0.891 (-3.264)***	-1.419 (-2.726)**	4.973 (12.079)***
C	38.698	-48.765	84.692
CointEq1 (α)	0.194 (0.050)	0.248 (-1.743)*	-0.153 (-0.196)
ρ_0	-0.041 (-1.259)	-0.106 (-0.880)	0.178 (0.631)
D1	-0.179 (-2.784)**	-0.121 (-1.641)*	-0.346 (-3.126)***
D2	-0.109 (-2.353)***	-0.144 (-3.634)***	-0.074 (-1.686)*
R-squared	0.859	0.816	0.828
Sum sq. residue	0.138	0.122	0.552
S.E. equation	0.067	0.071	0.127
Log likelihood	88.550	83.633	53.341
Error correction	*t*-1	*t*-2	*t*-3
Overall construction demand			
ΔCD	-0.491 (-2.941)***	-0.391 (-1.999)**	-0.004 (-0.032)
$\Delta CPPI$	3.515 (1.203)	-1.963 (-0.631)	-4.263 (-1.516)
ΔNI	-2.200 (-1.428)	-1.075 (-0.675)	-2.169(-1.382)
ΔHHE	0.641 (0.279)	6.679 (2.759)	-0.013 (-0.005)
ΔUR	0.7526 (1.381)	-0.157 (-0.321)	-0.217 (-0.547)
ΔPOP	13.749 (0.689)	39.717 (2.164)**	-6.155 (-0.227)
ΔIR	0.087 (0.402)	-0.024 (-0.115)	-0.508 (-2.626)***
ΔVOE	0.544 (1.112)	-0.379 (-0.743)	-0.030 (-0.057)
Residential construction demand			
ΔRCD	-0.034 (-0.117)	-0.108 (-0.063)	0.128 (0.932)
$\Delta CPPI$	6.619 (2.031)**	-1.393 (-0.407)	-3.151 (-0.819)
ΔNI	-0.639 (-0.381)	-1.335 (-0.843)	-2.956 (-1.857)**
ΔHHE	2.994 (1.325)*	4.765 (1.794)**	-1.251 (-0.493)
ΔUR	0.408 (0.699)	-0.749 (-1.503)	-0.784 (-1.801)
ΔPOP	33.418 (1.333)*	32.201 (1.430)*	-14.196 (-0.471)
ΔIR	0.327 (0.942)	0.017 (0.057)	-0.118 (-0.552)
ΔVOE	0.185 (0.369)	-0.548 (-1.069)	-0.513 (-0.853)
Non-residential construction demand			
$\Delta NRCD$	-0.011 (-0.028)	-0.269 (-0.820)	-0.065 (-0.249)
$\Delta CPPI$	0.333 (0.065)	-2.801 (-0.502)	1.756 (0.357)
ΔNI	-6.660 (-2.332)***	-1.054 (-0.365)	-2.339 (-0.839)
ΔHHE	-2.766 (-0.629)	7.278 (1.698)**	-1.261 (-0.279)
ΔUR	2.118 (2.344)***	0.591 (0.669)	0.838 (1.056)
ΔPOP	0.561 (0.018)	42.610 (1.175)**	-31.573 (-0.649)
ΔIR	0.364 (1.117)	0.600 (1.623)*	-0.649 (-1.842)**
ΔVOE	2.054 (1.527)*	0.593 (0.463)	2.333 (2.336)***

Table 2 Estimation results of the VEC models with intervention indicators for three types of construction demand

* denotes t-statistics significant at 0.1 level

** denotes t-statistics significant at 0.05 level

*** denotes t¯statistics significant at 0.01 level

The estimates of intervention indicators (e.g. the recent global financial crisis and seasonality) in these VEC models indicate whether these external impacts can affect these three types of construction demand in Australia. The estimation results suggest that overall, residential and non-residential construction demand have received significant negative impacts from both the late 2000s global financial crisis and seasonality. The VEC models for three types of construction demand were examined for their model fit based on values of R-squared, Sum square residue, S.E. equation and Log likelihood. These values suggest that the VEC models are stable and well-fitted. The R-square values in three VEC models indicate that above 81% of the variations in Australian construction demand could be captured by these models, and the results of model fit also suggest that the estimation for the overall construction demand is more accurate compared with the residential and non-residential construction demand.

Effects of the late 2000s Global Financial Crisis

The interventions of the recent global financial crisis on the overall, residential and non-residential construction demand were identified. The VEC models were solved with intervention indicators for the overall, residential and non-residential construction demand during the period of global financial crisis based on the pre-estimated coefficients of independent variables in Table 2. Through this procedure, the values of three types of construction demand with intervention analysis are estimated under two scenarios, each with different assumptions regarding the trend of the social and economic factors selected in estimating construction demand.

Scenario A

The change of construction demand is assumed to follow the trend of economic growth in the pre-crisis period. The VEC models of three types of construction demand are solved based on the dynamic solution technique, which solves the models according to the forecast series of selected macroeconomic variables during the event period. The solved results and actual values of three types of construction demand are shown in Figure 2.

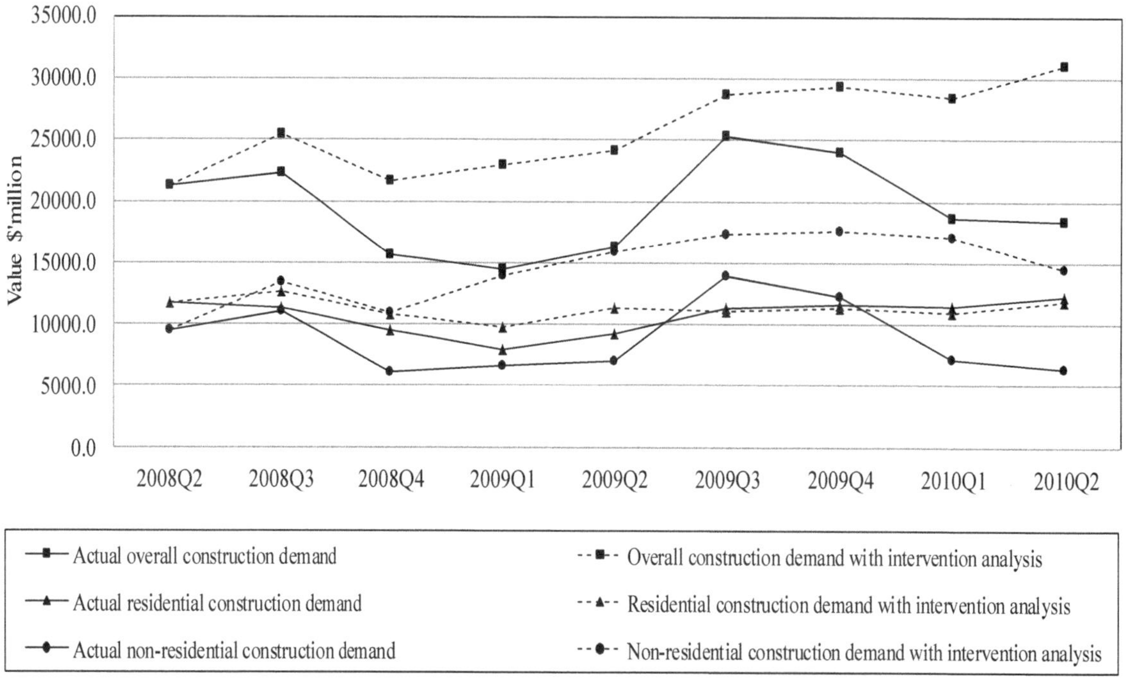

Figure 2 Comparison of different types of actual construction demand and construction demand with intervention analysis during the crisis (Scenario A)

The value of overall, residential and non-residential construction demand with intervention analysis are interpreted as the value of three types of construction demand changes from

2008Q3 to 2010Q2 without being shocked from the late 2000s global financial crisis. The gaps between values with intervention analysis and actual values can be seen as the net effects of the recent global financial tsunami on the level of demand for construction. The quarterly responses to the onset of the late 2000s global financial crisis, in terms of the change in the construction demand, are different in residential and non-residential industries. The non-residential construction industry was shocked deeper than the residential construction industry, showing a larger gap in the comparison.

The results of the impact of the late 2000s global financial crisis on the overall, residential and non-residential construction demand under Scenario A are summarised in Table 3. The total net effect of the current crisis on the residential construction demand was AUD −4971 million during 2008Q3 and 2010Q2, while the net effect of the crisis on the non-residential construction demand was AUD −50198.3 million. It should be noted that residential construction demand was positively impacted by the crisis between 2009Q4 and 2010Q2. This positive impact may be due to the series of effective stimulation strategies approved by the Australian Government, which resulted in a high demand in residential construction industry.

Scenario B

The change in construction demand is assumed to follow the trend of economic growth in the post-crisis era. A statistic solution technique is conducted to solve the value of overall, residential and non-residential construction demand, with intervention analysis based on the statistic data of selected macroeconomic variables. The results are compared with the actual value of construction demand in Figure 3. The comparison results also suggest that the impacts of the global financial crisis on the construction demand are different in the residential and non-residential industries. The non-residential construction industry is found to have been more negatively impacted than the residential construction industry during the crisis.

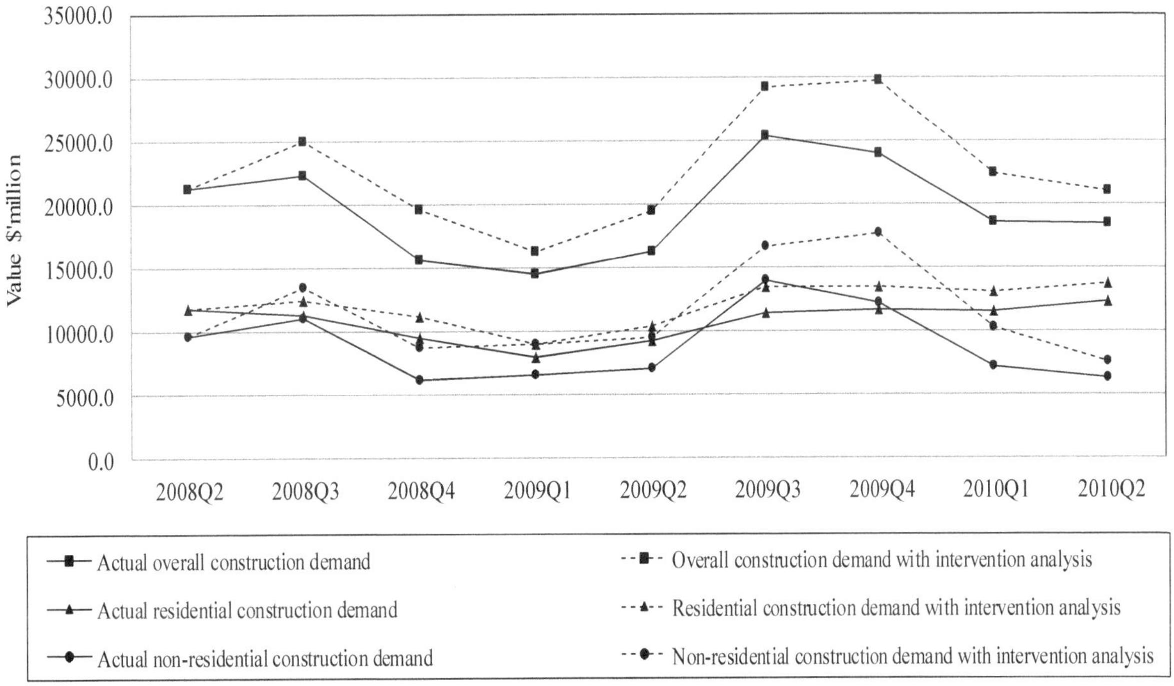

Figure 3 Comparison of different types of actual construction demand and construction demand with intervention analysis during the crisis (Scenario B)

Period	Overall construction demand			Residential construction demand			Non-residential construction demand		
	Actual	With analysis	Net effects	Actual	With analysis	Net effects	Actual	With analysis	Net effects
2008Q3	22333.6	25456.2	-3122.6	11281.4	12659.2	-1377.8	11052.2	13465.0	-2412.8
2008Q4	15662.0	21625.2	-5963.2	9529.8	10840.5	-1310.7	6132.2	10902.2	-4770.0
2009Q1	14476.2	22961.2	-8485.0	7887.9	9720.6	-1832.7	6588.3	13975.5	-7387.2
2009Q2	16241.9	24159.4	-7917.5	9213.2	11321.2	-2108.0	7028.7	15866.9	-8838.2
2009Q3	25349.7	28660.5	-3310.8	11345.6	11034.0	311.6	14004.1	17359.5	-3355.4
2009Q4	23973.9	29352.5	-5378.6	11652.3	11309.8	342.5	12190.0	17631.0	-5441.0
2010Q1	18612.9	28491.0	-9878.1	11518.2	10960.2	558.0	7177.0	17026.6	-9849.6
2010Q2	18428.6	31013.3	-12585.0	12298.9	11852.8	446.1	6312.1	14456.2	-8144.1
Total Net effects =			-56640.5			-4971.0			-50198.3

Table 3 Results of the impact of the late 2000s global financial crisis on the construction demand (Scenario A)

Note: the duration of the intervention period is from 2008Q3 to 2010Q2

Period	Overall construction demand			Residential construction demand			Non-residential construction demand		
	Actual	With analysis	Net effects	Actual	With analysis	Net effects	Actual	With analysis	Net effects
2008Q3	22333.6	24946.8	-2613.2	11281.4	12450.6	-1169.2	11052.2	13430.0	-2377.8
2008Q4	15662.0	19604.2	-3942.2	9529.8	11130.2	-1600.4	6132.2	8758.9	-2626.7
2009Q1	14476.2	16312.5	-1836.3	7887.9	8943.3	-1055.4	6588.3	9035.1	-2446.8
2009Q2	16241.9	19430.1	-3188.2	9213.2	10431.8	-1218.6	7028.7	9442.3	-2413.6
2009Q3	25349.7	29201.0	-3851.3	11345.6	13499.2	-2153.6	14004.1	16652.3	-2648.2
2009Q4	23973.9	29666.7	-5692.8	11652.3	13432.0	-1779.7	12190.0	17649.8	-5459.8
2010Q1	18612.9	22460.3	-3847.4	11518.2	13050.9	-1532.7	7177.0	10288.2	-3111.2
2010Q2	18428.6	21020.0	-2591.4	12298.9	13764.5	-1465.6	6312.1	7621.4	-1309.3
Total Net effects			-27562.8			-11975.2			-22393.4

Table 4 Results of the impact of the late 2000s global financial crisis on the construction demand (Scenario B)

Note: the duration of the intervention period is from 2008Q3 to 2010Q2

Based on Scenario B, the results of the impact of the late 2000s global financial crisis on the overall, residential and non-residential construction demand are summarized in Table 4. By comparing the net effect it can be seen that there is a significant decrease of AUD −22393.4 million in the value of demand for the non-residential construction industry, and a total negative effect with AUD −11975.2 million in the residential construction demand caused by the crisis. It is found that in 2009Q3, the residential construction industry received the greatest shock from the financial tsunami which caused a higher loss, with AUD −2153.6 million in the level of demand than other quarters. However, the level of demand in the non-residential construction industry was impacted the most, with AUD −5459.8 million by the crisis in 2009Q4. In regard to the results of the intervention analysis in Scenario B, the government policy makers should note that the actual value of residential and non-residential construction demand were AUD 1465.6 million and AUD 1309.3 million less than the estimated values in 2010Q2 respectively. In the latter quarter, a further boost strategy would be to increase investments on construction by these amounts to correct the effects of the late 2000s financial crisis on both the residential and non-residential industries.

Key economy stimulus policies, issued by the Australian government to boost the national economy from global economic downturn, are summarized in Table 5. This table indicates that a fund of nearly AUD 50 billion was announced, being allocated to the non-residential construction industry from 2008Q4 to 2009Q1; the First Home Owner Home Grant (established by the Australian government to encourage and assist residents to buy their first home) was tripled; and interest rates were cut from 7% to 3% within three quarters. These stimulation strategies have boosted both the national economy and the construction industry effectively. However, these strategies only stimulated investments in the construction industry in the short term, while approvals in the Australian construction industry reached the same level in the 2009Q3 as at the beginning of the global financial crisis. After that, the lagged effect of the crisis was enhanced, as no further boost strategies were applied by the Government from 2009Q2 to 2010Q2 and interest rates began to increase from October 2009. ABS (2012) reported a dramatic decline of greater than 25% in the value of construction work approved from 2009Q4 to 2010Q1. Furthermore, the results of Scenarios A and B indicate that the negative effects of the crisis on construction demand changed from strong to weak from 2008Q3 to 2009Q3, then reverted to strong during 2009Q4 and 2010Q2. This fluctuation in the loss of the value of demand may be largely due to the changes in Australian government policies in response to the financial crisis. The actual residential construction demand moves more closely with residential construction demand with intervention analysis when comparing the gaps between actual non-residential construction demand and its estimated value. This suggests that the residential construction industry reacts faster than the non-residential construction industry when facing a special global event.

Conclusions

Intervention analysis has become an effective technique of event estimation. The estimates of the effect of an event are based on the entire historical data of the variable concerned, and not simply on the comparison of a few quarters. Previous applications, of the intervention analysis in the construction industry, did not fully consider the dynamic relationships between the variable concerned and economic indicators. There were also problems with choosing an appropriate strategy to estimate the effect of the event. Furthermore, while previous research on intervention analysis discussed how the event affects the market or variables concerned, it did not consider the effect should the event not occur. The VEC model, with intervention indicators, was introduced to estimate the impact of the late 2000s global financial crisis on the construction demand in Australia. The dynamic relationships between construction demand and macroeconomic indicators were integrated into the estimation models. The estimation was conducted based on two scenarios, including that the change in construction demand is assumed to follow the trend of economic growth in

Dates of Announcements	Australia's responses to the global financial crisis
September 2008	The RBA announced plans to lower the cash rate by 25 basis points to 7 per cent.
October 2008	The RBA announced plans to lower the cash rate by 100 basis points to 6 per cent.
October 2008	The Australian government announced plans to guarantee all bank deposits and an economic stimulus package worth AUD 10.4 billion was announced. In this package, AUD 1.5 billion was allocated to support housing construction.
November 2008	The RBA announced plans to lower the cash rate by 75 basis points to 5.25 per cent.
December 2008	The Australian government announced that the First Home Buyer's grant doubled to AUD 14,000 for existing homes, and tripled to AUD 21,000 for new homes. A fund worth AUD 4.7 billion was allocated in commencement of large-scale infrastructure projects.
December 2008	The RBA announced plans to lower the cash rate by 100 basis points to 4.25 per cent.
February 2009	The Australian Government announced an AUD 42 billion stimulus package titled the Nation Building and Jobs Plan.
February 2009	The RBA announced plans to lower the cash rate by 100 basis points to 3.25 per cent.
April 2009	The RBA announced plans to lower the cash rate by 25 basis points to 3 per cent
May 2009 - August 2010	No further boost strategy was announced by the Government.
October 2009 - August 2010	The RBA announced plans to raise the cash rate from 3 per cent to 4.5 per cent

Table 5 Australia's response to the late 2000s global financial crisis

the pre-crisis period with no effects of the global financial crisis, and the change in construction demand is assumed to follow the trend of economic growth in the post-crisis era.

The estimates of the VEC models indicate that over 81% of variation in construction demand can be explained by the selected economic indicators. The lagged construction demand, the growth of population and the interest rate have significant roles in explaining the level of demand in the construction industry. Residential, non-residential and overall construction demands in Australian construction industries were affected significantly by both the recent global financial crisis and seasonality. Construction demand in the non-residential industry was affected more significantly than the residential industry by the crisis, shown by comparing the results of the net effect both in Scenarios A and B. It is also found that the actual residential construction demand is in line with the movements of the residential construction demand with intervention analysis, when compared with actual non-residential construction demand and its estimated value. This suggests that the residential construction industry was more reactive and was able to recover faster when experiencing a global crisis in comparison with the non-residential industry. The impact from the lagged effect of the crisis is worse than the onset of the crisis. Despite the effectiveness of the economy boost strategies issued by the Australian government, construction demand experienced a significant decline after the government removed the stimulation plan. According to the results of the intervention analysis, government policy makers should note that lagged negative effects on the non-residential construction demand became worse from 2009Q4 to 2010Q2, and a further boost strategy, such as increasing investments on construction, needs to be implemented to allow the Australian construction industry to recover from the late 2000s financial crisis..

This study focused on the impact of the recent global financial crisis and seasonality. Other global events and factors affecting the construction industry, such as the 1997 Asian Financial Crisis, the 2000 Sydney Olympic Games and the September 11 attacks, have not been considered in the scope of this study. . In particular, an abnormal variation of construction demand was observed in Australia during 2001 and 2002. This major depression was largely due to a change in Australian government policy regarding the new taxation law which came into effect in 2001. However, the VEC model with intervention indicators used in this study served as an experiment for an advanced econometrical method, which can be used to analyse the effect of special events and factors on construction and other industries.

References

ABS 2012, Australian economic indicators. Canberra, Australia Bureau of Statistics

Akintoye, A., Bowen, P. & Hardcastle, C. 1998, 'Macro-economic leading indicators of construction contract prices', *Construction Management and Economics*, **16** (2), 159-175

Akintoye, A. & Skitmore, M. 1994, 'Models of UK private sector quarterly construction demand', *Construction Management and Economics*, **12** (1), 3-13

Ball, M., Farshchi, M. & Grilli, M. 2000, 'Competition and the persistence of profits in the UK construction industry', *Construction Management and Economics*, **18** (7), 733-745

Berkmen, P., Gelos, R. G., Rennhack, R. & Walsh, J. 2009, The global financial crisis: Explaining cross-country differences in the output impact, International Monetary Fund Working Papers WP/09/280,available at http://www.imf.org/external/pubs/cat/longres.aspx?sk=23438.0 access on 15/08/2013.

Blankenburg, S. & Palma, J. G. 2009, 'Introduction: the global financial crisis', *Cambridge Journal of Economics*, **33** (4), 531-538.

Bonham, C., Gangnes, B. & Zhou, T. 2009, 'Modeling tourism: A fully identified VECM approach', *International Journal of Forecasting*, **25** (3), 531-549.

Box, G. E. P. & Tiao, G. C. 1975, 'Intervention analysis with applications to economic and environmental problems', *Journal of the American Statistical Association*, **70** (349), 70-79.

Dicky, D.A. & Fuller, W.A. 1979, 'Distribution of the estimators for autoregressive time series with a unit root', *Journal of the American Statistical Association*, **74** (336a), 427-431

Engle, R. F. & Granger, C. W. J. 1991, *Long-run economic relationships: readings in cointegration*, Oxford University Press, New York.

Fan, R. Y. C., Ng, S. T. & Wong, J. M. W. 2010, 'Reliability of the Box-Jenkins model for forecasting construction demand covering times of economic austerity', *Construction Management and Economics*, **28** (3), 241-254.

Hua, G. B. 1998, 'Forecasting residential construction demand in Singapore: a comparative study of the accuracy of time series, regression and artificial neural network techniques', *Engineering, Construction and Architectural Management*, **5** (3), 261-275.

Hua, G. B. 2005, 'The dynamic effects of the Asian financial crisis on construction demand and tender price levels in Singapore', *Building and Environment*, **40** (2), 267-276.

Hui, E. C. M. & Yue, S. 2006, 'Housing price bubbles in Hong Kong, Beijing and Shanghai: a comparetive study', The Journal of Real Estate Finance and Economics, 33 (4), 299-327.

Jiang, H. & Liu, C. 2011, 'Forecasting construction demand: A vector error correction model with dummy variables', *Construction Management and Economics*, **22** (9), 969-979.

Johansen, S. & Juselius, K. 1990, 'Maximum likelihood estimation and inference on cointegration -- with applications to the demand for money', *Oxford Bulletin of Economics and Statistics*, **52** (2), 169-210.

Li, Y. & Liu, C. 2012, 'Labour productivity measurement with variable returns to scale in Australia's construction industry', *Architectural Science Review*, **55** (2), 110-118.

Liu, C., Ma, L., Luo, Z. & Picken, D. 2009, 'An interdependence analysis of Australian house prices using variance decomposition', *International Journal of Housing Markets and Analysis*, **2** (3), 218-232.

Ma, L & Liu, C. 2013, 'Spatio-temporal analysis of house price convergence based on a demographical distance', *International Journla of Strategic Property managemnt*, **17**(3), 263-277.

Meikle, J. 2001, 'A review of recent trends in house construction and land prices in Great Britain', *Construction Management and Economics*, **19** (3), 259-265.

Murphy, L. 2011, 'The global financial crisis and the Australian and New Zealand housing markets', *Journal of Housing and the Built Environment*, **26** (3), 335-351.

Nayyar, D. 2011, 'The financial crisis, the great recession and the developing world', *Global Policy*, **2** (1), 20-32.

Ofori, G. 1990, *The construction industry: aspects of its economics and management*, Singapore University Press, Singapore.

Phillips, P. C. B. & Perron, P. 1988, 'Testing for a unit root in time series regression', *Biometrika*, **75** (2), 335-346.

Ramey, V. 1993, How important is the credit channel in the transmission of monetary policy?, *Carnegie-Rochester Conference Series on Public Policy*, North-Holland.

Song, Y. & Liu, C. 2006, 'The Australian construction linkages in the 1990s', *Architectural Science Review*, **49** (4), 408-417.

Tang, J. C. S., Karasudhi, P. & Tachopiyagoon, P. 1990, 'Thai construction industry: demand and projection', *Construction Management and Economics*, **8** (3), 249-257.

Yap, G. & Allen, D. 2011, 'Investigating other leading indicators influencing Australian domestic tourism demand', *Mathematics and Computers in Simulation*, **81** (7), 1365-1374.

Payment Mechanisms for Winter Road Maintenance Services

19

Adel Abdi, Hans Lind and Björn Birgisson, (KTH Royal Institute of Technology, Sweden)

Abstract

In countries with severe winters, a major portion of the annual budget for road maintenance is allocated to winter road maintenance. Thus it is important to identify an appropriate basis for the remuneration of the entrepreneurs who carry out the maintenance tasks, one that minimises or eliminates disputes and that satisfies both client organisations and contractors. The objective of this study is to investigate and evaluate the payment models applied in Sweden for winter road maintenance services and suggest possible improvements. Inadequate reimbursement models lead either to unnecessary cost overruns that affect the client's annual budget or cause cash flow problems for the contractor, which can result in safety issues. To solve the problems associated with paying for just-in-time road maintenance, cold region countries such as Sweden have developed various remuneration models, including some based on what is known as the Weather Index. The study uses a domestic questionnaire survey, analysis of a number of current contract documents, a series of meetings with project managers followed by an international benchmarking investigation. The study identified four winter maintenance remuneration models of which one is based on weather data. The study reveals that the payment model based on weather data statistics is applied only to roads with higher traffic flow and generates the most uncertainty about costs. Possible improvements should include more reliable weather data obtained from weather stations and bonuses related to customer satisfaction.

Keywords: Road maintenance, Winter maintenance services, Contract, Payment mechanism, Remuneration models

Introduction

Retaining the value of a country's existing infrastructure is extremely important for a healthy national economy. However, maintenance of infrastructure such as a road network is both difficult and costly because it is spread over a large geographical area and the tasks required are diverse and fragmented (Ng & Wong, 2007). It is even more difficult and expensive when the tasks are performed during winter. Yet any reduction in the budget for performing these tasks increases the vulnerability of the infrastructure (Jaarsma & van Dijk, 2002).

When outsourcing and contracting road infrastructure maintenance services, the main concept behind the payment mechanism is that contractors are compensated for services performed (Aziz, 2007). However, identifying the optimum payment method presents a challenge for road and transport authorities, which are subject to budget constraints. A fair remuneration model is a requirement if cost effective, quality services are to be performed in a timely manner. Such a model also acts as an incentive for contractors to tender for road maintenance projects. As Motawa & Kaka (2009) show, clear identification of an appropriate payment mechanism contributes to fair payment to contractors and improved work performance.

Most contractors engage subcontractors with the necessary skills for some of the contracted tasks to make up for the skills they lack. Subcontractors, in particular small and medium-sized contracting companies, consider an appropriate form of remuneration a vital factor in performing the tasks on schedule and to the required quality. In this connection, Judi & Rashid (2010) found that the financial capacity of most contractors is limited and that keeping their cash-flow in balance depends directly on how they get paid for work done, thus enabling them to continue operating.

Even medium-sized contractors in the supply chain need to hire service providers and suppliers to perform their contracted tasks. Improper payment methods can result in delays in performing the work and reduction of profitability, which also leads to traffic disruption and delay. Motawa & Kaka (2009) even show that many contractors try to minimise the outstanding balances owed by clients and increase project cash flow either by increasing efficiency (e.g. by instituting more effective management processes) or by resorting to unfair tactics such as over-measurement and delay in payment of subcontractors.

Although winter highway maintenance services seem limited and controllable tasks, the fact that these activities have serious consequences for society and are easily affected by unpredictable factors (such as improper payment models) had led to their being recognized as important, extensive and complex.

Winter maintenance of the state road network in Sweden is performed by the Swedish Transport Administration (STA) and integrated into a comprehensive road maintenance contract. A hybrid payment mechanism is applied in the contract, partly for performance of year-round services and partly for winter services. When contracting road maintenance services, STA is interested in just-in-time service delivery before the contractor is paid. The payment model for winter maintenance is based particularly on weather characteristics that are addressed in the contract documents.

The objective of this study is to describe and analyse different dimensions of the payment system for winter road maintenance, both from a theoretical perspective and in relation to practical feasibility. The focus is on the current Swedish system and how it can be improved.

Theoretical Background

When developing the basic structure of a payment system, four principles should be addressed. The first basic principle is the measurability of the contracted and the performed services (i.e. the performed services should be measurable both qualitatively and quantitatively). A third party (e.g. a court) must be able to evaluate whether the work has been carried out as agreed. For example, according to the Irish Department of the Environment and Local Government Note 12 (2000) performance-based payment should be based upon achieving standards that are practically measurable during the whole period of the contract. Mattsson & Lind (2009) indicate that problems can arise in contracts where payments are subject to the results achieved or subject to keeping a road in a certain condition. There is often disagreement about whether the specific standards were met. This can be an especially large problem in winter maintenance contracts as road conditions can change very quickly.

The second basic principle, also underlined by the Irish authority (2000) is that the payment system has to be simple enough to be understandable. The third basic principle is the inclusion of proper incentives within a specific payment mechanism. Should there be a fixed payment for the services to be performed, or should there be a payment for a specific action carried out? As will be discussed below, a fixed payment can lead to contractors doing too little, while paying according to work actually carried out can lead to contractors doing too much. In this context, Motawa & Kaka (2008, 2009) state that both payment mechanisms have advantages and disadvantages.

The fourth basic principle in designing a payment mechanism is the allocation of risk. Winter conditions vary from year to year. In some years a lot of work is needed to guarantee certain results for each type of road, while in other years much less work is needed. A fixed payment puts this risk on the contractor, while payment in relation to the work carried out puts the risk on the client. As discussed in Mattsson & Lind (2009), putting greater risk on the contractor can lead to reduced competition as smaller and risk-averse companies may not be willing to bid for such a contract.

As Aziz (2007) states, in order to develop payment systems, existing payment systems should be studied and analysed so that the information can be used to create new payment mechanisms. Råsled (1992) claims that the difficulty level of a project, product or service has a crucial and decisive influence on whether a certain payment model (e.g. unit-price) is advisable and feasible, or whether payment should be based on cost accounting.

Payment Framework in Sweden

The Legal Framework

While all public procurement and contracting within the construction industry in Sweden is governed by European Union (EU) directive 2004/18/EC incorporated into the Swedish *Act on Public Procurement* (see Abdi, Lind & Birgisson, 2013b), the payment mechanism for the total contract is governed by the domestic agreement *General Condition of Contract for Design and Construct Contracts for Building, Civil Engineering and Installation Works – ABT 06*. The agreement is a legal document that all the actors in the Swedish construction industry regard as the basis for how payment should be implemented. Road maintenance services in Sweden are performed according to a so-called *total contract*, described as a *turnkey contract for the execution of the operation and maintenance of public roads.* This means the payment system for this type of contract is also regulated by this legal document. International contractors who are awarded a contract to perform projects in Sweden have to follow this agreement. An important question is how flexible or rigid this document is, but that is beyond the scope of this study.

Contract Type

In Sweden, road maintenance is divided into routine and structural maintenance. Routine or corrective maintenance is the work required to maintain the functionality of the road network all year long to specific road standards. Structural or preventive maintenance is the work required to maintain the whole road asset for long term usage by maintaining and improving the asset's values. In order to facilitate road maintenance services, the whole Swedish road network is divided into 115 maintenance areas (1st September 2012) within STA's six regional administrative directorates. A routine maintenance contract is procured area-wide (i.e. one contract per maintenance area). Winter road maintenance services are bundled with other road maintenance services into Area Maintenance Contracts (AMC). Abdi, Lind & Birgisson (2013b) discuss whether this kind of bundled contract is the best option.

Empirical Investigations – Method and Results

Earlier studies (Abdi, 2011; Abdi, Lind & Birgisson, 2013c) revealed diffuse and widespread dissatisfaction with the procurement procedure and contracts for current Swedish winter maintenance services. The outcome of these studies motivated additional study to gain insight into the payment system and how it is linked to the whole maintenance performance procedure.

This research started with a domestic questionnaire survey and continued by collecting a number of current AMC documents, with the support of Swedish Transport Administration (STA), for analysis and evaluation. The questionnaire for the domestic survey was designed with a total of 70 questions divided into four themes. Totally 41 questionnaires were sent to a number of selected experienced staff at 18 Swedish client organisations and contracting companies. The core theme for this part of the study was the payment system used in road maintenance contracts including winter road maintenance. A series of meetings with a number of project managers was also conducted to collect their insight into the current payment system. An international benchmarking investigation was also carried out by requesting information from a number of Road and Transport Authorities (RTA) around the world with climates similar to Sweden. The selected RTA were from Washington DC, USA; Ontario, Canada; Finland; Norway; Iceland; Denmark; Estonia; Lithuania; Latvia; England; Austria and Japan. Six of the twelve selected RTA responded to the request. The results of each avenue of research are described in the corresponding subsections below.

Selected Results of the Domestic Survey

The responses to the questionnaire revealed that there are two basic forms of remuneration for all road maintenance tasks performed year round, which thus include winter maintenance: performance-based and prescribed payments. Both forms are based on a fixed pricing model for one to two years, depending on the maintenance area and the contract length. After the end of the initial year (or two years) indexing is applied to the remaining contract period. Every month the contractor is entitled to payments partly on a performance basis and partly according to a prescribed payment plan. The performance basis is intended to adjust for quantities (fixed unit-price multiplied by actual agreed quantities). The prescribed payment plan is set up annually by the clients. This latter form of compensation is intended for non-adjustable quantities (fixed costs) according to the bill of quantities. Usually, the prescribed fixed payment is intended for the contractor's winter organisation and cannot be counted as an adjustable amount. The results from examining current AMC documents confirmed this after reviewing and analysing five randomly chosen maintenance areas across Sweden. The analysis of the contract documents also revealed that there are advance payment rules in some AMC, but not in others. Figure 1 gives an overview of the payment mechanism for road maintenance services in Sweden where AMC stands for Area Maintenance Contract.

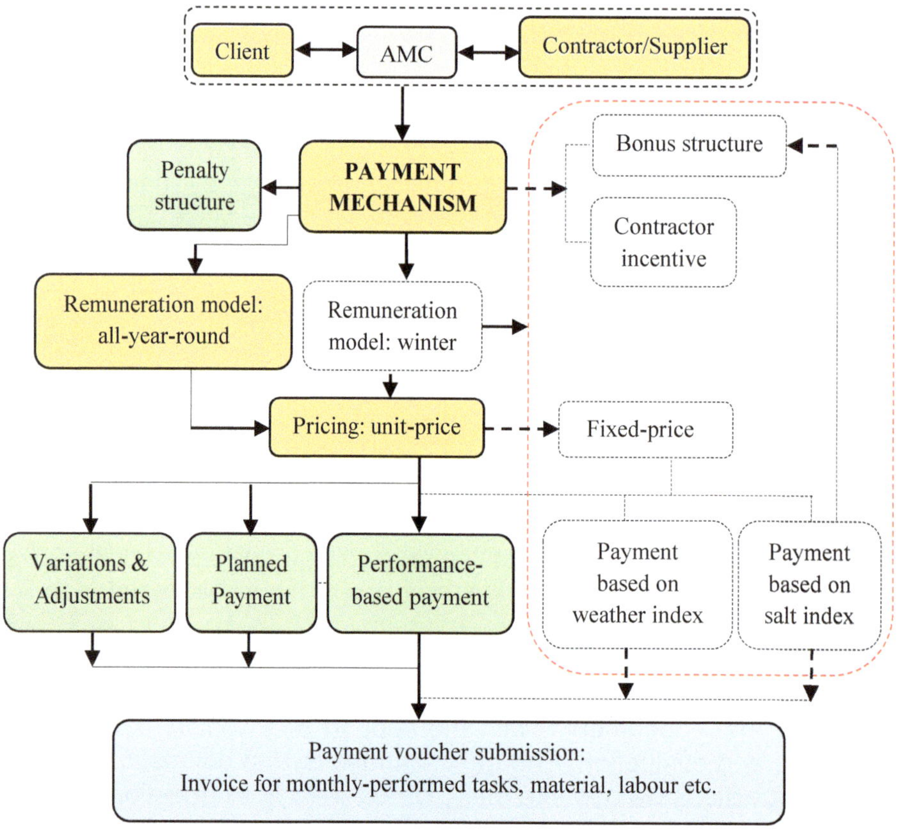

Figure 1 Flow chart of the Swedish payment mechanism for road AMC

Payment System for Winter Road Maintenance

Although winter road maintenance services are integrated into the AMC (as mentioned earlier), additional rules apply in the payment system for performance of these services. These rules vary from maintenance area to maintenance area due to different regional and area-specific circumstances. Some contracts that were awarded some years ago are still valid and are driven by older rules, while other contracts are newer with revised rules. Currently, there are a number of different remuneration models for winter maintenance adapted to different maintenance contracts. Winter models generally consist of fixed and variable components that are similar to the remuneration of the other road maintenance models. However, the mechanism for payment of winter road maintenance differs based on weather data and outcomes, making them rather complicated and in some cases unclear and uncertain.

Weather situation		Weather outcome and condition
Group H1		**Number of weather outcomes of type HR1, HR2 and HS**
HR1	Slippery road surface due to moderate hoar frost[1]	Road surface temperature is at least 0.5°C lower than dew point temperature or road surface temperature is lower than +1.0 °C
HR2	Slippery road surface due to heavy hoar frost	Road surface temperature is at least 2.0°C lower than dew point temperature or road surface temperature is lower than +1.0 °C
HS	Slippery road surface due to light snowfall	Amount of accumulated snow is between 0.01 and 0.50 cm at a temperature between +1.0°C and -10.0 °C during a snow period of highest 4h
Group H2		**Number of weather outcomes of type HT and HN**
HT	Slippery road surface due to moist /wet surface freezing	Wet road surface temperature drops below +1.0 °C due rain, snow or sleet under 12 hours
HN	Slippery road surface due to rain or sleet on cold surface	Road surface temperature is lower than +1.0 °C due to rainfall or sleet
Group S1		**Number of weather outcomes of type S and D both with snow amount >0.51 cm per interval of highest 4 hours.**
S	Snowfall	Precipitation is snow
D	Drifting snow	Wind speed at least 6m/s under at least 4 h in average and if drift-capable snow is presented
Group S2		**Number of weather outcomes of type D with amount of snow ≤0.50 cm per interval of highest 4 hours**
D	Drifting snow	Wind speed at least 6m/s under at least 4 h in average and if drift-capable snow is presented

Table 1 Weather situations generated by WI from basic data of RWIS and the output groups (Source: STA, 2012)

Information about weather is obtained from a *Weather Index* (WI)[2] developed by the STA. The WI is the basis on which contractors to plan and perform their tasks, and then receive payments for the work done during winter. This index is also a compensation model through which the payment for winter road maintenance services is calculated. The WI for each area is calculated partly by a number of climate factors. Basic raw data is obtained from weather stations within a maintenance area connected to the STA's Road Weather Information System (RWIS), and partly from scaled Mesoscale Analysis System (MESAN scaled) weather data obtained from the Swedish Meteorological and Hydrological Institute (SMHI). The raw data from RWIS consists of air temperature, road surface temperature, dew point temperature, relative humidity, type and amount of precipitation, and wind speed. MESAN

[1] A deposit of interlocking ice crystals formed by direct sublimation on a surface, sometimes referred to as Rime Frost (EU Cost 344, 2002)

[2] WI is a further development of an earlier software program used for winter remuneration model 'Weather Compensation Combination'(Swedish: VädErsKombi)

analysis is a compilation of all conceivable meteorological observations around the clock. The data obtained from MESAN are average values over a 22×22 km wide area, while the RWIS data are area-specific values. Each kind of obtained data is weighted depending on its quality. Different weather situations are generated as items by WI and are divided into four different groups of weather outcomes that are presented in the output of the WI software program (Table 1).

In addition to the above weather situations, two other weather situations have to be accessed by entrepreneurs, namely special weather SV1 and SV2. Special weather SV1 is a weather situation where the wind speed is at least 10 m/sec for 6 hours and there is risk of snowdrifts. Special weather SV2 is a weather situation with snow intensity of at least 1.5 cm/h for 15 hours. In both cases, entrepreneurs have to use all available resources for organising their maintenance tasks. The remuneration in such severe situations is on a time and materials basis.

The way outcomes in Table 1 are compensated for depends on the amount of each item. A severe winter may generate eight HR1s and four HR2s monthly, while a mild winter may generate fewer HRs per month. Payment for all work performed is made after the average of the amount of items in the H and S groups has been calculated. The unit-price for each item in each group has already been priced in the bill of quantities.

The Swedish road network is divided into five classes (Table 2) according to traffic flow (Annual Average Daily Traffic – AADT), and the compensation of contractors for the winter tasks varies between classes.

Traffic flow *(AADT)*	**>16000**	**8000-16000**	**2000-7999**	**500-1999**	**<500**
Road class	1	2	3	4	5

Table 2 Road classification in Sweden (Source: STA 2012)

There are two models for the variable remuneration part of the payment mechanism for winter measures. The first model is based on weather outcomes from WI and applies only to road classes 1 to 3. The second one is intended for the implemented measures per km road and applies only to road classes 4 to 5. The variable compensation for road classes 1 to 3 is priced in the bill of quantities per weather outcomes. The amount is adjustable for the actual number of weather outcomes according to the WI. The amount in the bill of quantities is calculated based on the statistics of the past five years. The variable compensation for road classes 4 to 5 is priced in the bill of quantities per road class and road kilometre (km) performed services, excluding de-icing materials. De-icing materials for road class 4 to 5 are reimbursed separately according to actual quantity used and according to the client's specified unit-price in the bill of quantities.

The fixed remuneration covers the fixed costs of the contractor's winter organisation that are weather-dependent. The payment system based on weather data statistics is common to all 115 existing AMC for performed winter road services. In addition to the weather index, a *salt index* is applied in the contracts. This describes the actual salt consumption compared to the client-recommended use of salt for each type of weather situation. While the structure includes a penalty in the payment mechanism for performance of both winter and other road maintenance services, the bonus structure is intended to create incentives for contractors to reduce salt consumption during winter. Thus the payment mechanism is influenced by a bonus system which only aims to reduce salt consumption during winter, nothing else.

Meeting with the Project Managers

A series of meetings were conducted with a number of the client organisation's project managers to confirm the obtained data by analysing contract documents and to understand other possible concerns with the current payment systems. Almost all the managers who work with the client organisation agreed that the payment mechanism is far from perfect, particularly during severe winters. They also stated that the current remuneration model based on WI is very complicated and there are a number of uncertainties in the model. Moreover a lack of skilled personnel using the WI also affects payments. In some cases, the complexity of the current model has led to disputes. A number of managers stated that the model works in certain areas, but not in others.

Some project managers said that before a contractor is compensated for work done, the performed tasks have to be followed up and verified by a Global Positioning System (GPS). This verification method is intended for road classes 4 to 5. This method does not function within all maintenance areas. Follow-up and verification of performed measures for km-road by GPS is not always easy to sort out.

The survey recognised at least four remuneration models in the payment mechanism for winter road maintenance services, summarised as follows:

- Remuneration with variable costs, where payment depends on how long the measures have taken, that is, the contractor is reimbursed by SEK/h (Swedish kronor per hour)
- Remuneration with variable costs with time (h) incentives, where the contractor gets paid per hour for work done during winter season. After the end of the winter season, payments are reconciled with the contractor according to the weather outcomes (WI) model
- Remuneration with variable costs with road length (SEK/km) incentives, where the contractor gets paid per km-road and reconciliation is made with the contractor after the season
- Remuneration according to the weather outcomes (WI) where the unit-price is already priced in the bill of quantities for different weather situations (Table 1)

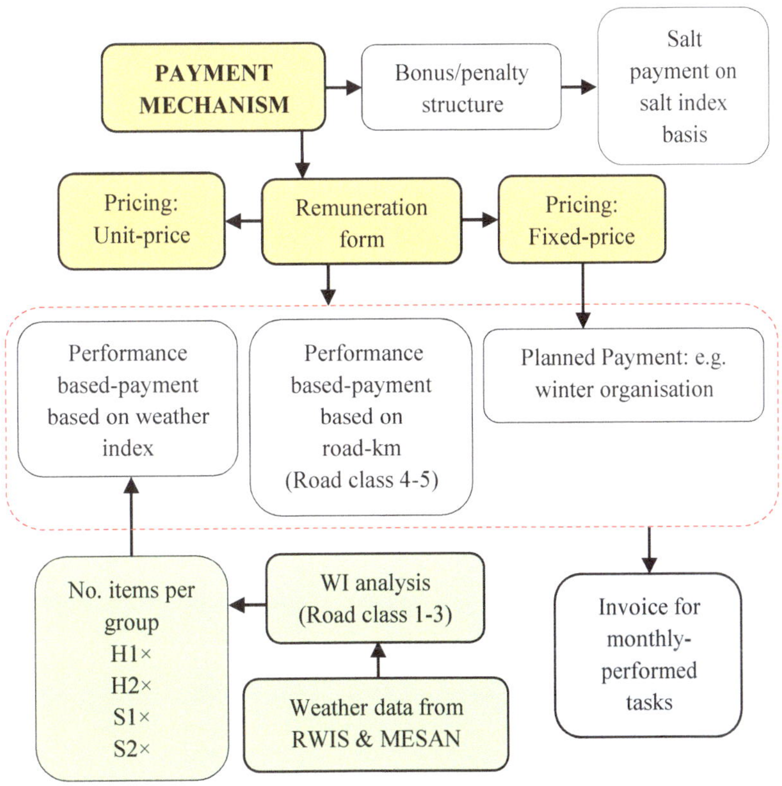

Figure 2 Flow chart of the Swedish payment mechanism for winter road maintenance

Because contracts are different in different areas with different contract durations, the models differ from one area to another. The two last models are newer ones that are applied in newly procured contracts, while the former two have already been applied in some Area Maintenance Contracts (AMC). Figure 2 illustrates how the contractor receives payment for winter road services.

Selected Results from the International Survey

Concerning payment method for winter road maintenance, the obtained responses from six of twelve responding Road and Transport Authorities (RTA) in Canada; Ontario, Denmark, Finland, Iceland, Lithuania and Estonia show that winter road maintenance is often organised differently from one RTA to another. The level of knowledge and experience in selecting and applying appropriate payment mechanisms also varies between RTA in different countries. According to the responses submitted, winter maintenance services are integrated into one road maintenance contract (i.e. bundled with other road maintenance services) in all the responding RTA except one. Concerning forms of remuneration, there is a clear difference between the six responding RTA. While variable payment related to the amount of work performed is used in Demark and Lithuania, fixed-price payment is used in Estonia and Ontario, Canada. Finland and Iceland apply both the fixed-price and variable model. Figure 3 is a compilation of the benchmarking study and gives an overview of the payment mechanism for road maintenance services in the responding countries' RTA. Note that the boxes enclosed by dashed lines show the components that are not included in the payment mechanism in the specific country listed at the top of the column.

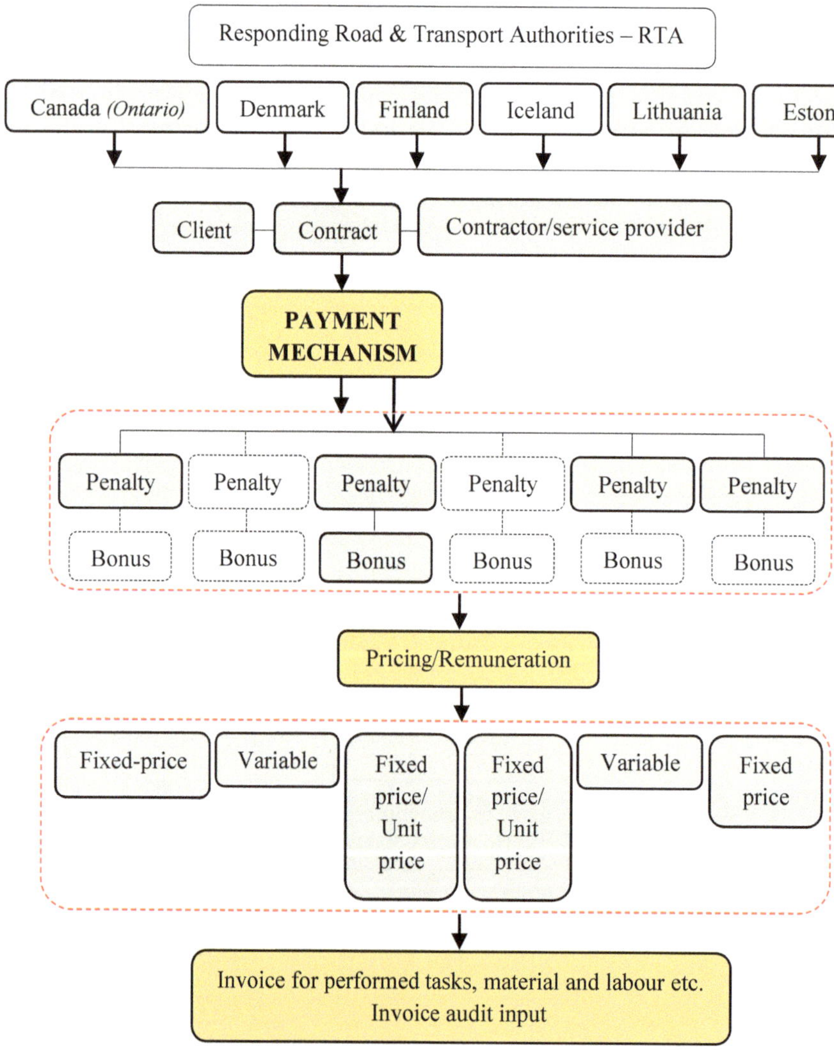

Figure 3 Schematic representation of the payment mechanism in the studied countries

As the information from the responding Road and Transport Authorities (RTA) shows, there are similarities to Sweden's area-wide contracts with different durations being applied in some responding countries like Finland and Ontario, Canada.

Concerning the inclusion of penalties and bonuses in the contract, three responding RTA apply only penalties (Ontario, Lithuania and Estonia), two of them apply neither penalties nor bonuses (Denmark and Iceland), and only one RTA applies both penalties and bonuses (Finland). The penalty and bonus components in the payment mechanism cover not only winter maintenance but also other road maintenance services (see Figure 3). The way in which penalties are applied is as follows:

- Penalties are applied if the quality of performed work is poor after inspection. If the work is finished ahead of schedule with the required quality, extra payment will be available (Finland).
- Penalties are applied if the quality does not meet ISO standards. No bonus is available (Ontario, Canada).
- Penalties, as determined in the contract, are applied if the requirements for road conditions are not met. There are no bonuses (Estonia).
- Although the contractors are state-owned contracting companies, they still face penalties if they do not follow the requirements in the contract (Lithuania).

When asked '*If the performed work is paid by variable payment related to the amount of work performed, there is risk that you have to pay for more work than is actually carried out*', only one responding RTA, which applies the variable remuneration model, totally agreed with the statement. Three of them partly agreed and one totally disagreed. Three commented as follows:

- In Denmark, the performed work is logged and followed up through the winter management system, otherwise there is risk of paying more for less performed work (agree with question)
- In Iceland, the service vehicles are tracked, thus there is no risk of paying more for less work done (disagree)
- In Lithuania, there is full control over the situation by monthly follow-up of the amount of performed work according to the contract, which is clearly stated in the contract (disagree)

When asked '*If the performed work is paid by fixed-price, there is risk that all the measures are not performed correctly*', only one responding RTA, which applies the fixed-price remuneration model, totally disagrees with the statement. Three of them partly agreed and one totally agreed with this statement. Additional comments were received from two RTA:

- Ontario, Canada, disagreed with the statement and commented that the contract states clearly what performance levels are to be met. Penalties are applied if performance levels are not met.
- Denmark agreed with the statement and commented that they follow the performed work through their winter system management just as they followed for the variable payment method.

Note that in Denmark road maintenance services are performed partly in-house and partly outsourced.

Four of the six responding countries use WI to facilitate organisational and payment mechanisms for winter tasks just as Sweden does. WI is a remuneration model that is served by data from RWIS (mentioned earlier). As the survey shows, the road networks in all the responding countries are equipped with the RWIS. The crucial components of a payment mechanism for remuneration of performed winter maintenance tasks can be highlighted and summarised in Table 3.

Country	Winter maintenance services		Contract period *(year)*	Option in contract	RWIS maintenance		WI access for cost control	Payment mechanism			
								Remuneration		Penalty	Bonus
	Bundled in road contract	Separate contract			Bundled in road contract	Separate contract		Variable* *(Unit-price)*	Fixed-price		
Canada *(Ontario)*	X	-	≥10	-	-	X	Yes	-	X	Yes	No
Denmark	X	-	4	2	-	X	Yes	X	-	No	No
Finland	X	-	5-7	No	X	-	Yes	X	X	Yes	Yes
Iceland	-	X	3-5	2 (1+1)	-	X	Yes	X	X	No	No
Lithuania	X	-	1	No	-	X	No	X	-	Yes	No
Estonia	X	-	5-8	No	-	X	No	-	X	No	No
Sweden	X	-	4-6	2 (1+1)	-	X	Yes	X	X	Yes**	Yes**

**Variable remuneration model, related to the amount of work performed according to the unit-price in the bill of quantities*
*** Only for salt consumption control during winter*

Table 3 International benchmarking of winter road maintenance contracts

Due to the limited response to the benchmarking survey, it is uncertain to what extent WI is used in other countries.

Discussion

Two fundamentally different main forms of payment appear in the AMC, namely fixed price and variable payment. The second form is related to the amount of work performed on the basis of unit-price. Jacobson (2004) states that these remuneration forms rarely appear completely separate from each other. This is partly because an assignment can very rarely be specified with full certainty in advance, and partly because the activities are often associated with greater risks in certain areas than in others.

One of the payment models uses the Winter Index in which different kinds of registered weather situations within the main groups of slippery and snow are already priced in the bill of quantities. The contractor will receive payment in accordance with unit-price for each situation (Table 1). As our investigation revealed, one of the uncertainties with the model is that the selection of the WI outcome determines the payment, not the contractor's real performed measures. Another uncertainty with the model is the way in which the number of weather outcomes is measured and noted in the procurement specification. Weather data statistics from past years are usually used for new contracts. There is also uncertainty due to problematic weather conditions, such as fine rain that in the presence of low temperatures can cause slippery road conditions that are difficult to detect. The compensation model can in some cases create uncertainty and confuse the contractor when selecting the correct weather outcome (i.e. whether the selected weather outcome is compensable or not). In such cases, the contractor has to wait for a long time to select actions that are entitled to any payment. This implies a risk of conditions remaining slippery with snow on the road surface because lesser measures are taken, which in turn jeopardises road safety.

Another aspect of uncertainty is the measurability of the performed work in order to be paid fairly. Milgrom & Roberts (1992, p. 369) state that this is one reason for using fixed pay when performance and productivity are difficult to measure with any accuracy or even to describe in sufficient detail. According to the survey, the outcome of the WI system is strongly influenced by a number of factors (e.g. climate factor). Wesström & Böhm (2008) found that some of these factors can be attributed to the local characteristics of a maintenance area. As this current study reveals, there are disagreements between the client and contractor concerning types of weather situations or events (e.g. the amount of HR1, HR2 or SV1 or SV2). This is due to the definitions and measurement methods that are not optimum.

As mentioned earlier, RWIS is a system of weather stations for collecting weather data to help contractors plan winter road maintenance (see Abdi, Lind & Birgisson, 2012) and then get paid for the tasks performed. This system needs to be maintained and serviced in order to provide accurate weather data in real time. Operation and maintenance of this system is performed according to a separate contract called a framework agreement. Earlier studies (Abdi, Lind & Birgisson, 2012) revealed that the maintenance of this system has not kept up with the performance of those winter tasks that are stated in the AMC. Inaccurate weather data collected in the WI system leads to inaccurate outcomes of the WI. This, in turn, affects the accuracy of the amount of performed winter tasks, culminating in the creation of fuzzy invoices.

Analysis of the contract documents showed that when weather data is missing due to the failure of one or more weather stations, the weather outcome has to be calculated as follows to get invoices as accurate as possible: If a combination of the values from any of the selected MESAN and RWIS stations is available, then the average of the combination is used during the failure period. If values from all combinations are missing, then real resource capacity is reimbursed according to the legal agreement ABT 06 chapter 6, while the stations are repaired. From a broader perspective, Milgrom & Roberts (1992, p. 369) state, 'Attempting to pay for performance in these circumstances may actually amount to little more than a random payment policy'. This statement can be interpreted to mean that compensating a contractor according to the real resource capacity of the entrepreneur implies a more reliable payment mechanism than the payment according to WI outcome. Payment is made monthly, but a final reconciliation of the number of weather outcomes is done when the winter season ends.

In one of the investigated maintenance areas, both the contractor and the client were very satisfied with RWIS-based remuneration for the past winter season. However, the same survey showed that a number of weather stations in the maintenance area had errors and repair took so long that winter was over before they were operational. The study also shows that the assigned RWIS contractor had delayed repair and maintenance of weather stations due to delays in previous payments by the client.

Conclusion and Recommendations

This study, which is the concluding part of a broader research project undertaken in Sweden, focuses on the forms of remuneration in road maintenance contracts. Its main focus is to provide deeper knowledge and understanding of the payment mechanism for winter maintenance tasks in Sweden that are integrated in more comprehensive contracts and bundled with other road maintenance services. As the earlier studies revealed (Abdi, Lind & Birgisson, 2013b), several road maintenance services are bundled in a questionable manner in current Swedish road maintenance contracts. Focusing on the payment mechanism and the possibilities of improving it, the general conclusion and recommendations are as follows: From a risk allocation perspective, the combination of a fixed part and a weather dependent part seems to be the rational choice. A large client organisation with a number of AMC can pool the weather risks over the whole country, whereas small contractors may have problems if they enter into a fixed price contract and encounter a severe winter.

The weather dependent part in the Swedish contract is to a large extent also a fixed price contract, in the sense that there is no relation to the actions carried out by the contractor. The exceptions are when there are very special weather conditions. As both theory and the international questionnaire suggest, a fixed price contract creates a risk that the contractor does too little, especially when there are vague stipulations in the contract.

The issue of monitoring actual road conditions in relation to the demands in the contract is very important for applying proper road maintenance measures and payment methods. Monitoring of work performance is, however, costly and will always be imperfect when there are rapidly changing conditions. This means that it is very important to think in terms of creating long-term incentives for the contractor not to use loopholes in the contract and in the payment mechanism. One way to do this is bonuses related to general performance. The Swedish bonus model is currently only applied for salt consumption and control, but it would be very interesting to test the effect of bonuses in relation to customer satisfaction. Customer satisfaction could be measured by questionnaires given to road users. The bonuses can be related to absolute levels of customer satisfaction and to relative customer satisfaction (i.e. how satisfied they are in this district compared to other districts). Bonuses related to internal satisfaction could also be tested and could, for example, be based on questionnaires to the client's staff that has been working with the contractor in question.

As argued in another study (Abdi, Lind & Birgisson, 2013a), the use of a partnering concept in winter maintenance contracts might be rational, as this creates more flexibility. If such a contract contributes to establishing trust between the parties, it would also reduce the incentives for the contractor to exploit imperfections in the contract.

Another important aspect of creating long term incentives for the contractor is discussed in another study (Abdi, Lind & Birgisson, 2013b) and relates to the procurement process for contractors. Incentives are strengthened if contractor performance in the current period is an important determinant of whether the contractor will get the contract for the coming period. This can be achieved both by options in the current contract and by giving large weight to earlier performance when contractors are selected. Documented customer and client satisfaction can be the foundation for such an evaluation.

Predicting the effect of a certain payment mechanism is difficult and therefore systematic experimentation is important. A specific mechanism can then be tested and evaluated in a smaller number of AMC before using it on a larger scale.

References

Abdi, A., Lind, H. & Birgisson, B. (2013a) 'Effective winter highway maintenance through application of partnering concept', *International Journal of Economics and Management Engineering*, **3** (3), 112-125, DOI:10.5963/IJEME0303003

Abdi, A., Lind, H. & Birgisson, B. (2013b) 'Public procurement of winter road maintenance services based on EU procurement directive: Lessons from Sweden', *Journal of Investment and Management*, **2** (4), 70-86, doi:10.11648/j.jim.20130204.11

Abdi, A., Lind, H. & Birgisson, B. (2013c) 'Designing appropriate contract for achieving efficient winter road and railway maintenance with high performance quality: A survey of the state of practice in Sweden', *Accepted for publishing in the International Journal of Quality and Service Sciences, 1st April 2013*

Abdi, A., Lind, H. & Birgisson, B. (2012) 'Use of Road Weather Information System (RWIS) as assistive tool for effective winter road maintenance: Technical and contractual interactions', *International Journal of Engineering and Technology*, **2** (12), 2002-2012

Abdi, A. (2012) '*Winter road maintenance on contractual basis: An international benchmarking*', Manuscript for Research Rep., Royal Institute of Technology – KTH, Stockholm, Sweden

Abdi, A. (2011) '*Technical and contractual aspects in winter highway and railway operation and maintenance: A survey of current technical systems and contract forms in Sweden*', Licentiate thesis, TRITA-VBT 11:08, ISSN 1650-867X, Royal Institute of Technology – KTH, Stockholm, Sweden

Aziz, A. M. A. (2007) 'A survey of the payment mechanisms for transportation DBFO projects in British Columbia', *Construction Management and Economics*, **25** (5), 529-543

Department of the Environment and Local Government (2000) '*Payment Mechanisms*', Public Private Guidance Note 12, 14 April 2000, Dublin, Rep. of Ireland

European COST 344 Action, Improvements to Snow and Ice Control (2002) '*WINTERTERM - Dictionary of Terms for Winter Maintenance on European Roads and Bridges*', Durth Roos Consulting GmbH, Germany

Jaarsma, C. F. & Dijk, T. van. (2002) 'Financing local rural road maintenance. Who should pay what share and why?', *Transport Research Part A*, **36**, 507-524

Judi, S. S. & Rashid, R. A. (2010) 'Contractor's right of action for late or non-payment by the employer', *Journal of Surveying, Construction & Property*, **1** (1), 65-95

Laurinavicius, A. & Jukneviciute-Zilinskiene, L. (2011) 'Eleven years of RWIS operation in Lithuania: Possibilities for the use of the data collected', *8th International Environmental Engineering Conference,* 19-20 May, Vilnius, Lithuania

Mattsson, H-Å. & Lind, H. (2009) 'Experiences from procurement of integrated bridge maintenance in Sweden', *European Journal of Transport and Infrastructure Research*, **9** (2), 143-163

Milgrom, P. & Roberts, J. (1992) *'Economics, Organization and Management',* Prentice Hall, New Jersey, USA

Motawa, I. & Kaka, A. (2008) 'Payment mechanisms for integrated teams in construction', *Australasian Journal of Construction Economics and Building'*, **8** (2), 1-10

Motawa, I. & Kaka, A. (2009) 'Modelling payment mechanisms for supply chain in construction', *Engineering, Construction and Architectural Management*', **16** (4), 325-336

Ng, T. S. & Wong, Y. M. W. (2007) 'Payment and audits mechanisms for non-private-funded PPP-based infrastructure maintenance projects', *Construction Management and Economics*, **25**(9), 915-923

Ramachandra, T. & Rotimi, J. O. (2011) 'The nature of payment problems in the New Zealand construction industry', *Australasian Journal of Construction Economics and Building*, **11** (2), 22-33

Råsled, B. (1992) '*Building and Civil Engineering Contracting*', Byggförlaget, Stockholm, Sweden

Stenbeck, T. (2004) '*Incentives to innovations in road and rail maintenance and operations*', Licentiate thesis, TRITA-INFRA 04-038, ISSN 1651-0216, Royal Institute of Technology – KTH, Stockholm, Sweden

Stenbeck, T. (2009) 'Budgeting performance-based winter maintenance: snow influence on highway maintenance cost', *Journal of infrastructure systems.* **15** (3), 251-260

Strong, C., Shvetsov, Y. & Sharp, J. (2005) '*Development of roadway weather severity index*', Final Technical Report, Collage of engineering, Montana State University, Montana, USA

Swedish Construction Federation (2006) '*General Conditions of Contract for Design and Construct Contract for Building, Civil Engineering and Installation Works – ABT 06*', (Allmänna Bestämmelser för Totalentreprenader avseende byggnads-, anläggnings- och installationsarbeten), Stockholm, Sweden

Swedish Transport Administration – STA (2005) '*Road maintenance contract*', Contract documents to Edsby maintenance area, Gävle, Sweden

Swedish Transport Administration – STA (2008) '*Road maintenance contract*', Contract documents to Bollnäs maintenance area, Gävle, Sweden

Swedish Transport Administration – STA (2009) '*Road maintenance contract*', Contract documents to Borlänge maintenance area, Gävle, Sweden

Swedish Transport Administration – STA (2009) '*Road maintenance contract*', Contract documents to Borås maintenance area, Gothenburg, Sweden

Swedish Transport Administration – STA (2009) *'Rules on measuring and payment for basic maintenance package'*, General regulations, Borlänge, Sweden

Swedish Transport Administration – STA (2012) *'Road maintenance contract'*, Contract documents to Väsby maintenance area, Stockholm, Sweden

Swedish Transport Administration – STA (2012) *'Rules on measuring and payment'*, Contract enclosure to Väsby maintenance area contract, Stockholm, Sweden

Swedish Transport Administration – STA (2013) *'Winter road maintenance'*, Working material, Borlänge, Sweden

Tighe, S. L., Mills, B., Haas, C. T. & Baiz, S. (2007) *'Using Road Weather Information Systems (RWIS) to control load restrictions on gravel and surface-treated highways'*, Technical Report, Ministry of Transportation, Engineering Standards Branch Report, Ontario, Canada

Venäläinen, A. & Kangas, M. (2003) 'Estimation of winter road maintenance costs using climate data', *Meteorol. Appl.* **10**, 69-73, DOI: 10.1017/S1350482703005073

Wesström, D. & Böhm, H. (2008) *'Follow-up of the Skanska's operation activities: Winter road maintenance during 2007'*, MSc thesis, Lund Institute of Technology with Lund University, Lund, Sweden

World Road Association, (2010) *'Snow and Ice databook'*, Technical committee B5, Winter service, ISBN 978-2-550-57709-6, Québec, Canada

Permissions

All chapters in this book were first published in AJCEB, by UTS ePress; hereby published with permission under the Creative Commons Attribution License or equivalent. Every chapter published in this book has been scrutinized by our experts. Their significance has been extensively debated. The topics covered herein carry significant findings which will fuel the growth of the discipline. They may even be implemented as practical applications or may be referred to as a beginning point for another development.

The contributors of this book come from diverse backgrounds, making this book a truly international effort. This book will bring forth new frontiers with its revolutionizing research information and detailed analysis of the nascent developments around the world.

We would like to thank all the contributing authors for lending their expertise to make the book truly unique. They have played a crucial role in the development of this book. Without their invaluable contributions this book wouldn't have been possible. They have made vital efforts to compile up to date information on the varied aspects of this subject to make this book a valuable addition to the collection of many professionals and students.

This book was conceptualized with the vision of imparting up-to-date information and advanced data in this field. To ensure the same, a matchless editorial board was set up. Every individual on the board went through rigorous rounds of assessment to prove their worth. After which they invested a large part of their time researching and compiling the most relevant data for our readers.

The editorial board has been involved in producing this book since its inception. They have spent rigorous hours researching and exploring the diverse topics which have resulted in the successful publishing of this book. They have passed on their knowledge of decades through this book. To expedite this challenging task, the publisher supported the team at every step. A small team of assistant editors was also appointed to further simplify the editing procedure and attain best results for the readers.

Apart from the editorial board, the designing team has also invested a significant amount of their time in understanding the subject and creating the most relevant covers. They scrutinized every image to scout for the most suitable representation of the subject and create an appropriate cover for the book.

The publishing team has been an ardent support to the editorial, designing and production team. Their endless efforts to recruit the best for this project, has resulted in the accomplishment of this book. They are a veteran in the field of academics and their pool of knowledge is as vast as their experience in printing. Their expertise and guidance has proved useful at every step. Their uncompromising quality standards have made this book an exceptional effort. Their encouragement from time to time has been an inspiration for everyone.

The publisher and the editorial board hope that this book will prove to be a valuable piece of knowledge for researchers, students, practitioners and scholars across the globe.

List of Contributors

Mastura Jaafar
Universiti Sains Malaysia, Malaysia

Nurhidayah Mohd Radzi
Universiti Sains Malaysia, Malaysia

Imriyas Kamardeen
University of New South Wales, Australia

Aneurin Grant
University of West Florida, USA

Jimmie Hinze
University of Florida, USA

David Finnie
Otago Polytechnic, New Zealand

Paul H K Ho
City University of Hong Kong, Hong Kong

Elizabeth Grant
The University of Adelaide, Australia

Alex Albert
University of Colorado at Boulder, USA

Matthew R Hallowel
University of Colorado at Boulder, USA

Ximing Ruan
University of the West of England, UK

Edward Ochieng
Liverpool John Moores University, UK

Andrew Price
Loughborough University, UK

Charles Egbu
University of Salford, UK

Lisa Robyn Barnes
University of Newcastle, Australia

Norman Croker
University of Newcastle, Australia

Abimbola Windapo
University of Cape Town, South Africa

Jimoh Richard
Federal University of Technology Minna, Nigeria

Jasper van Wyk
Nelson Mandela Metropolitan University, South Africa

Ayedh Alqahtani
Curtin University, Australia

Andrew Whyte
Curtin University, Australia

Wai Yee Betty Chiu
The University of Hong Kong, Hong Kong

Fung Fai Ng
The University of Hong Kong, Hong Kong

Riza Yosia Sunindijo
University of New South Wales, Australia

Patrick X. W. Zou
University of Canberra, Canberra, Australia

Jennifer MacDonald
University of Technology Sydney, Australia

Julie Mills
University of South Australia, Australia

Raufdeen Rameezdeen
University of South Australia, Australia

Anushi Rodrigo
University of Northumbria, UK

Aki Pekuri
University of Oulu, Finland

Laura Pekuri
University of Oulu, Finland

Harri Haapasalo
University of Oulu, Finland

Heng Jiang
Deakin University, Geelong, Australia

Xiao-Hua Jin
University of Western Sydney, Australia

Chunlu Liu
Deakin University, Australia

Adel Abdi
KTH Royal Institute of Technology, Sweden

Hans Lind
KTH Royal Institute of Technology, Sweden

Björn Birgisson
KTH Royal Institute of Technology, Sweden